Living on an ACTIVE Earth

Perspectives on Earthquake Science

Committee on the Science of Earthquakes

Board on Earth Sciences and Resources

Division on Earth and Life Studies

NATIONAL RESEARCH COUNCIL
OF THE NATIONAL ACADEMIES

THE NATIONAL ACADEMIES PRESS
Washington, D.C.
www.nap.edu

THE NATIONAL ACADEMIES PRESS • 500 Fifth Street, N.W. • Washington, D.C. 20001

This study was supported through endowment funds provided by the National Academy of Sciences. Any opinions, findings, conclusions, or recommendations expressed in this publication are those of the author(s) and do not necessarily reflect the views of the organizations or agencies that provided support for the project.

Library of Congress Cataloging-in-Publication Data

Living on an active earth : perspectives on earthquake science /
Committee on the Science of Earthquakes, Board on Earth Sciences and
Resources, Division on Earth and Life Studies, National Research Council
of the National Academies.
p. cm.
Includes bibliographical references and index.
ISBN 0-309-06562-3 (Book)
ISBN 0309-50631-X (PDF)
1. Seismology—Research. 2. Earthquake hazard analysis. I. National
Research Council (U.S.). Committee on the Science of Earthquakes.
QE539 .L58 2002
551.22'07'2073—dc21a

2002151540

Additional copies of this report are available from the National Academies Press, 500 Fifth Street, N.W., Lockbox 285, Washington, DC 20055; (800) 624-6242 or (202) 334-3313 (in the Washington metropolitan area); Internet, http://www.nap.edu

Cover: Different perspectives on earthquake science. *Left:* Global tectonic map, generated from digital ocean bathymetry and land topography data. SOURCE: NOAA National Geophysical Data Center. *Upper right:* Geologists examining fault slip from the 1954 Dixie Valley-Fairview Peaks, Nevada, earthquake. SOURCE: Photograph by Karl Steinbrugge, Steinbrugge Collection, Earthquake Engineering Research Center, University of California, Berkeley. *Center right:* Synthetic aperture radar interferometry image of deformation caused by the 1999 Hector Mine earthquake. SOURCE: D. Sandwell, L. Sichoix, A. Jacobs, R. Scharroo, B. Minster, Y. Bock, P. Jamason, E. Price, and H. Zebker, Institute of Geophysics and Planetary Physics, La Jolla, Calif. *Bottom right:* Rubble of collapsed buildings in the town of Golcuk, Turkey, resulting from the 1999 Izmit earthquake. SOURCE: Photograph by Enric Marti. Copyright (1999) Associated Press.

Printed in the United States of America

THE NATIONAL ACADEMIES

Advisers to the Nation on Science, Engineering, and Medicine

The **National Academy of Sciences** is a private, nonprofit, self-perpetuating society of distinguished scholars engaged in scientific and engineering research, dedicated to the furtherance of science and technology and to their use for the general welfare. Upon the authority of the charter granted to it by the Congress in 1863, the Academy has a mandate that requires it to advise the federal government on scientific and technical matters. Dr. Bruce M. Alberts is president of the National Academy of Sciences.

The **National Academy of Engineering** was established in 1964, under the charter of the National Academy of Sciences, as a parallel organization of outstanding engineers. It is autonomous in its administration and in the selection of its members, sharing with the National Academy of Sciences the responsibility for advising the federal government. The National Academy of Engineering also sponsors engineering programs aimed at meeting national needs, encourages education and research, and recognizes the superior achievements of engineers. Dr. Wm. A. Wulf is president of the National Academy of Engineering.

The **Institute of Medicine** was established in 1970 by the National Academy of Sciences to secure the services of eminent members of appropriate professions in the examination of policy matters pertaining to the health of the public. The Institute acts under the responsibility given to the National Academy of Sciences by its congressional charter to be an adviser to the federal government and, upon its own initiative, to identify issues of medical care, research, and education. Dr. Harvey V. Fineberg is president of the Institute of Medicine.

The **National Research Council** was organized by the National Academy of Sciences in 1916 to associate the broad community of science and technology with the Academy's purposes of furthering knowledge and advising the federal government. Functioning in accordance with general policies determined by the Academy, the Council has become the principal operating agency of both the National Academy of Sciences and the National Academy of Engineering in providing services to the government, the public, and the scientific and engineering communities. The Council is administered jointly by both Academies and the Institute of Medicine. Dr. Bruce M. Alberts and Dr. Wm. A. Wulf are chair and vice chair, respectively, of the National Research Council.

www.national-academies.org

COMMITTEE ON THE SCIENCE OF EARTHQUAKES

THOMAS H. JORDAN, *Chair*, University of Southern California, Los Angeles
GREGORY BEROZA, Stanford University, Stanford, California
C. ALLIN CORNELL, Stanford University, Portola Valley, California
C. B. CROUSE, URS Corporation, Seattle, Washington
JAMES DIETERICH, U.S. Geological Survey, Menlo Park, California
ARTHUR FRANKEL, U.S. Geological Survey, Denver, Colorado
DAVID D. JACKSON, University of California, Los Angeles
ARCH JOHNSTON, University of Memphis, Tennessee
HIROO KANAMORI, California Institute of Technology, Pasadena
JAMES S. LANGER, University of California, Santa Barbara
MARCIA K. MCNUTT, Monterey Bay Aquarium Research Institute, Moss Landing, California
JAMES R. RICE, Harvard University, Cambridge, Massachusetts
BARBARA A. ROMANOWICZ, University of California, Berkeley
KERRY SIEH, California Institute of Technology, Pasadena
PAUL G. SOMERVILLE, URS Corporation, Pasadena, California

National Research Council Staff

Anne M. Linn, Study Director (from September 2000)
Tamara L. Dickinson, Study Director (May to November 1998)
Ellen Kappel, Study Director (July 1999 to January 2000)
Charles Meade, Study Director (until January 1998)
Monica Lipscomb, Research Assistant (from October 2001)
Verna J. Bowen, Administrative Assistant (from June 1998)
Steven Shannon, Project Assistant (until October 1996)
Susan Sherwin, Project Assistant (October 1996 to June 1998)

Acknowledgments

This report has been reviewed in draft form by individuals chosen for their diverse perspectives and technical expertise, in accordance with procedures approved by the National Research Council's Report Review Committee. The purpose of this independent review is to provide candid and critical comments that will assist the institution in making its published report as sound as possible and to ensure that the report meets institutional standards for objectivity, evidence, and responsiveness to the study charge. The review comments and draft manuscript remain confidential to protect the integrity of the deliberative process. We wish to thank the following individuals for their review of this report:

Brian Atwater, University of Washington, Seattle
Bruce A. Bolt, University of California, Berkeley
Adam M. Dziewonski, Harvard University, Cambridge, Massachusetts
J. Freeman Gilbert, University of California, San Diego
James E. Monsees, Parsons Brinckerhoff, Inc., Orange, California
Stuart Nishenko, PG&E, San Francisco, California
Terry Tullis, Brown University, Providence, Rhode Island

Although the reviewers listed above have provided many constructive comments and suggestions, they were not asked to endorse the conclusions or recommendations nor did they see the final draft of the report before its release. The review of this report was overseen by Clarence R. Allen, professor emeritus, California Institute of Technology, Pasadena.

Appointed by the National Research Council, he was responsible for making certain that an independent examination of this report was carried out in accordance with institutional procedures and that all review comments were carefully considered. Responsibility for the final content of this report rests entirely with the authoring committee and the institution.

Preface

Recent earthquakes in California, Japan, Taiwan, and Turkey have demonstrated the devastating consequences of seismicity in large urban areas. Vulnerability to these seismic hazards can be reduced through scientific research in support of mitigation. However, applying this research to mitigation of seismic hazards has been a challenge. The study presented in this report was motivated by questions surrounding the effectiveness of the "knowledge-based" strategy taken by the National Earthquake Hazard Reduction Program (NEHRP). A series of critiques in the early to mid 1990s, including a 1995 report by the congressional Office of Technology Assessment (*1*), concluded that the NEHRP approach short-changed practical measures for mitigating earthquake losses, creating an "implementation gap" in which risk-reduction efforts lagged far behind the knowledge base created by basic research. A preliminary review by the National Research Council (NRC) Committee on Seismology indicated that the debate over the structure of a national program for earthquake risk reduction lacked adequate information about the prospects for, and potential payoffs from, long-term basic research and the relationship of current research activities, both basic and applied, to mitigation efforts.

Based on this assessment, a proposal was submitted to the National Academies, which established the Committee on the Science of Earthquakes and provided a grant from its endowment funds to

> assess the current scientific understanding of earthquake processes. In its work, the study committee will prepare a comprehensive summary of the multidisciplinary research throughout the earth and physical sciences on the origins, properties, and consequences of earthquakes, assess

> the research goals for the field of earthquake science, particularly as they support engineering and policy efforts to improve seismic mitigation strategies, and identify strategies to improve the communication of earthquake science to engineers, policy makers, and the general public.

This report is meant to provide a technical reference for scientists, engineers, and policy makers concerned with understanding earthquakes and reducing society's vulnerability to seismic hazards.

To gather information for the study the committee met with representatives from government agencies, nongovernmental organizations, and private sector companies concerned with earthquake research and engineering, hazard mitigation, and earthquake insurance. Briefings were provided by AXA Reinsurance, the Department of Defense, Department of Energy, Federal Emergency Management Agency, Incorporated Research Institutions for Seismology, National Aeronautics and Space Administration, National Institute of Standards and Technology, National Science Foundation, Office of Science and Technology Policy, and U.S. Geological Survey. Input from the broader Earth science community was solicited through an editorial in *Seismological Research Letters* (2) and the committee's web site, which included a series of white papers on different earthquake topics. Altogether the committee met eight times to gather information and prepare its report.

The committee thanks the following individuals for making presentations or providing background material, figures, or other input: Cliff Astill, Aykut Barka, Harley Benz, Yehuda Ben-Zion, Margaret Boettcher, Steve Bratt, Robert Bucknam, George Budd, Rhett Butler, Wuchen Chi, Rob Clayton, Paula Davidson, James Dolan, Douglas Dreger, John Filson, Chris Goldfinger, Tom Henyey, Steve Hickman, Ken Hudnut, Jeff Kimball, Richard Krimm, Michael Mahoney, Jeff McGuire, Robert Nadeau, Amos Nur, Elaine Padovani, Robert Page, Gilles Peltzer, Eliza Richardson, Charles Rubin, Ronald Sack, Charles Sammis, Kaye Shedlock, David Simpson, Shyam Sunder, Louis Walter, Jim Whitcomb, Cecily Wolfe, Nicholas Woodward, Richard Wright, Howard Zebker, and Mark Zoback. The committee also thanks the staff of the National Research Council for their support of this project. Charles Meade obtained funding for the study and helped the committee generate most of the raw material for this report before he left the NRC to pursue a new career. Tammy Dickinson and Ellen Kappel provided figures and interim support to the committee. The committee is particularly grateful to Anne Linn, whose outstanding efforts brought this study to a successful conclusion.

Thomas H. Jordan
Chair

NOTES

1. Office of Technology Assessment, *Reducing Earthquake Losses,* OTA-ETI-623, U.S. Government Printing Office, Washington, D.C., 162 pp., 1995.
2. T.H. Jordan, Is the study of earthquakes a basic science? *Seis. Res. Lett.,* **68**, 259-261, 1997.

Contents

1

The Challenge of Earthquake Science

Few natural events disrupt human society more than large earthquakes. Civilization depends on a massive "built environment" anchored in the Earth's active crust, framing a habitat extremely vulnerable to seismic shaking. To cope with the all-too-frequent destruction caused by earthquakes, people have long sought to improve their practical knowledge about where and when such events might occur, and what happens when they do. Science has shown that seismic activity can be understood in terms of a basic machinery of deformation that shapes the face of the planet. Consequently, the pragmatic inquiry into the causes and effects of earthquakes has become increasingly fused with the quest for a more fundamental understanding of the geologically active Earth.

This report surveys all aspects of earthquake science, basic and applied, from ancient times to the present day (*1*). Despite rapid progress in the latter part of the twentieth century, the study of earthquakes, like the science of many other complex natural systems, is still in its juvenile stages of exploration and discovery. The brittleness and opacity of the deforming crust have made headway arduous, slower in some respects than in the study of the Earth's oceans and atmosphere. However, in just the last decade, new instrumental networks for recording seismic waves and geodetic motions have mushroomed across the planet, and new methods for deciphering the geological record of earthquakes have been applied to active faults in many tectonic environments. High-performance computing is now furnishing the means to process massive streams of observations and, through numerical simulation, to quantify many aspects of earthquake behavior that are completely resistant to theoretical manipulation and manual calculations. These new research capabilities are transforming the field from a haphazard collection of disciplinary activities to a more coordinated "system-level" science—one that seeks to describe seismic activity not just in terms of individual events, but as an evolutionary process involving dynamic interactions within networks of faults.

The scientific challenge is to leverage these advances into an understanding of earthquake phenomena that is both profound and practical. The research needed to move toward these objectives is the focus of this report. The National Earthquake Hazard Reduction Program, the mainstay for federal earthquake research over the past 25 years (Appendix A), has opened many areas of fruitful inquiry. New possibilities are arising from the system-level approach that now organizes the study of active faulting and crustal deformation. In appraising research opportunities, the Committee on the Science of Earthquakes has sought to keep in focus the rationale for future earthquake research from four complementary perspectives: (1) the need to improve seismic safety and performance of the built environment, especially in highly exposed urban areas; (2) the requirements for disseminating information rapidly during earthquake crises; (3) the fresh opportunities for exciting basic science, particularly in the context of current research on complex natural systems; and (4) the responsibility for educating people at all levels of society about the active Earth on which they live. These perspectives are summarized below.

1.1 SEISMIC SAFETY AND PERFORMANCE

Earthquakes are hazards primarily because strong ground shaking destroys things that people have constructed—buildings, transportation lifelines, and communication systems (Figure 1.1). Earthquakes are also responsible for secondary, though often very damaging, effects such as soil liquefaction, landslides, tsunamis, and fires. Over the last century, earthquakes have caused an average of 10,000 deaths per year worldwide and hundreds of billions of dollars in economic losses. The United States has seen less seismic devastation than many other countries, primarily owing to its lower population density and superior building construction (2). Nevertheless, the annualized long-term loss due to U.S. earthquakes is currently estimated at $4.4 billion per year (3), and this figure appears to be rising rapidly, despite continuing improvements in building codes and structural design (Figure 1.2). California leads with the highest risk, but the problem is truly national: 38 other states face substantial earthquake hazards, including 46 million people in metropolitan areas at moderate to high risk outside of California (Figure 1.3).

FIGURE 1.1 Scene of destruction in Kobe, Japan, from the 1995 Hyogo-ken Nanbu earthquake (magnitude 6.9), showing a collapsed section of the Hanshin Expressway. This event killed at least 5500 people, injured more than 26,000, and was responsible for approximately $200 billion in direct economic losses. SOURCE: Pan-Asia Newspaper Alliance.

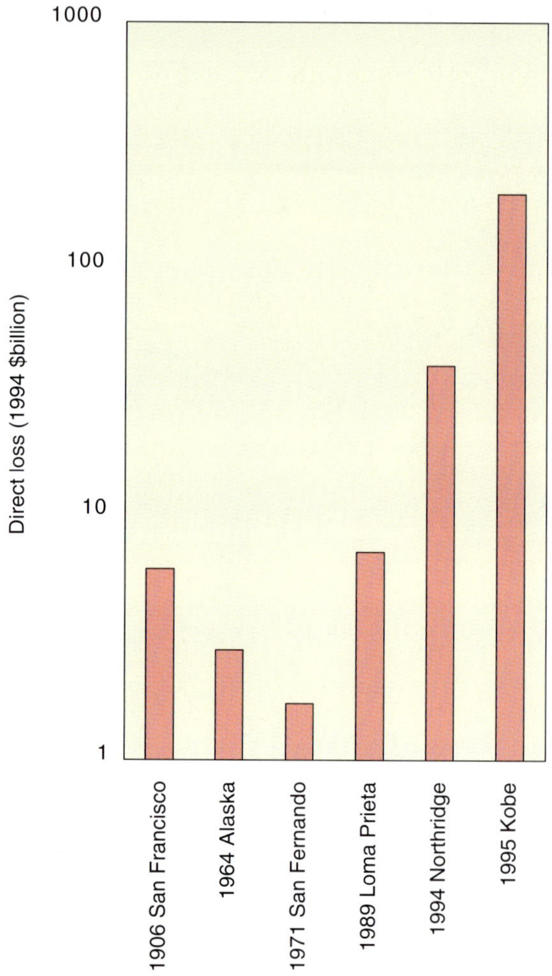

FIGURE 1.2 Direct economic losses, given in inflation-adjusted 1994 dollars on a logarithmic scale, from some major earthquakes in the United States and Japan. The plot shows a near-exponential rise in the losses caused by urban earthquakes of approximately equal size from 1971 to 1995. SOURCE: Compiled from Office of Technology Assessment, *Reducing Earthquake Losses*, OTA-ETI-623, U.S. Government Printing Office, Washington, D.C., 162 pp., 1995; R.T. Eguchi, J.D. Goltz, C.E. Taylor, S.E. Chang, P.J. Flores, L.A. Johnson, H.A. Seligson, and N.C. Blais, Direct economic losses in the Northridge earthquake: A three-year post-event perspective, *Earthquake Spectra*, 14, 245-264, 1998; National Institute of Standards and Technology, *January 17, 1995 Hyogoken-Nanbu (Kobe) Earthquake: Performance of Structures, Lifelines, and Fire Protection Systems*, NIST Special Publication 901, Washington, D.C., 573 pp., July 1996.

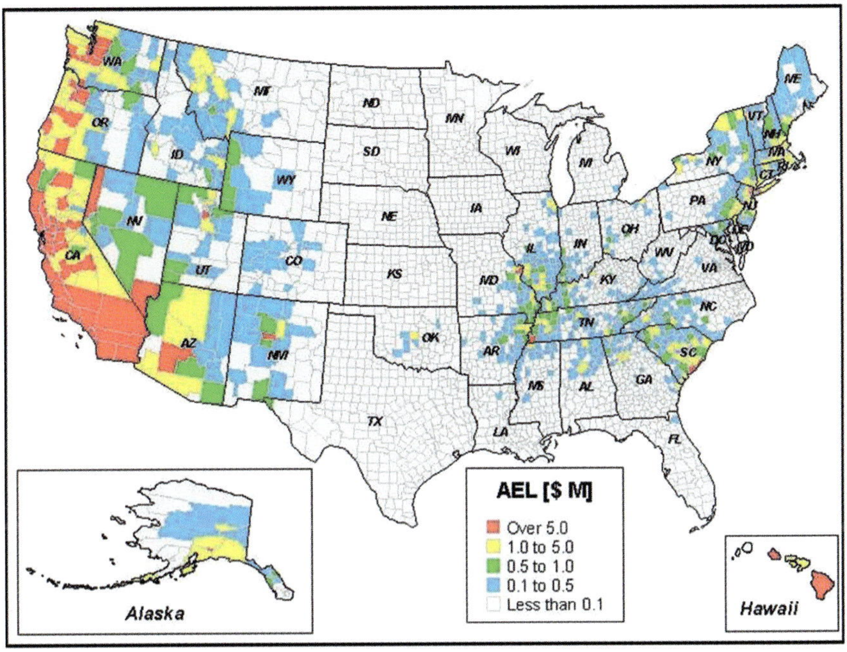

FIGURE 1.3 Current annualized earthquake losses (AEL) in millions of dollars, estimated by the Federal Emergency Management Agency (FEMA) on a county-by-county basis using the HAZUS method. Twenty-four states have an AEL greater than $10 million. The total AEL estimated for the entire United States is about $4.4 billion. SOURCE: FEMA, *HAZUS 99 Estimated Annualized Earthquake Losses for the United States*, FEMA Report 366, Washington, D.C., 33 pp., February 2001.

The prediction of losses from future natural disasters is notoriously uncertain (4), but more accurate projections are being established with better technical input from earthquake science, engineering, and economics. Scenarios constructed using loss estimation tools have begun to quantify the magnitude of the risk that now faces large population centers in earthquake-prone regions. According to a 1995 report (5), a repeat of the 1906 San Francisco earthquake would likely result in a total loss of $170 billion to $225 billion (in 1994 dollars). The comparable loss for the 1994 Northridge earthquake, the costliest U.S. disaster on record, was about five times lower. The direct losses in a repeat of the 1923 Kanto earthquake, near Tokyo, would truly be staggering—$2 trillion to $3 trillion—and the indirect economic costs could be much higher (6). Nevertheless, much of the Pacific Rim and other earthquake-prone regions are urbanizing rapidly (Figure 1.4) and will face risks similar to those of existing megacities, such as Los Angeles, Istanbul, and Singapore. One expert has speculated that the urbanization in earthquake-prone areas of the developing world may result in a four- to tenfold increase in the annual fatality rate over the next 30 years, reversing a long-standing trend (*7*).

The losses expected from future events (*risk*) depend on the population and amount of infrastructure concentrated in a given area (*exposure*) and the vulnerability of the built environment (*fragility*), as well as the hazard itself. The seismic hazard levels in Alaska and California are both high, but California's exposure is much greater, which yields a much larger total risk (Figure 1.3). The growth in losses charted in Figure 1.2 comes primarily from the increased exposure, especially in urbanizing regions (*8*). Exposure can be lowered by a judicious choice of building sites and careful land-use planning; but only by so much—in seismically active regions, all sites face significant hazard. Earthquake risk reduction must thus rely on lowering vulnerability through earthquake-resilient design of new structures and retrofitting or rehabilitating inadequate older structures to improve their seismic safety. Provisions for earthquake design are now an integral part of building codes in most seismically active parts of the United States, although some states with moderate earthquake risk do not have state seismic codes (see Chapter 2). Code improvements and related problems of implementation continue to require scientific guidance to take into account regional and local variations in seismic hazards, as well as uncertainties and improvements in hazard estimates (*9*).

The fundamental question of how much to invest in seismic safety has become a pressing problem for the national economy. Several hundred billion dollars are currently spent per annum on new construction in seismically active areas of the United States, and about 1 percent of this overall investment is associated with seismic reinforcement (*10*). Seismic retrofitting is much more expensive; the costs for unreinforced masonry buildings are typically about 20 percent of new construction, and the values for other building types are comparably high. An alarming example can be found in Los Angeles County, where a recent study showed that more than half of the hospitals are vulnerable to collapse in a strong earthquake. The cost for upgrading these facilities to state-mandated levels of seismic safety has been estimated conservatively at $7 billion to $8 billion, which exceeds the entire assessed value of all hospital property in the county and comes at a time when nearly all hospitals are losing money (*11*). Resolving such economic and political issues is at least as difficult as the engineering and science issues, and it illustrates the need for coordinated planning across all aspects of seismic safety. To facilitate planning, understandable earthquake information must be made avail-

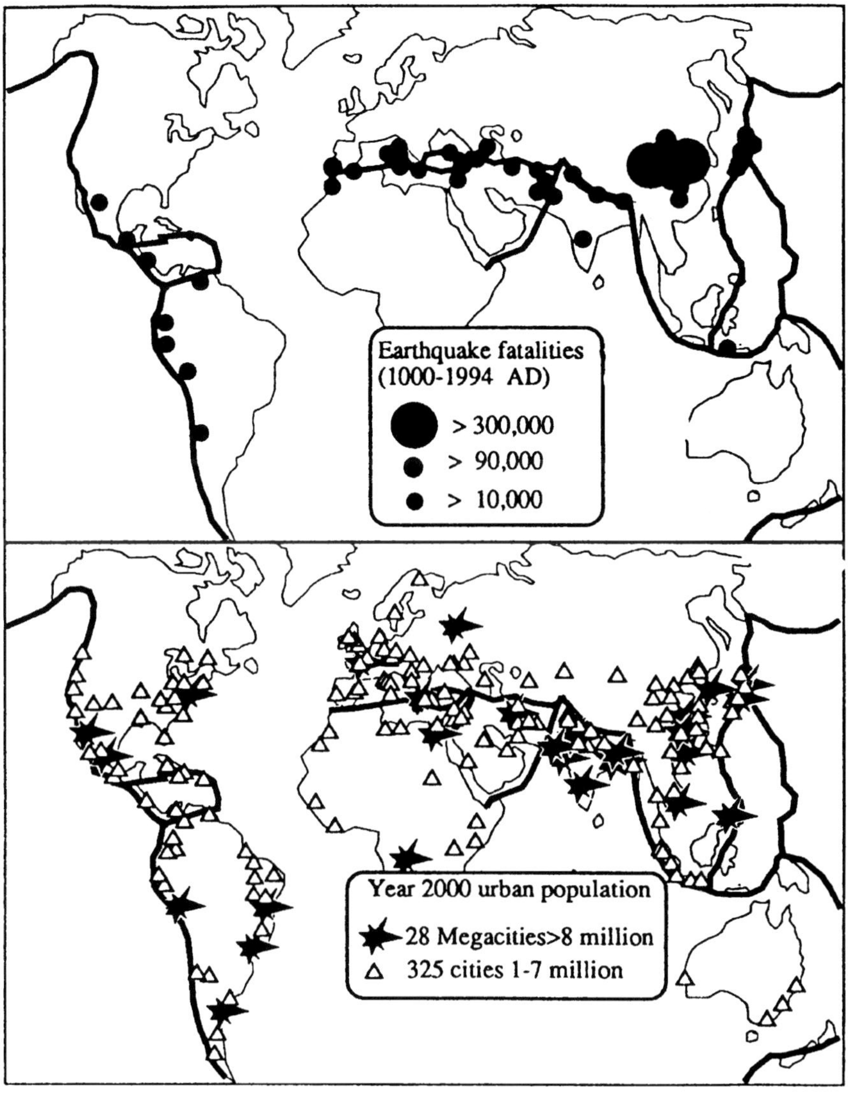

FIGURE 1.4 Major cities (1 million to 7 million inhabitants, open triangles) and megacities (more than 8 million inhabitants, solid stars) in the year 2000, relative to the major earthquake belts (shaded regions). These cities, which house approximately 20 percent of the global population, tend to be concentrated in regions of high seismic risk. SOURCE: R. Bilham, Global fatalities in the past 2000 years: Prognosis for the next 30, in *Reduction and Predictability of Natural Disasters*, J. Rundle, F. Klein, and D. Turcotte, eds., Santa Fe Institute Studies in the Sciences of Complexity, vol. 25, Addison-Wesley, Boston, Mass., pp. 19-31, 1995. Copyright 1995 by Westview Press. Reprinted with permission of Westview Press, a member of Perseus Books, L.L.C.

able to a wide range of professionals, as well as people at large, who ultimately decide what level of public safety is acceptable.

Requirements for a built environment that can withstand seismic shaking have motivated much research on specialized construction materials and advanced engineering methods. As these efforts have matured, engineers have begun to employ more detailed characterization of strong ground motions in structural design and testing. In the process, the coupling between earthquake engineering and science has been strengthened. Engineers are now interested in going beyond the basic life-safety requirement of preventing structural collapse. Demand is growing for techniques to design buildings that can retain specified levels of functionality after earthquake shaking of a specified probability of occurrence (see Chapter 3). The success of "performance-based engineering" will depend on two related developments: (1) the formulation of new, more diagnostic measures of structural damage (primarily an engineering task) and (2) more sophisticated treatments of ground shaking, including parameterizations of seismograms that are better suited to the probabilistic prediction of damage states than peak ground acceleration and other classical intensity measures (primarily a seismological task). Thus, collaboration between the earthquake science and engineering communities is essential.

1.2 SEISMIC INFORMATION FOR EMERGENCY RESPONSE

Science and technology can do nothing to prevent or control large earthquakes, and as yet no known method can reliably predict when, where, and how big future tremors will be. However, once an earthquake occurs, advanced *seismic information systems* can transmit signals from dense networks of seismometers to central processing facilities and, in a fraction of a minute, pinpoint the initial fault rupture (hypocenter) and determine other diagnostic features (e.g., fault orientation). If equipped with strong-motion sensors that accurately record the most violent shaking (when velocity reaches a meter per second and acceleration sometimes exceeds gravity), these automated systems can also deliver accurate maps in nearly real time of where the ground shaking was strong enough to cause significant damage (Figure 1.5). This information about ground shaking can be incorporated into earthquake loss estimation programs such as HAZUS, which provide assessments of potential damage (*12*). Such information can be crucial in helping emergency managers and other officials deploy equipment and personnel as quickly as possible to save people trapped in rubble and to reduce further property losses from fires and other secondary effects. Bulletins about the magnitude and boundaries of the shaking can also be channeled through the mass media, reduc-

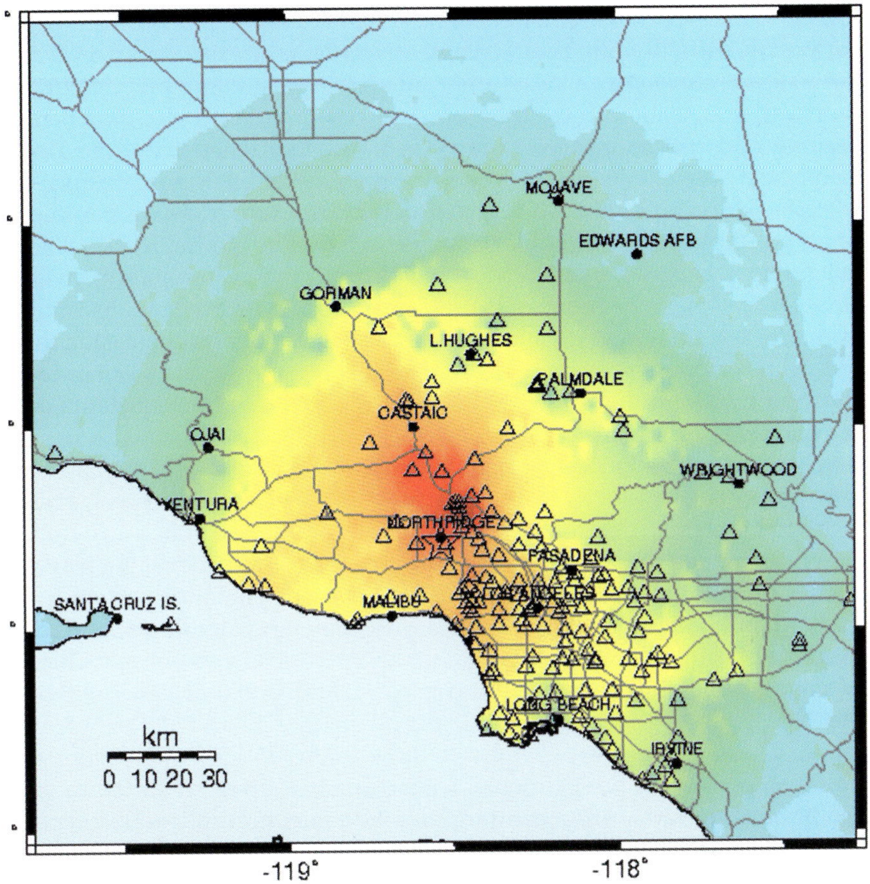

FIGURE 1.5 A map of the shaking intensity for the 1994 Northridge earthquake derived from peak ground accelerations measured by strong-motion instruments (triangles) in the Los Angeles area. In regions with advanced seismic information systems, maps such as this can be broadcast to emergency management agencies within a few minutes, providing critical information for organizing emergency response. SOURCE: U.S. Geological Survey.

ing public confusion during disasters and allaying fears aroused by minor tremors.

Seismic information systems with the capabilities described above are now operational or under development in high-risk areas such as California, the Pacific Northwest, and the intermountain West, but the delivery of rapid information following earthquakes poses many technological and scientific challenges. The experience with recording large earthquakes is still fairly thin; with rare exceptions, areas of more moderate risk are currently serviced only by sparse seismographic networks with antiquated instrumentation and uneven capabilities for digital recording and processing. To remedy some of the deficiencies, the U.S. Geological Survey has begun to deploy the Advanced National Seismic System, which will improve regional networks, expand the distribution of strong-motion and building response sensors in the high-risk urban areas, and package and deliver earthquake information automatically (see Chapter 4). Research is needed in many aspects of post-event analysis, for example, assimilating data into strong-motion predictions, increasing the reliability of aftershock predictions, and identifying areas of enhanced short-term risk through the development of models of how earthquakes transfer stresses from one fault to another.

With appropriate technology, seismic information systems can be used as *earthquake warning systems*. Because electronic signals travel much faster than seismic disturbances, it is possible to notify regions away from the epicenter that an earthquake is in progress before any damaging waves arrive. Suboceanic earthquakes sometimes generate tsunamis (sea waves) that can inundate shorelines thousands of kilometers from the source. For example, the great 1964 Alaska earthquake generated tsunamis that killed 17 people along the Oregon-California coast, and tsunamis generated by the 1960 Chile earthquake killed 61 people in Hawaii and 122 people in Japan. These waves travel relatively slowly (500 to 700 kilometers per hour), so post-event predictions of tsunami arrival times and amplitudes can be used to warn coastal communities soon enough to allow for preparation and evacuation. Agencies such as the National Oceanic and Atmospheric Administration operate tsunami-warning networks that depend on precise seismic information to function properly (*13*). However, how tsunamis are generated by suboceanic earthquakes and how they run up along coastlines are still poorly understood.

Advance warnings of strong motions caused by the fast-moving (2 to 8 kilometers per second) ground waves are more problematic than for tsunamis, because there is so little time for an event to be evaluated and an advisory broadcast through civil defense or other warning systems. The damage zones of large earthquakes usually have radii of 200 kilometers or less, and it takes only 60 to 100 seconds for the most damaging waves (shear and surface waves) to propagate to this distance. Nevertheless, the time is adequate to issue electronic warnings that can initiate emergency shutdowns and other protective actions within power generation, transportation, and computer systems, provided that decisions can be automated reliably (see Chapter 3). The implementation of this capability will require seismic information systems that are robust with respect to regional disruptions in power and communications and methods for making (and aborting) decisions under stressful conditions, in addition to research on the basic problems of rupture dynamics, wave propagation, and site response.

1.3 BASIC GEOSCIENCE

In its broadest sense, the science of earthquakes seeks a comprehensive, physics-based understanding of seismic behavior and associated geological phenomena. While plate tectonics has provided a simple and precise kinematic framework for relating individual earthquakes to geological deformations on a global scale (see Chapter 2), its extension to a fully dynamic theory of Earth deformation remains a work in progress. Among the fundamental questions that still lack satisfactory answers are: Why do crustal strains on the Earth tend to localize on major faults, rather than be distributed over broad zones of more continuous deformation, as has been inferred for the surface of Venus? How does brittle crustal deformation couple to the ductile motion of the convecting solid mantle? What physical mechanisms explain the large earthquakes that occur in the descending limbs of mantle convection down to depths of nearly 700 kilometers? How predictable are large earthquakes? Such questions connect the study of earthquakes to many basic aspects of solid-Earth research.

Correspondingly, many disciplines—seismology, tectonic geodesy, earthquake geology, rock mechanics, complex systems theory, and information technology—are producing the conceptual innovations needed for the practical issues of seismic hazard analysis, risk reduction, and rapid emergency response. Fundamental work on elastic wave scattering in fractal media has been applied to prediction of the strong ground motions that damage buildings. The ultraprecise techniques of space geodesy have measured the deformation across the broad zone of faulting that extends from the Wasatch Range in Utah to the Pacific coast, improving constraints on probabilistic earthquake forecasting. The need to predict where earthquakes will occur has inspired novel techniques for deciphering where and when earthquakes have occurred in prehistory. This practical focus has provided new opportunities for basic geological research in the youthful fields of paleoseismology and neotectonics.

Other applications of earthquake science include the study of volcanic seismicity, which has led to improvements in predicting volcanic eruptions, and the detection of nuclear explosions, a strategic capability for the verification of international arms control treaties. The coupling of seismological and archaeological studies in the Middle East has shed light on the role of earthquakes in the rise and fall of ancient urban centers and political systems. Laboratory studies of fault friction and earthquake fracture mechanics have led to constitutive relations and dynamical models that have found applications in materials engineering. The power-law scaling relations between earthquake frequency and size, combined with the recognition that nearly all of the Earth's crust may be close to its critical state of failure, have stimulated theories of nonlinear system dynamics (*14*).

The fundamental interactions that govern active faulting are distributed over an enormous range of spatial and temporal scales—sequences of great earthquakes on thousand-kilometer faults over hundreds of years are coupled dynamically to deformation processes operating in milliseconds over millimeters. These processes, difficult to study directly in the field, can be replicated in the laboratory only at the lower end of this range. Sampling and in situ observation by trenching, tunneling, and drilling are confined to the outer few kilometers. Understanding the physical processes at the much greater depths where earthquakes typically nucleate depends on the ability to construct models that combine surface observations with seismic imaging and other remote-sensing data.

Earthquakes in nearly all tectonic environments share similar scaling laws (e.g., frequency-magnitude statistics, aftershock decay rates, stress drops), which suggests that some of the most basic aspects of earthquake behavior are universal and not sensitive to the details of deformation processes. However, differing theories have sparked considerable controversy about how small-scale processes such as rock damage and fluid flow are involved in active faulting. A comprehensive theory must integrate earthquake behavior across all dimensions of the problem. This challenge is a primary motivation for the National Science Foundation's ambitious EarthScope initiative, which will employ four new technologies for observing active deformation in the United States over a wide range of scales (*15*).

1.4 EDUCATION

Educating people about earthquakes can effectively reduce human and economic losses during seismic disasters. The pedagogical responsibilities of earthquake science are expanding as concerns about the vulnerability of the built environment increase. The issues encompass the delivery of earthquake information to the public; earthquake-oriented curricula in schools at all levels; the career focus of young researchers, and the transfer of knowledge to engineers, emergency managers, and government officials.

The Internet has greatly enhanced the capability for delivering a wide variety of earthquake information. Many public and private organizations maintain web sites that host a variety of earthquake information services. These sites display up-to-the-minute maps of earthquake epicenters and strong motions, describe seismic hazards and damage, offer tips to homeowners about retrofitting and insurance, and make available curricular material. Regional seismic networks and earthquake response organizations update their web sites regularly. The Hector Mine (California) earthquake of October 16, 1999, and the Nisqually (Washington) earthquake of February 28, 2001, were among the first U.S. "cyber-quakes" in the sense that the Internet became the dominant medium for exchanging data and posting results in the minutes and hours following the ruptures (*16*).

Information and communication technologies have irrevocably altered how scientists interact with the public in several subtle ways. New tools for digital representation and visualization are available to animate scientific descriptions of earthquakes and present research results in more attractive and intelligible formats. Renderings of numerical simulations now allow people to visualize more readily the complex physical processes on space-time scales too large or small, or places too remote, to be observed directly. Worldwide improvement in communication systems is feeding the public's interest in global problems. When devastating earthquakes occurred in El Salvador and the Indian province of Gujarat in January 2001, coverage by U.S. broadcast and print media, as well as through the Internet, was quicker and more comprehensive than it had been for most previous, remote earthquakes.

Every effort should be made to educate the public that earthquakes are natural phenomena that cannot be controlled (*17*) but can be understood. The goal is to teach people enough about earthquake science for them to become smart users of information and know how to prepare for and react to seismic events. Earthquakes furnish compelling examples of physics and mathematics in real-world action, and research on earthquakes illustrates the process of empirical inquiry and the scientific method in many interesting ways. The silver lining of seismic disruption is perhaps the "teachable moment" when a major earthquake captures public attention and the media are filled with technical explanations. Properly prepared students can learn a lot about natural science from these intense periods of reporting.

The content and methods for teaching earthquake science in primary and secondary schools remain pressing subjects that should engage more scientists involved in basic research. National science education standards recommend that Earth science be taught at all grade levels (*18*). For example, students should learn about fossils and rock and soil properties in kindergarten through fourth grade; geologic history and the structure of the Earth in grades 5 through 8; and the origin and evolution of the Earth system in grades 9 through 12. A number of states have adopted these standards, which provide many opportunities for giving students a better understanding of how the Earth works and how its internal forces are acting to shape their environment.

Researchers tend to be more concerned with higher education. Many four-year colleges and universities offer coursework in Earth science that satisfies the general science requirements for nonmajors. Most of these classes include material on earthquake processes and active tectonics, and they can be quite popular. However, fewer college students are majoring in Earth science (*19*). The baccalaureate in this field is rarely an adequate degree, and its status as a pre-professional degree (e.g., premedical or prelaw) has never been high, except as preparation for graduate studies in Earth science (*20*). Graduate education, especially at the Ph.D. level, is research oriented and therefore sustained by the migration of young geoscientists into teaching careers at research universities. Earth science needs a healthy program of earthquake research to attract the best candidates.

1.5 PREDICTIVE UNDERSTANDING

The main goal of earthquake research is to learn how to predict the behavior of earthquake systems. This goal drives the science for two distinct reasons. First, society has a compelling strategic need to anticipate earthquake devastation, which is becoming more costly as large urban centers expand in tectonically active regions. In this context, prediction has come to mean the accurate forecasting of the time, place, and size of specific large earthquakes, ideally in a short enough time to allow nearby communities to prepare for a calamity. Three decades ago, many geoscientists thought this type of short-term, event-specific earthquake prediction was right around the corner. However, proposed schemes for short-term earthquake prediction have not proven successful. In particular, no unambiguous signals precursory to large earthquakes have been identified, even in areas such as Japan and California where monitoring capability has improved substantially. Although a categorical statement about the theoretical infeasibilty of event-specific prediction appears to be premature, few seismologists remain optimistic that short-term, event-specific prediction will be feasible in the foreseeable future.

It is true, nevertheless, that many aspects of earthquake behavior can be anticipated with enough precision to be useful in mitigating risk. The long-term potential of near-surface faults to cause future large earthquakes can be assessed by combining geological field studies of previous slippage with seismic and geodetic monitoring of current activity. Seismologists are learning how geological complexity controls the strong ground motion during large earthquakes, and engineers are learning how to predict the effects of seismic waves on buildings, lifelines, and critical facilities such as large bridges and nuclear power plants. Together they have quantified the long-term expectations for potentially destructive shaking in the form of seismic hazard maps, and they are striving to improve these forecasts by understanding how past earthquakes influence the likelihood of future events. Consideration of slip models along faults is now leading to predictions of site-specific ground motions, needed by engineers for the design of large urban structures that can withstand seismic shaking.

The second reason for seeking a predictive understanding is epistemological and generic to fundamental research on a variety of geosystems, be they localized—like volcanoes, petroleum reservoirs, and groundwater systems—or global—like the oceans and atmosphere. These geosystems are so complex and their underlying physical and chemical processes are so difficult to characterize that the traditional reductionist approach, based on the elucidation of fundamental laws, is incomplete as an investigative method. In the face of such complexity, the ability to routinely extrapolate observable behavior becomes an essential measure of how well a system is understood. Model-based prediction plays an integral role in this type of empiricism as the first step in an iterated cycle of data gathering and analysis, hypothesis testing, and model improvement.

1.6 ORGANIZATION OF THE REPORT

This report summarizes progress in earthquake science and assesses the opportunities for advancement. Although not intended to be exhaustive, the report documents major scientific achievements through extensive references and technical notes. Chapter 2 charts the rise of earthquake science and engineering, introducing the technical concepts and terminology in their historical context. Chapter 3 gives the current view of seismic hazards on a national and global scale and shows how an improved characterization of these hazards can reduce earthquake losses by lessening risk and speeding response. Chapter 4 describes the observational activities and research methods in four major disciplines—seismology, tectonic geodesy, earthquake geology, and rock mechanics—and discusses the advanced technologies being employed to gather new information on the detailed workings of active fault systems. The integration of this information into a predictive, physics-based framework is the subject of Chapter 5, which lays out key scientific questions in five areas of interdisciplinary research—fault systems, fault-zone processes, rupture dynamics, wave propagation, and seismic hazard analysis. Chapter 6 examines the goals of earthquake research over the next decade and the resources and technological investments needed to achieve these goals in nine major areas of interest. Federal programs supporting earthquake research and engineering are summarized in Appendix A. Finally, a list of acronyms is given in Appendix B.

NOTES

1. Some scientists prefer to follow Robert Mallet, who first employed (in 1862) the term "seismology" as a generic name for the study of earthquakes (e.g., B.A. Bolt, *Seis. Res. Lett.*, **68**, 714, 1997). However, modern usage has tended to restrict seismology to the study of waves and other motions of the Earth produced by earthquakes, explosions, and other large energy sources (see K. Aki and P.G. Richards, *Quantitative Seismology: Theory and Methods*, Vol. I, p. 1, W.H. Freeman, San Francisco, 1980). The committee has chosen the less elegant but more ecumenical title of "earthquake science" because it emphasizes the multidisciplinary nature of a field that has drawn practitioners from geology, geodesy, rock mechanics, and the physics of complex systems, in addition to seismologists.

2. Averaged over the twentieth century, the U.S. population comprised about 6 percent of the global population but suffered only 0.1 percent of the human toll in earthquake disasters. The reasons for the amount of earthquake damage and number of deaths observed in a particular country during a particular earthquake vary, but in general are due to (1) the location of the event relative to population centers, (2) inadequate seismic codes or poor enforcement of them, (3) poor design practice, (4) inferior construction materials, (5) lack of inspection, (6) large inventory of older structures and insufficient funds to retrofit them, and (7) corruption. For example, many of the buildings subjected to the 1999 Turkey earthquakes did not meet strength and detailing requirements in the applicable building codes. In addition, landslides, subsidence, and lateral spreading due to soil failure contributed to the heavy damage observed in the coastal area of the Bay of Izmit (1999 Kocaeli, Turkey, Earthquake Reconnaissance Report, *Earthquake Spectra*, **16**, Suppl. A, December 2000). On the other hand, most buildings affected by the 2001 India earthquake did not comply with the seismic code provisions because the codes are not mandatory (Preliminary Observations on the Origin and Effects of the January 26, 2001, Bhuj (Gujarat, India) Earthquake, *EERI Newsletter*, **35**, April 2001).

3. Federal Emergency Management Agency, *HAZUS®99 Estimated Annualized Earthquake Losses for the United States*, FEMA Report 366, Washington, D.C., 33 pp., February 2001.

4. National Research Council, *Estimating Losses from Future Earthquakes*, National Academy Press, Washington, D.C., 231 pp., 1989.

5. Risk Management Solutions, Inc., *What if the 1906 Earthquake Strikes Again? A San Francisco Bay Scenario*, Menlo Park, Calif., 81 pp., May 1995. Estimated losses are assumed to follow a beta distribution, with the range corresponding to the 50th and 90th percentiles (i.e., the probability that the actual losses will be less than the lower bound is 50 percent, and the probability that the losses will be greater than the upper bound is 10 percent). Direct economic losses do not include expenses for debris removal and emergency relief or indirect losses from firms that depend on earthquake-damaged businesses, lower tax revenues due to reduced business revenues, loss of jobs, or decreased valuations in national and world markets.

6. Risk Management Solutions, Inc., *What if the 1923 Earthquake Strikes Again? A Five-Prefecture Tokyo Region Scenario*, Menlo Park, Calif., 97 pp., November 1995. Estimates quoted in the text are in 1994 dollars. The report goes on to speculate: "This [direct] loss represents 44-70% of the gross domestic product of Japan for 1994. The region is a vital link in the world economy, and the ripple effects of its losses would be felt around the world. . . . To finance the enormous sum needed to begin reconstruction after the scenario earthquake,

the Japanese government and private corporations would need to sell foreign assets and/or issue new securities [which] would upset the supply and demand balance at the bond markets and push up long-term interest rates." For further comments on indirect losses, see T. Uchida (Thoughts about the great Tokyo Earthquake, *By the Way*, May/June 1995).

7. R. Bilham, Global fatalities in the past 2000 years: Prognosis for the next 30, in *Reduction and Predictability of Natural Disasters*, J. Rundle, F. Klein, and D. Turcotte, eds., Santa Fe Institute Studies in the Sciences of Complexity, vol. 25, Addison-Wesley, Boston, Mass., pp. 19-31, 1995. Over the past 400 years, the annualized global mean fatality rate from earthquakes has fallen by more than an order of magnitude, from 10 to 100 per million before 1600 to about 1 per million during the last half of the twentieth century. Yet Bilham raises the specter of more than 1 million fatalities from a single event if a major earthquake occurs near one of the dozens of supercities in developing countries within seismically active zones.

8. Increases in economic exposure are not limited to urban centers, however. The moderate (M 5.2) Yountville earthquake of September 3, 2000, in a rural area 10 miles northwest of Napa, California, resulted in a direct loss of at least $50 million; 25 people were injured—2 critically—and building inspectors red-tagged 16 buildings and yellow-tagged 168 others. The governor declared a state of emergency and obtained federal disaster relief funds (<http://quake.wr.usgs.gov/recent/reports/napa>).

9. California Seismic Safety Commission, *California at Risk, 1994 Status Report*, SSC 94-01, Sacramento, Calif., p. 1, 1994.

10. *Current Construction Reports, Value of Construction Put in Place*, C30/01-5, U.S. Department of Commerce, U.S. Census Bureau, Washington, D.C., 21 pp., 2001; Office of Technology Assessment, *Reducing Earthquake Losses*, OTA-ETI-623, U.S. Government Printing Office, Washington, D.C., 162 pp., 1995.

11. *Summary of Hospital Performance Ratings*, Office of Statewide Health Planning and Development, Sacramento, Calif., 20 pp., March 2001; Study Finds High Quake Risk for Hospitals, *Los Angeles Times*, March 29, 2001.

12. HAZUS is a public-domain software package and database that estimates losses from ground motions. (See Chapter 3, Section 3.4.) HAZUS was run by the Federal Emergency Management Agency immediately following the 2001 Nisqually, Washington, earthquake and played a role in the decision for a Presidential Disaster Declaration (FEMA Disaster Federal Register Notice 1361, March 2001, <http://www.appl.fema.gov/library/dfrn/2001/d1361in.htm>).

13. Earthquake information is provided by seismic stations operated by tsunami warning centers, the U.S. Geological Survey National Earthquake Information Center, and international sources. If the location and magnitude of an earthquake meet the known criteria for generating a tsunami, a warning is issued to emergency managers and the general public. The warning includes predicted tsunami arrival times at selected coastal communities within the geographic area defined by the maximum distance the tsunami could travel in a few hours. See <http://www.geophys.washington.edu/tsunami/general/warning/warning.html>.

14. P. Bak, *How Nature Works: The Science of Self-Organized Criticality*, Springer-Verlag, New York, 212 pp., 1996; J.S. Langer, J.M. Carlson, C.R. Myers, and B.E. Shaw, Slip complexity in dynamic models of earthquake faults, *Proc. Natl. Acad. Sci.*, **93**, 3825-3829, 1996; B.E. Shaw and J.R. Rice, Existence of continuum complexity in the elastodynamics of repeated fault ruptures, *J. Geophys. Res.*, **105**, 791-810, 2000.

15. The EarthScope program will deploy four technologies for observing the active tectonics and structure of the North American continent: (1) *USArray*, for high-resolution seismological imaging of the structure of the continental crust and upper mantle beneath the conterminous United States, Alaska, and adjacent regions; (2) *San Andreas Fault Observatory*

at Depth (SAFOD), for probing and monitoring the San Andreas fault at seismogenic depths; (3) *Plate Boundary Observatory* (PBO), for measuring deformations of the western United States using strainmeters and ultraprecise geodesy of the Global Positioning System; and (4) *Interferometric Synthetic Aperture Radar (InSAR) Initiative*, for using satellite-based InSAR to map surface deformations, especially the deformation fields associated with active faults and volcanoes (<http://www.earthscope.org>).

16. For example, see the web-based clearinghouse for the Nisqually earthquake set up by the University of Washington at <http://maximus.ce.washington.edu/~nisqually/>.

17. In the 1960s, studies of seismicity induced by reservoir loading (D. Simpson, W. Leith, and C. Scholz, Two types of reservoir-induced seismicity, *Bull. Seis. Soc. Am.*, **78**, 2025-2040, 1988) and injection of fluids into rock masses (J. Healy, W. Rubey, D. Griggs, and C.B. Raleigh, The Denver earthquakes, Colorado, USA, *Science*, **161**, 1301-1310, 1968) stimulated speculations that similar methods might be adaptable to relieving tectonic stresses prior to large earthquakes. In 1969, a National Research Council report put forward the argument not only for earthquake prediction, but also for earthquake control: "Prevention or control of destructive earthquakes must rank as a major goal of seismology" (National Research Council, *Seismology: Responsibilities and Requirements of a Growing Science, Part 1: Summary and Recommendations*, National Academy Press, Washington, D.C., p. 15, 1969). This can-do attitude was quickly replaced by the recognition that the stresses and strains in the brittle crust are such complex functions of space and time and so poorly known that any program aimed at beneficial control would be expensive, ineffective, and potentially dangerous, especially in regions of high seismic activity.

18. National Research Council, *National Science Education Standards*, National Academy Press, Washington, D.C., 262 pp., 1996.

19. Over the past two decades, undergraduate enrollment in the geosciences has declined from a high of 36,893 in 1983 to 10,454 in 2000. See American Geological Institute's survey of historical enrollment and degree information at <http://www.agiweb.org/career/enroll.html>.

20. According to the Bureau of Labor Statistics, a bachelor's degree in geology or geophysics is adequate for some entry-level jobs, but more job opportunities and better jobs with good advancement potential usually require at least a master's degree. See U.S. Department of Labor, Occupational Outlook Handbook, 2000-2001 edition, Bulletin 2520, <http://www.bls.gov/oco/ocos050.htm>.

2

Rise of Earthquake Science

Earthquakes have engaged human inquiry since ancient times, but the scientific study of earthquakes is a fairly recent endeavor. Instrumental recordings of earthquakes were not made until the last quarter of the nineteenth century, and the primary mechanism for the generation of earthquake waves—the release of accumulated strain by sudden slippage on a fault—was not widely recognized until the beginning of the twentieth century. The rise of earthquake science during the last hundred years illustrates how the field has progressed through a deep interplay among the disciplines of geology, physics, and engineering (*1*). This chapter presents a historical narrative of the development of the basic concepts of earthquake science that sets the stage for later parts of the report, and it concludes with some historical lessons applicable to future research.

2.1 EARLY SPECULATIONS

Ancient societies often developed religious and animistic explanations of earthquakes. Hellenic mythology attributed the phenomenon to Poseidon, the god of the sea, perhaps because of the association of seismic shaking with tsunamis, which are common in the northeastern Mediterranean (Figure 2.1). Elsewhere, earthquakes were connected with the movements of animals: a spider or catfish (Japan), a mole or elephant (India), an ox (Turkey), a hog (Mongolia), and a tortoise (Native America). The Norse attributed earthquakes to subterranean writhing of the imprisoned god Loki in his vain attempt to avoid venom dripping from a serpent's tooth.

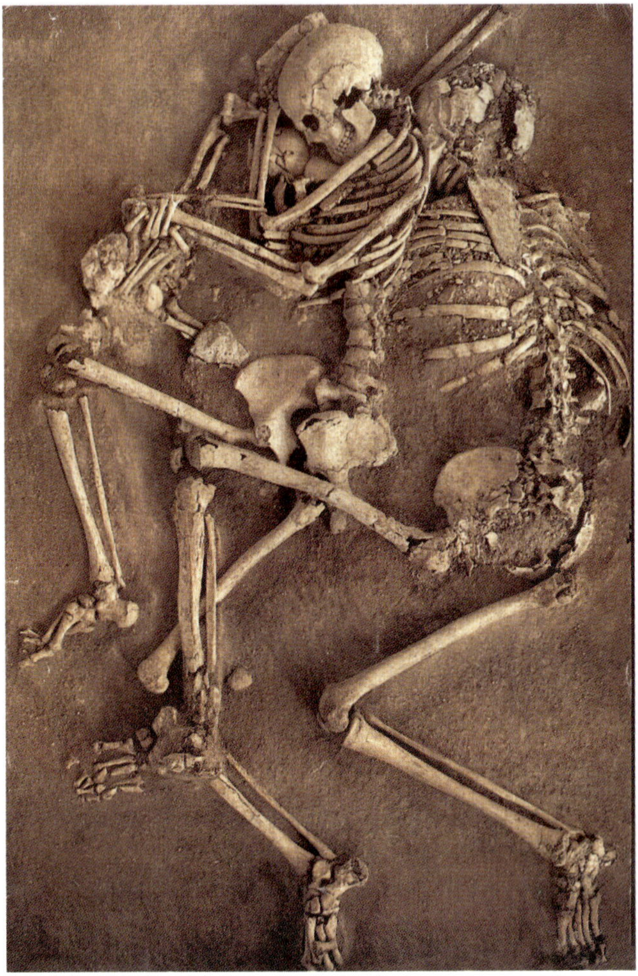

FIGURE 2.2 The remains of a family—a man, woman, and child (small skull visible next to the woman's skull)—crushed to death in A.D. 365 in the city of Kourian on the island of Cyprus when their dwelling collapsed on top of them during an earthquake. SOURCE: D. Soren, The day the world ended at Kourion, *National Geographic*, 30-53, July 1988; Copyright Martha Cooper.

zontal and whether they released vapor from the ground. He noted that "places whose subsoil is poor are shaken more because of the large amount of the wind they absorb." The correlation he observed between the intensity of the ground motions and the weakness of the rocks on which structures are built remains central to seismic hazard analysis.

2.2 DISCOVERY OF SEISMIC FAULTING

Aristotle's ideas and their variants persisted well into the nineteenth century (3). In the early 1800s, geology was a new scientific discipline, and most of its practitioners believed that volcanism caused earthquakes, both of which are common in geologically active regions. A vigorous adherent to the volcanic theory was the Irish engineer Robert Mallet, who coined the term seismology in his quantitative study of the 1857 earthquake in southern Italy (4). By this time, however, evidence had been accumulating that earthquakes are incremental episodes in the building of mountain belts and other large crustal structures, a process that geologists named tectonics. Charles Lyell, in the fifth edition of his seminal book *The Principles of Geology* (1837), was among the first to recognize that large earthquakes sometimes accompany abrupt changes in the ground surface (5). He based this conclusion on reports of the 1819 Rann of Cutch (Kachchh) earthquake in western India—near the disastrous January 26, 2001, Bhuj earthquake—and, in later editions, on the Wairarapa, New Zealand, earthquake of 1855. A protégé of Lyell's, Charles Darwin, experienced a great earthquake while visiting Chile in 1835 during his voyages on the *H.M.S. Beagle*. Following the earthquake, he and Captain FitzRoy noticed that in many places the coastline had risen several meters, causing barnacles to die because of prolonged exposure to air. He also noticed marine fossils in sediments hundreds of meters above the sea and concluded that seismic uplift was the mechanism by which the mountains of the coast had risen. Darwin applied James Hutton's principle of uniformitarianism—"the present is the key to the past"—and inferred that the mountain range had been uplifted incrementally by many earthquakes over many millennia (6).

Fault Slippage as the Geological Cause of Earthquakes

The leap from these observations to the conclusion that earthquakes result from slippage on geological faults was not a small one. The vast majority of earthquakes are accompanied by no surface faulting, and even when such ruptures had been found, questions arose as to whether the ground breaking shook the Earth or the Earth shaking broke the ground. Moreover, the methodology for mapping fault displacements and understanding their relationships to geological deformations, the discipline of structural geology, had not yet been systematized. A series of field studies—by G.K. Gilbert in California (1872), A. McKay in New Zealand (1888), B. Koto in Japan (1891), and C.L. Griesbach in Baluchistan (1892)—demonstrated that fault motion generates earthquakes, thereby documenting that the surface faulting associated with each of these earthquakes was consistent with the long-term, regional tectonic deformation that geologists had mapped (Figure 2.3).

FIGURE 2.3 Photograph of the 1891 Nobi (Mino-Owari) earthquake scarp at Midori, taken by B. Koto, a professor of geology at the Imperial University of Tokyo. Based on his geological investigations, Koto concluded, "The sudden elevations, depressions, or lateral shiftings of large tracts of country that take place at the time of destructive earthquakes are usually considered as the effects rather than the cause of subterranean commotion; but in my opinion it can be confidently asserted that the sudden formation of the 'great fault of Neo' was the actual cause of the great earthquake." This photograph appeared in *The Great Earthquake in Japan, 1891*, published by the Seismological Society of Japan, which was one of the first comprehensive scientific reports of an earthquake. The damage caused by the Nobi earthquake motivated Japan to create an Earthquake Investigation Committee, which set up the first government-sponsored research program on the causes and effects of earthquakes. SOURCE: J. Milne and W.K. Burton, *The Great Earthquake in Japan, 1891*, 2nd ed., Lane, Crawford & Co., Yokohama, Japan, 69 pp. + 30 plates, 1892.

Among the geological investigations of this early phase of tectonics, Gilbert's studies in the western United States were seminal for earthquake science. From the new fault scarps of the 1872 Owens Valley earthquake, he observed that the Sierra Nevada, bounding the west side of the valley, had moved upward and away from the valley floor. This type of faulting was consistent with his theory that the Basin and Range Province between the Sierra Nevada and the Wasatch Mountains of Utah had been formed by tectonic extension (7). He also recognized the similarity of the

FIGURE 2.4 Aerial view, Salt Lake City, showing the scarps of active normal faults of the Wasatch Front, first recognized by G.K. Gilbert. SOURCE: Utah Geological Survey.

Owens Valley break to a series of similar piedmont scarps along the Wasatch Front near Salt Lake City (Figure 2.4). By careful geological analysis, he documented that the Wasatch scarps were probably caused by individual fault movements during the recent geological past. This work laid the foundation for paleoseismology, the subdiscipline of geology that employs features of the geological record to deduce the fault displacement and age of individual, prehistoric earthquakes (*8*).

Geological studies were supplemented by the new techniques of geodesy, which provide precise data on crustal deformations. Geodesy grew out of two practical arts, astronomical positioning and land surveying, and became established as a field of scientific study in the mid-nineteenth century. One of the first earthquakes to be measured geodetically was the Tapanuli earthquake of May 17, 1892, in Sumatra, which happened during a triangulation survey by the Dutch Geodetic Survey. The surveyor in charge, J.J.A. Müller, discovered that the angles between the survey monuments had changed during the earthquake, and he concluded that a horizontal displacement of at least 2 meters had occurred along a structure later recognized to be a branch of the Great Sumatran fault. R.D. Oldham of the Geological Survey of India inferred that the changes in survey angles and elevations following the great Assam earthquake of June 12, 1897, were due to co-seismic tectonic movements. C.S. Middlemiss reached the same conclusion for the Kangra earthquake of April 4, 1905, also in the well-surveyed foothills of the Himalaya (*9*).

Mechanical Theories of Faulting

The notion that earthquakes result from fault movements linked the geophysical disciplines of seismology and geodesy directly to structural geology and tectonics, whose practitioners sought to explain the form, arrangement, and interrelationships among the rock structures in the upper part of the Earth's crust. Although Hutton, Lyell, and the other founders of the discipline of geology had investigated the great vertical deformations required by the rise of mountain belts, the association of these deformations with large horizontal movements was not established until the latter part of the nineteenth century (*10*). Geological mapping showed that some horizontal movements could be accommodated by the ductile folding of sedimentary strata and plastic distortion of igneous rocks, but that much of the deformation takes place as *cataclastic flow* (i.e., as slippage in thin zones of failure in the brittle materials that make up the outer layers of the crust). Planes of failure on the larger geological scales are referred to as faults, classified as *normal*, *reverse*, or *strike-slip* according to their orientation and the direction of slip (Figure 2.5).

In 1905, E.M. Anderson (*11*) developed a successful theory of these faulting types, based on the premises that one of the principal compressive stresses is oriented vertically and that failure is initiated according to a rule published in 1781 by the French engineer and mathematician Charles Augustin de Coulomb. The Coulomb criterion states that slippage occurs when the shear stress on a plane reaches a critical value τ_c that depends linearly on the effective normal stress σ_n^{eff} acting across that plane:

$$\tau_c = \tau_0 + \mu\sigma_n^{eff}, \tag{2.1}$$

where τ_0 is the (zero-pressure) *cohesive strength* of the rock and μ is a dimensionless number called the *coefficient of internal friction*, which usually lies between 0.5 and 1.0. Anderson's theory made quantitative predictions about the angles of shallow faulting that fit the observations rather well (except in regions where fault planes were controlled by strength anisotropy like sedimentary layering). However, it could not explain the existence of large, nearly horizontal thrust sheets that formed at deeper structural levels in many mountain belts. Owing to the large lithostatic load, the total normal stress σ_n acting on such fault planes was

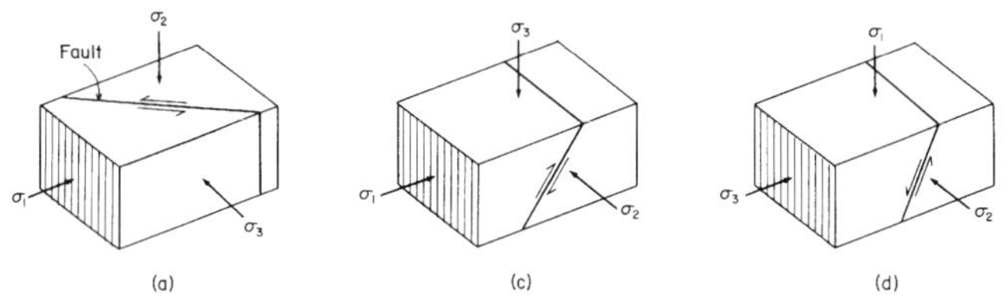

FIGURE 2.5 Fault types showing principal stress axes and Coulomb angles. For a typical coefficient of friction in rocks ($\mu \approx 0.6$), the Coulomb criterion (Equation 2.1) implies that a homogeneous, isotropic material should fail under triaxial stress ($\sigma_1 > \sigma_2 > \sigma_3$) along a plane that contains the σ_2 axis and lies at an angle about 30° to the σ_1 direction. According to Anderson's theory, normal faults should therefore occur where the vertical stress σ_V is the maximum principal stress σ_1, and the initial dips of these extensional faults should be steep (about 60°); reverse faults ($\sigma_V = \sigma_3$) should initiate as thrusts with shallow dips of about 30°, and strike-slip faults ($\sigma_V = \sigma_2$) should develop as vertical planes striking at about 30° to the σ_1 direction. SOURCE: Reprinted from K. Mogi, *Earthquake Prediction*, Academic Press, Tokyo, 355 pp., 1985, Copyright 1985 with permission from Elsevier Science.

much greater than any plausible tectonic shear stress, so it was difficult to see how failure could happen. M.K. Hubbert and W.W. Rubey resolved this quandary in 1959 (*12*) by recognizing that the effective normal stress in the Coulomb criterion should be the difference between σ_n and the fluid pressure P_f:

$$\sigma_n^{eff} = \sigma_n - P_f. \tag{2.2}$$

They proposed that overthrust zones were *overpressurized*; that is P_f in these zones was substantially greater than the pressure expected for hydrostatic equilibrium and could approach lithostatic values (*13*). Hence, σ_n^{eff} could be much smaller than σ_n. Overpressurization may explain why some faults, such as California's San Andreas, appear to be exceptionally weak.

Elastic Rebound Model

When a 470-kilometer segment of the newly recognized San Andreas rift ruptured across northern California in 1906 (Box 2.2, Figures 2.6 and 2.7), both geologists and engineers jumped at the opportunity to observe first-hand the effects of a major earthquake. Three days after the earthquake, while the fires of San Francisco were still smoldering, California Governor George C. Pardee appointed a State Earthquake Investigation

BOX 2.2 San Francisco, California, 1906

At approximately 5:12 a.m. local time on April 18, 1906, a small fracture nucleated on the San Andreas fault at a depth of about 10 kilometers beneath the Golden Gate (*20*). The rupture expanded outward, quickly reaching its terminal velocity of about 2.5 kilometers per second (5600 miles per hour). Its upper front broke through the ground surface at the epicenter within a few seconds, and its lower front decelerated as it spread downward into the more ductile levels of the middle crust, while the two sides continued to propagate in opposite directions along the San Andreas. Near the epicenter, the rupture displaced the opposite sides of the fault rightward by an average of about 4 meters (a right-lateral strike-slip). On the southeastern branch, the total slip diminished as the rupture traveled down the San Francisco peninsula and vanished 100 kilometers away from the epicenter. To the northwest, the fracture ripped across the neck of the Point Reyes peninsula and entered Tomales Bay, where the total slip increased to 7 meters, sending out seismic waves that damaged Santa Rosa, Fort Ross, and other towns of the northern Coast Ranges. The rupture continued up the coast to Point Arena, where it went offshore, eventually stopping near a major bend in the fault at Cape Mendocino (Figure 2.6). At least 700 people were killed, perhaps as many as 3000, and many buildings were severely damaged.[1] In San Francisco, the quake ignited at least 60 separate fires, which burned unabated for three days, consuming 42,000 buildings and destroying a considerable fraction of the West Coast's largest city.

[1] G. Hansen and E. Condon, *Denial of Disaster: The Untold Story of the San Francisco Earthquake and Fire of 1906*, Cameron and Co., San Francisco, 160 pp., 1989. These authors present evidence that the scale of the 1906 disaster, in terms of both property destroyed and lives lost, was deliberately underreported to protect economic interests. They also argue that the damage directly caused by the earthquake was preferentially reported as fire damage, because the latter was more likely to be covered by insurance.

Commission, headed by Berkeley Professor Andrew C. Lawson, to coordinate a wide-ranging set of scientific and engineering studies (*14*). The first volume of the Lawson Report (1908) compiled reports by more than 20 specialists on a variety of observations: the geological setting of the San Andreas; the fault displacements inferred from field observations and geodetic measurements; reports of the arrival time, duration, and intensity of the seismic waves; seismographic recordings from around the world; and detailed surveys of the damage to structures throughout Northern California. The latter demonstrated that the destruction was closely related to building design and construction, as well as to local geology. The intensity maps of San Francisco clearly show that some of the strongest shaking occurred in the soft sediment of China Basin and in the present Marina district, two San Francisco neighborhoods that would be severely damaged in the Loma Prieta earthquake some 83 years later (*15*). This interdiscipli-

FIGURE 2.6 San Andreas fault system in California, showing the extent of the surface rupture, damage area, and felt area of the 1906 earthquake. SOURCE: T.H. Jordan and J.B. Minster, Measuring crustal deformation in the American west, *Sci. Am.*, **256**, 48-58, 1988. Illustration by Hank Iken.

FIGURE 2.7 Panoramic view of the ruins of San Francisco after the April 1906 earthquake and fire, viewed from the Stanford Mansion site. SOURCE: Lester Guensey, Library of Congress, Prints and Photographs Division, [Lc-USZ62-123408 DLC].

nary synthesis is still being mined for information about the 1906 earthquake and its implications for future seismic activity (*16*).

Professor Henry Fielding Reid of Johns Hopkins University wrote the second volume of the Lawson Report (1910), presenting his celebrated elastic rebound hypothesis. Reid's 1911 follow-up paper (*17*) summarized his theory in five propositions:

- The fracture of the rocks, which causes a tectonic earthquake, is the result of elastic strains, greater than the strength of the rock can withstand, produced by the relative displacements of neighboring portions of the earth's crust.
- These relative displacements are not produced suddenly at the time of the fracture, but attain their maximum amounts gradually during a more or less long period of time.
- The only mass movements that occur at the time of the earthquake are the sudden elastic rebounds of the sides of the fracture towards positions of no elastic strain; and these movements extend to distances of only a few miles from the fracture.
- The earthquake vibrations originate in the surface of the fracture; the surface from which they start is at first a very small area, which may quickly become very large, but at a rate not greater than the velocity of compressional elastic waves in rock.
- The energy liberated at the time of an earthquake was, immediately before the rupture, in the form of energy of elastic strain of the rock.

Today all of these propositions are accepted with only minor modifications (*18*). Although some geologists, for at least the latter half of the nineteenth century, had considered the notion that most large earthquakes result from fault slippage, Reid's hypothesis was boldly revolutionary. The horizontal tectonic displacements he postulated had no well-estab-

lished geologic basis, for example, and they would remain mysterious until the plate-tectonic revolution of the 1960s (*19*).

2.3 SEISMOMETRY AND THE QUANTIFICATION OF EARTHQUAKES

In 1883, the English mining engineer John Milne suggested that "it is not unlikely that every large earthquake might with proper appliances be recorded at any point of the globe." His vision was fulfilled six years later when Ernst von Rebeur-Paschwitz recorded seismic waves on delicate horizontal pendulums at Potsdam and Wilhemshaven in Germany from the April 17, 1889, earthquake in Tokyo, Japan. By the turn of the century, the British Association for the Advancement of Science was sponsoring a global network of more than 40 stations, most equipped with instruments of Milne's design (*21*); other deployments followed, expanding the coverage and density of seismographic recordings (*22*). Working with records of the great Assam earthquake of June 12, 1897, Oldham identified three basic wave types: the small primary (*P* or compressional) and secondary (*S* or shear) waves that traveled through the body of the Earth and the "large" (*L*) waves that propagated across its outer surface (*23*).

Hypocentral Locations and Earth Structure

Milne investigated the velocities of the *P*, *S*, and *L* waves by plotting their travel times as a function of distance for earthquakes whose location had been fixed by local observations. From curves fit to these travel times, he could then determine the distance from the observing stations to an event with an unknown epicenter, and he could fix its location from the intersection of arcs drawn at the estimated distance from three or more such stations. By applying this simple technique, he and others began to compile catalogs of instrumentally determined earthquake epicenters (*24*).

Improved locations meant that seismologists could use the travel time of the seismic waves to develop better models of the variations of wave velocities with depth, which in turn could be used to improve the location of the earthquake's initial radiation (hypocenter), as well as its origin time. This cycle of iterative refinement of Earth models and earthquake locations, along with advances in the distribution and quality of the seismometer networks, steadily decreased the uncertainties in both. It also led to some major discoveries. In 1906, Oldham presented the first seismological evidence that the Earth had a central core, and in 1914, Beno Gutenberg obtained a relatively precise depth (about 2900 kilometers) to the boundary between the core and the solid-rock shell or mantle (German for coat) surrounding it. From regional recordings of the 1909 Croatian earthquake, the Serbian seismologist Andriji Mohorovicic discovered the sharp increase in seismic velocities that bears his name, often abbreviated the Moho, which separates the lighter, more silica-rich crust from the ultramafic (iron- and magnesium-rich) mantle.

After Milne's death in 1913, H.H. Turner, an Oxford professor, took over the determination of earthquake hypocenters and origin times. Turner's efforts to compile earthquake data systematically led, after the First World War, to the founding of the International Seismological Summary (ISS) (*25*). While preparing the ISS bulletins, Turner (1922) noticed some events with anomalous travel times, which he proposed had hypocenters much deeper than that of typical earthquakes. In 1928, Kiyoo Wadati established the reality of such "deep-focus" earthquakes as much as 700 kilometers beneath volcanic arcs such as Japan and the Marianas, and he subsequently delineated planar regions of seismicity (now called Wadati-Benioff zones) extending from the ocean trenches at the face of the arcs down to these deep events. The Danish seismologist Inge Lehmann discovered the Earth's inner core in 1936; this "planet within a planet" has since been shown to be a solid metallic sphere two-thirds the size of the Moon at the center of the liquid iron-nickel outer core. By the time Harold Jeffreys and Keith Bullen finalized their travel-time tables in 1940, the Earth's internal structure was known well enough to estimate the hypocenter of large earthquakes with a standard error often less than 10 kilometers and origin time with a standard error of less than 2 seconds (*26*).

Earthquake Magnitude and Energy

The next important step in the development of instrumental seismology was the quantification of earthquake size. Maps of seismic damage were made in Italy as early as the late eighteenth century. In the 1880s, M.S. Rossi of Italy and F. Forel of Switzerland defined standards for grading qualitative observations by integer values that increase with the amount of shaking and disruption. Versions of their "intensity scale," as modified by G. Mercalli and others, are still used to map intensity after strong events (*27*), but they do not measure the intrinsic size of an earthquake, nor can they be applied to events that humans have not felt and observed (i.e., almost all earthquakes). The availability of instrumental recordings and the desire to standardize the seismological bulletins motivated seismologists to estimate the intrinsic size of earthquakes by measuring the amplitude of the seismic waves at a station and correcting them for propagation effects, such as the spreading out of wave energy and its attenuation by internal friction. Several such scales were developed, including one by Wadati in 1931, but the most popular and successful schemes were based on the standard magnitude scale that Charles Richter of Caltech published in 1935.

Richter recognized that seismographic amplitude provides a first-order measure of the radiated energy but that these data are highly variable depending on the type of seismograph, distance to the earthquake, and local site conditions. To normalize for these factors, he considered only southern California earthquakes recorded on Caltech's standardized network of Wood-Anderson torsion seismometers (*28*). He defined the local magnitude scale for such events by the formula

$$M_L = \log A - \log A_0, \tag{2.3}$$

where A is the maximum amplitude of the seismic trace on the standard seismogram; A_0 is the amplitude at that same distance for a reference earthquake with $M_L = 0$; and all logarithms are base 10. He fixed the reference level A_0 by specifying a magnitude-zero earthquake as an event with an amplitude of 1 micron on a standard Wood-Anderson seismogram at a distance of 100 kilometers (*29*). An earthquake of magnitude 3.0 thus had an amplitude of 1 millimeter at 100 kilometers, which was about the smallest level measurable on this type of pen-written seismogram (*30*). Corrections for recordings made at other distances were determined empirically and incorporated into a simple graphic procedure.

During the next decade, Richter and Gutenberg refined and extended the methodology to include earthquakes recorded by various instrument types and at teleseismic distances. Gutenberg published a series of papers in 1945 detailing the construction of magnitude scales based on the maximum amplitude of long-period surface waves (M_S), which could be applied to shallow earthquakes at any distance, and teleseismic body waves (m_b), which could be applied to earthquakes too deep to excite ordinary surface waves. To the extent possible, these scales were calibrated to agree with Richter's definition of magnitude, although various discrepancies became apparent as experience accumulated (*31*). In 1956, Gutenberg and Richter used surface-wave magnitudes as the basis for an energy formula (with E in joules):

$$\log E = 1.5M_S + 4.8. \tag{2.4}$$

This relationship implies that earthquake energies vary over at least 12 orders of magnitude, a much larger range than previously supposed. It also allows comparison with a new source of seismic energy, the atomic bomb. Seismic signals were recorded by regional stations from the first Trinity test in 1945 (*32*) and an underwater explosion at Bikini atoll in July of 1946, the Baker test; both generated compressional waves observed at teleseismic distances. The energy released from Baker, a Hiroshima-type device, was about 8×10^{13} joules. Assuming a 1 percent seismic efficiency, Gutenberg and Richter calculated a body-wave magnitude of 5.1 from their revised energy formulas, which agreed reasonably well with their observed value of 5.3 (*33*). Seismology thus embarked on a new mission, the detection and measurement of nuclear explosions. By 1959, the reliable identification of small underground nuclear explosions had become the primary technical issue confronting the verification of a comprehensive nuclear test ban treaty, and the resulting U.S. program in nuclear explosion seismology, Project Vela Uniform, motivated important developments in earthquake science (*34*).

Seismicity of the Earth

Observational and theoretical research in Japan, North America, and Europe during the 1930s markedly improved seismogram interpretation. Seismographic readings from an increasingly dense global network of stations were compiled and published regularly in the International Seismological Summary, an invaluable source of data for refining event locations. By 1940, the ability to locate earthquakes was sufficiently advanced to allow the systematic analysis of global seismicity. Gutenberg and Richter produced their first synthesis in 1941, based on their relocation of hypocenters and estimation of magnitudes (*35*). They used focal depth to formalize the nomenclature of shallow (less than 70 kilometers), intermediate (70 to 300 kilometers), and deep (greater than 300 kilometers) earthquakes; they confirmed that Wadati's depth of 300 kilometers for the transition from intermediate focus to deep focus was a minimum in earthquake occurrence rate, and they showed a sharp cutoff in global seismicity at about 700 kilometers. Their classic treatise *Seismicity of the Earth* documented a number of observations about the geographic distribution of seismicity that helped to establish the plate-tectonic theory: (1) most large earthquakes occur in narrow belts that outline a set of stable blocks, the largest comprising the central and western Pacific basin; (2) nearly all intermediate and deep seismicity is associated with planar zones that dip beneath volcanic island arcs and arc-like orogenic (mountain-building) structures; and (3) seismicity in the ocean basins is concentrated near the crest of the oceanic ridges and rises.

Gutenberg and Richter also discussed a series of issues related to the size distribution and energy release of earthquakes. They found that the total number of earthquakes N greater than some magnitude M in a fixed time interval obeyed the relationship (*36*)

$$\log N = a - bM, \tag{2.5}$$

where a and b are empirical constants. Equation 2.5 is equivalent to $N = N_0 10^{-bM}$; in this form, $N_0 = 10^a$ is the total number of earthquakes whose magnitude exceeds zero. This is an extrinsic parameter that depends on the temporal interval and spatial volume considered, whereas b describes an exponential fall-off in seismicity with magnitude, a parameter more intrinsic to the faulting process. For a global distribution of shallow shocks, they estimated $b \approx 0.9$, so that a decrease in one unit of magnitude gives an eightfold increase in frequency. Subsequent studies have confirmed that regional seismicity typically follows these Gutenberg-Richter statistics, with b values ranging from 0.5 to 2.0. Because spatial extent and energy release grow exponentially with magnitude, Gutenberg-Richter statistics imply a power-law scaling between frequency and size (*37*).

Gutenberg and Richter noted that even though small earthquakes are much more common than large events, the big ones dominate the energy distribution. According to their energy formula (Equation 2.4), an increase by one magnitude unit gives a 32-fold increase in energy, so that a summation over all events still implies that the total energy increases about a factor of 4 per unit magnitude. They used this type of calculation to dispel the popular notion that minor shocks can function as a "safety valve" to delay a great earthquake. They found that the total annual energy release from all earthquakes was only a fraction of the heat flow from the solid Earth, estimated a few years earlier by the British geophysicist Edward Bullard (*38*). This calculation was consistent with the idea that earthquakes were a form of work done by a thermodynamically inefficient heat engine operating in the Earth's interior.

Earthquakes as Dislocations

Although it was known that earthquakes usually originate from sudden movements across a fault, the actual mechanics of the rupture process remained obscure (*39*), and a quantitative theory of how this dislocation forms and generates elastic waves through the spontaneous action of material failure was completely lacking.

Progress toward a dynamic description of faulting began in Japan, where the high density of seismic stations allowed seismologists to recognize coherent geographic patterns in the seismic radiation. They mapped the first-arriving *P*-wave pulses into regions of compression (first motion up) and dilatation (first motion down), separated by nodal lines where the initial arrival was very weak (*40*). Stimulated by these observations, H. Nakano formulated, in 1923, the problem of deducing the orientation of the faulting from the pattern of first motions (*41*). He expressed the radiation from an instantaneous event in terms of a system of dipolar forces at the earthquake hypocenter. The results appeared to be ambiguous, because the observed "beachball" radiation pattern of *P* waves (Figure 2.8) could be explained either by a single couple of such forces or by a double couple. A 40-year controversy ensued regarding which of these models is physically correct, until understanding began to grow in the 1960s of the definitive theoretical conclusion that a fault dislocation is equivalent to a double couple (*42*).

The dislocation model also shed light on the dynamic coupling between the brittle, seismogenic layer and its ductile, aseismic substrate. Geodetic data from the 1906 earthquake had shown that the process of strain accumulation and release was concentrated near the fault. In 1961, Michael Chinnery (*43*) showed that the displacement from a uniform vertical dislocation decays to half its maximum value at a horizontal distance equal to the depth of faulting, and he applied this result to estimate a rupture depth of 2 to 6 kilometers for the 1906 earthquake. Later workers used Chinnery's model to provide a physical model for Reid's elastic rebound theory, arguing that the deformation before the 1906 earthquake was due to nearly steady slip at depth on the San Andreas fault, while the shallow part of the fault slipped enough in the earthquake itself to catch up, at least approximately, with the lower fault surface.

2.4 PLATE TECTONICS

Alfred Wegener, a German meteorologist, first put forward his theory of continental drift in 1912 (*44*). He marshaled geological arguments that the continents had once been joined as a supercontinent he named Pangea, but he imagined that they moved apart at very rapid rates—tens of meters per year (*45*)—like buoyant, granitic ships plowing through a denser, basaltic sea of oceanic crust. Jeffreys showed in 1924 that this idea, as well as the dynamic mechanisms Wegener proposed for causing drift (e.g., westward drag on the continents by lunar and solar tidal forces), were

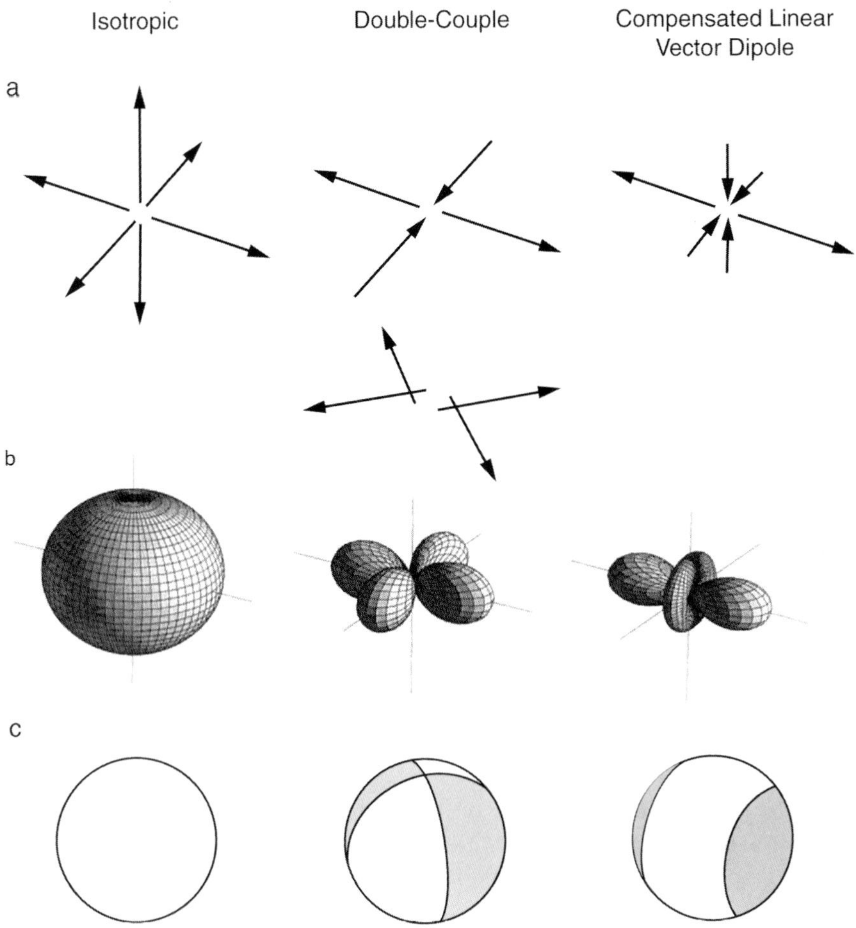

FIGURE 2.8 Graphical representations of three basic types of seismic point sources (*left to right*): isotropic, double couple, and compensated linear vector dipole. (a) Principal axis coordinates of equivalent force systems. (b) Compressional wave radiation patterns. (c) Curves of intersection of nodal surfaces with the focal sphere. Each of these source types is a specialization of the seismic moment tensor **M**. SOURCE: Modified from B.R. Julian, A.D. Miller, and G.R. Foulger, Non-double-couple earthquakes: 1. Theory, *Rev. Geophys.*, **36**, 525-549, 1998. Copyright 1998 American Geophysical Union. Modified by permission of American Geophysical Union.

physically untenable, and most of the geological community discredited Wegener's hypothesis (*46*). In the 1930s, however, the South African geologist A.L. du Toit assembled an impressive set of additional geologic data that supported continental drift beginning in the Mesozoic Era, and the empirical case in its favor was further strengthened when E. Irving and S.K. Runcorn published their compilations of paleomagnetic pole positions in 1956. The paleomagnetic data indicated drifting rates on the order of centimeters per year, several orders of magnitude slower than Wegener had hypothesized. Within the next 10 years, the key elements of plate tectonics were put in place. The main conceptual breakthrough was the recognition that on a global scale, the amount of new basaltic crust generated by *seafloor spreading*—the bilateral separation of the seafloor along the mid-ocean ridge axis—is balanced by *subduction*—the thrusting of basaltic crust into the mantle at the oceanic trenches.

Seafloor Spreading and Transform Faults

Submarine mountain ranges, mapped in the 1870s, came into focus as a world-encircling system of extensional tectonics after the Second World War. Marine geologists Maurice Ewing and Bruce Heezen, based at Columbia University, mapped a narrow, nearly continuous "median valley" along the ridge crests in the Atlantic, Indian, and Antarctic Oceans, which they inferred to be a locus of active rifting and the source of the mid-ocean seismicity that Gutenberg and Richter had documented (*47*). In the early 1960s, Harry H. Hess of Princeton University and Robert S. Dietz of the Scripps Institution of Oceanography advanced the concept of seafloor spreading to account for observations of such phenomena as the paucity of deep-sea sediments and the tendency for oceanic islands to subside with time (*48*). In his famous 1960 "geopoetry" preprint, Hess noted that crustal creation at the Mid-Atlantic Ridge implies a more plausible mechanism for continental drift than the type originally envisaged by Wegener: "The continents do not plow through oceanic crust impelled by unknown forces; rather they ride passively on mantle material as it comes to the surface at the crest of the ridge and then moves laterally away from it."

Two distinct predictions based on the theory of seafloor spreading were confirmed in 1966. The first involved the striped patterns of magnetic anomalies being mapped on the flanks of the mid-ocean ridges. In 1963, F. Vine and D. Matthews suggested that such anomalies record the reversals of the Earth's magnetic field through remnant magnetization frozen into the oceanic rocks as they diverge and cool away from the ridge axis. These geomagnetic "tape recordings" were shown to be symmetric about this axis and consistent with the time scale of geomagnetic reversals worked out from lava flows on land; moreover, the spreading speed measured from the magnetometer profiles in the Atlantic was found to be nearly constant and in agreement with the average opening rate obtained from the paleomagnetic data on continental rocks (*49*).

The second confirmation came from the study of earthquakes on the mid-ocean ridges. Horizontal displacements as large as several hundred kilometers had been documented for strike-slip faults on land, by H.W. Wellman for the Alpine fault in New Zealand and by M. Hill and T.W. Dibblee for the San Andreas (*50*), but even larger displacements—greater than 1000 kilometers—could be inferred from the offsets of magnetic anomalies observed across fracture zones in the Pacific Ocean (*51*). In a 1965 paper that laid out the basic ideas of the plate theory, the Canadian geophysicist J. Tuzo Wilson recognized that fracture zones were relics of faulting that was active only along those portions connecting two segments of a spreading ridge, which he called *transform faults* (*52*). His model implied that the sense of motion across a transform fault would be the opposite to the ridge axis offset. The seismologist Lynn Sykes of Columbia University verified this prediction in an investigation of the focal mechanisms from transform-fault earthquakes (Figure 2.9).

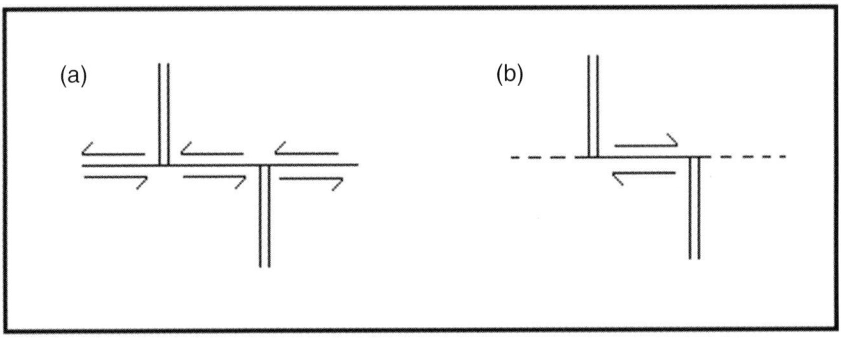

FIGURE 2.9 Two interpretations of two ridge segments offset by a fault. (a) In the pre-plate-tectonic interpretation, the two ridge segments (double lines) would have been offset in a sinistral (left-lateral) sense along the fault (solid line). Earthquakes should occur along the entire fault line. (b) According to the plate-tectonic theory, the two ridge segments were never one continuous feature. Spreading of the seafloor away from the ridges causes dextral (right-lateral) motions only along the section of the fault between the two ridge segments (the transform fault). The extensions of the faults beyond the ridge segments, the fracture zones (dashed lines), are aseismic. Earthquake observations conclusively demonstrated the validity of interpretation (b) for the mid-ocean ridges. SOURCE: Modified from L. Sykes, Mechanism of earthquakes and nature of faulting on mid-ocean ridges, *J. Geophys. Res.*, **72**, 2131-2153, 1967. Copyright 1967 American Geophysical Union. Reproduced by permission of American Geophysical Union.

Sykes's study was facilitated by the rapidly accumulating collection of seismograms, readily available on photomicrofiche, from the new World Wide Standardized Seismographic Network (WWSSN) set up under Project Vela Uniform. These high-quality seismometers had good timing systems, fairly broad bandwidth, and a nearly uniform response to ground motions, and they were installed and permanently staffed around the world at recording sites with relatively low background noise levels (*53*). The high density of stations allowed smaller events to be located precisely and their focal mechanisms to be determined more rapidly and accurately than ever before. One result was much more accurate maps of global seismicity, which clearly delineated the major plate boundaries, as well as the Wadati-Benioff zones of deep seismicity (Figure 2.10).

Subduction of Oceanic Lithosphere

If the Earth's surface area is to remain constant, then the creation of new oceanic crust at the ridge crests necessarily implies that some old crust is being recycled back into the mantle. This inference was consistent with the theories of mantle convection that attributed the volcanic arcs and linear zones of compressive orogenesis to convective downwellings (*54*), which David Griggs had discussed as early as 1939, calling it "a convection cell covering the whole of the Pacific basin, comprising sinking peripheral currents localizing the circum-Pacific mountains and rising currents in the center" (*55*). Griggs belonged to a growing group of "mobilists" who espoused the view that the Earth's solid mantle is actively convecting like a fluid heated from below, causing large horizontal displacements of the crust, including continental drift (*56*). The alternative, expanding-Earth hypothesis states that the planetary radius is increasing, perhaps owing to radioactive heating or possibly to a universal decrease in gravitational strength with time, and that seafloor spreading accommodates the associated increase in surface area (*57*). Thus, new oceanic crust created at the spreading centers does not have to be balanced by the sinking of old crust back into the mantle.

Because of this controversy, as well as the geologic complexity of the problem, subduction was the last piece of the plate-tectonic puzzle to fall into place (*58*). While the system of oceanic ridges and transform faults fit neatly together in seafloor spreading, the compressional arcs and mountain belts juxtaposed all types of active faulting, which continued to baffle geologists. Benioff had pointed out the asymmetric polarity of the island arcs, correctly proposing that the deep oceanic trenches are surface expressions of giant reverse faults (*59*). Robert Coats used this idea to account for the initial formation of island arcs such as the Aleutians and the geochemical data bearing on the development of the

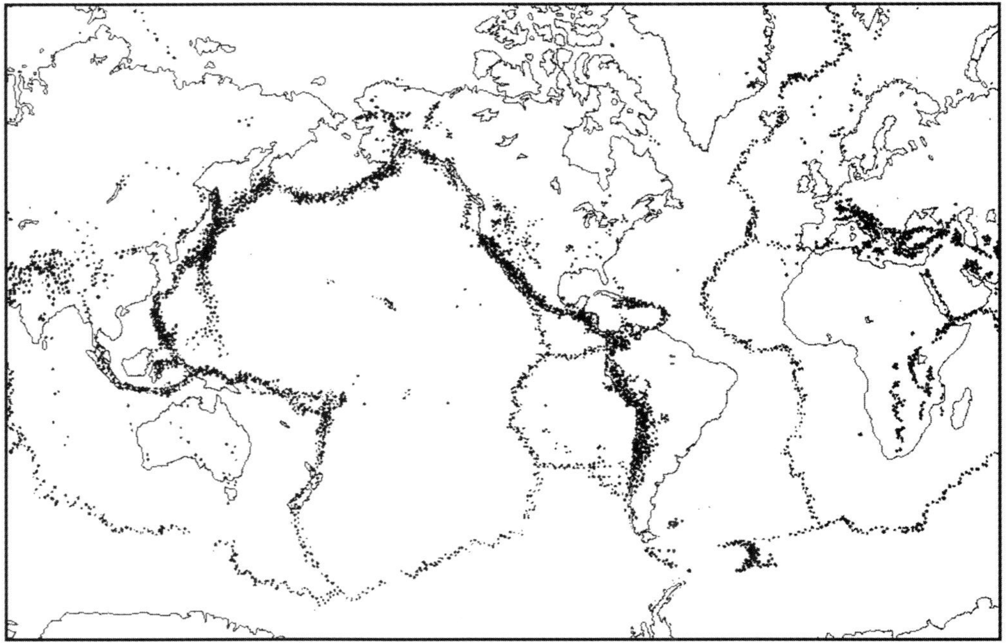

FIGURE 2.10 Locations of earthquake epicenters with body-wave magnitudes greater than 4.5 for the period 1960-1967, which incorporated the improved data of the WWSSN. Computer-generated epicenter maps such as this one first became available in about 1964 and were used to delineate the major plate boundaries and the descending slabs of cold oceanic lithosphere. Dots are epicenters reported by the U.S. Coast and Geodetic Survey. SOURCE: B.L. Isacks, J. Oliver, and L.R. Sykes, Seismology and the new global tectonics, *J. Geophys. Res.*, **73**, 5855-5899, 1968. Copyright 1968 American Geophysical Union. Reproduced by permission of American Geophysical Union.

andesitic stratovolcanoes characteristic of these arcs (*60*). Benioff's model was based on several misconceptions, however, including the assumption that intermediate- and deep-focus seismicity could be explained by extrapolating trench-type reverse faulting into the mid-mantle transition zone. In fact, the focal mechanism of most earthquakes with hypocenters deeper than 70 kilometers does not agree with Benioff's model of reverse faulting (*61*).

The definitive evidence for "thrust tectonics" finally arrived in the form of the great 1964 Alaska earthquake (Box 2.3). The enormous energy released in this event (~3×10^{18} joules) set the Earth to ringing like a bell and allowed precise studies of the terrestrial free oscillations, whose period might be as long as 54 minutes (*62*). A permanent strain of 10^{-8} was recorded by the Benioff strainmeter on Oahu, more than 4000 kilometers away, consistent with a fault-dislocation model of the earthquake (*63*). However, the high-amplitude waves drove most of the pendulum seismometers offscale (*64*). Moreover, field geologists could not find the fault; all ground breaks were ascribable to secondary effects. What they did observe was a systematic pattern of large vertical motions—uplifts as high as 12 meters and depressions as deep as 2.3 meters, which could easily be mapped along the rugged coastlines by observing the displacement of beaches and the stranded colonies of sessile marine organisms such as barnacles (just as Darwin had done for the 1835 Chile earthquake). By combining this pattern with the seismological and geodetic data, they inferred that the rupture represented the slippage of the Pacific Ocean crust beneath the continental margin of southern Alaska along a huge thrust fault. Geologist George Plafker concluded that "arc structures are sites of down-welling mantle convection currents and that planar seismic zones dipping beneath them mark the zone of shearing produced by downward-moving material thrust against a less mobile block of the crust and upper mantle" (*65*). By connecting the Alaska megathrust with the more steeply inclined plane of deeper seismicity under the Aleutian Arc, Plafker articulated one of the central tenets of plate tectonics.

Plafker's conclusions were bolstered by more accurate sets of focal mechanisms that William Stauder and his colleagues at St. Louis University derived (*66*). Dan McKenzie and Robert Parker took the next major step toward completion of the plate theory in 1967, when they showed that slip vectors from Stauder's mechanisms of Alaskan earthquakes could be combined with the azimuth of the San Andreas fault to compute a consistent pole of instantaneous rotation for the Pacific and North American plates (*67*). At the same time, Jason Morgan's analysis of seafloor spreading rates and transform-fault azimuths demonstrated the global consistency of plate kinematics (*68*).

Clarity came with the realization that the plate is a cold mechanical boundary layer that can act as a mechanical stress guide, capable of transmitting forces for thousands of kilometers from one boundary to another (*69*). The essential elements of the subduction process were brought together in a 1968 paper by seismologists Brian Isacks, Jack Oliver, and Lynn Sykes (*70*). In addition to obtaining improved data on earthquake locations and focal mechanisms, they delineated a dipping slab of mantle material with distinctively high seismic velocity and low attenuation, which coincided with the Wadati-Benioff planes of deep seismicity (*71*). They found that they could account for their results, as well as most of the other data on plate tectonics, in terms of three mechanical layers, which J. Barrell and R.A. Daly had postulated earlier in the century to explain the vertical motions associated with isostatic compensation. A cold, strong *lithosphere* was generated by seafloor spreading at the ridge axis and subsequent conductive cooling of the oceanic crust and upper mantle, attaining a thickness of about 100 kilometers. It slid over and eventually subducted back into a hot, weak *asthenosphere*. Earthquakes of the Wadati-Benioff zones were generated primarily by stresses internal to the descending slab of oceanic lithosphere when it encountered a stronger, interior *mesosphere* at a depth of about 700 kilometers.

Deformation of the Continents

Plate tectonics was astounding in its simplicity and the economy with which it explained so many previously disparate geological observations. In the late 1960s and 1970s, geological data were reappraised in the light of the "new global tectonics," leading to some important extensions of the basic plate theory. However, a major problem was the obvious contrast in mechanical behavior of the oceanic and continental lithospheres. Geophysical surveys in the ocean basins revealed much narrower plate boundaries than observed on land. The volcanic rifts of active crust formation along the mid-ocean ridges were found to be only a few kilometers wide, for example, whereas volcanic activity in continental rifts could be mapped over tens to hundreds of kilometers. Similar differences were observed for transform faults; in the oceans, the active slip is confined to very narrow zones, in marked contrast to the broad belts of continental strike-slip tectonics, which often involve many distributed, interdependent fault systems. For example, only about two-thirds of the relative motion between the Pacific and North American plates turned out to be accommodated along the infamous San Andreas fault; the remainder is taken up on subsidiary faults and by oblique extension in the Basin and Range Province (see Section 3.2).

In 1970, Tanya Atwater (*72*) explained the geological evolution of western North America over the last 30 million years as the consequence

BOX 2.3 Prince William Sound, Alaska, 1964

The earthquake nucleated beneath Prince William Sound at about 5:36 p.m. on Good Friday, March 27, 1964. As the rupture spread outward, its progress to the north and east was stopped at the tectonic transition beneath the Chugach Mountains, behind the port of Valdez, Alaska, but to the southwest it continued unimpeded at 3 kilometers per second down the Alaska coastline, paralleling the axis of the Aleutian Trench for more than 700 kilometers, to beyond Kodiak Island. The district geologist of Valdez, Ralph G. Migliaccio, filed the following report:[1]

> Within seconds of the initial tremors, it was apparent to eyewitnesses that something violent was occurring in the area of the Valdez waterfront . . . Men, women, and children were seen staggering around the dock, looking for something to hold onto. None had time to escape, since the failure was so sudden and violent. Some 300 feet of dock disappeared. Almost immediately a large wave rose up, smashing everything in its path. . . . Several people stated the wave was 30 to 40 feet high, or more. . . . This wave crossed the waterfront and, in some areas reached beyond McKinley Street. . . . Approximately 10 minutes after the initial wave receded, a second wave or surge crossed the waterfront carrying large amounts of wreckage, etc. . . . There followed a lull of approximately 5 or 6 hours during which time search parties were able to search the waterfront area for possible survivors. There were none.

The height of the tsunami measured 9.1 meters at Valdez, but 24.2 meters at Blackstone Bay on the outer coast of the Kodiak Island group and 27.4 meters at Chenega on the Kenai Peninsula. The city of Anchorage, 100 kilometers west of the epicenter, was shielded from the big tsunami, but it experienced considerable damage, especially in the low-lying regions of unconsolidated sediment that became liquefied by the shaking. Robert B. Atwood, editor of the *Anchorage Daily Times*, who lived in the Turnagain Heights residential section, described his experiences during the landslide:

> I had just started to practice playing the trumpet when the earthquake occurred. In a few short moments it was obvious that this earthquake was no minor one. . . . I headed for the door . . . Tall trees were falling in our yard. I moved to a spot where I thought it would be safe, but, as I moved, I saw cracks appear in the earth. Pieces of the ground in jigsaw-puzzle shapes moved up and down, tilted at all angles. I tried to move away, but more appeared in every direction. . . . Table-top pieces of earth moved upward, standing like toadstools with great overhangs, some were turned at crazy angles. A chasm opened beneath me. I tumbled down . . . Then my neighbor's house collapsed and slid into the chasm. For a time it threatened to come down on top of me, but the earth was still moving, and the chasm opened to receive the house.

Migliaccio and Atwood had witnessed the second largest earthquake of the twentieth century. The plane of the rupture inferred from the dimensions of the aftershock zone was the size of Iowa (800 kilometers by 200 kilometers), and geodetic data showed that the offset along the fault averaged more than 10 meters. The product of these three numbers, which is proportional to a measure of earthquake size called the seismic moment (Equation 2.6), was thus 2000 cubic kilometers, about 100 times greater than the 1906 San Francisco earthquake. Among instrumentally recorded earthquakes, only the Chilean earthquake of 1960, which occurred in a similar tectonic setting, was bigger (by a factor of about 3). Both of these great earthquakes engendered tsunamis of large amplitude that propagated across the Pacific Ocean basin and caused damage and death thousands of kilometers from their source. Along the Oregon-California coast, 16 people were killed by the Alaska tsunami. In Crescent City, California, a series of large tsunamis inundated the harbor, beginning at four and a half hours, with the third and fourth wave causing the most damage. After the first two had struck, seven people returned to a seaside tavern to recover their valuables. Since the ocean seemed to have returned to normal, they remained to have a drink and were caught by the third wave, which killed five of them.[2]

[1] National Research Council, *The Great Alaska Earthquake of 1964*, National Academy Press, Washington, D.C., 15 volumes, 1972-1973.
[2] B. Bolt, *Earthquakes and Geological Discovery*, W.H. Freeman, New York, p. 155, 1963.

of the North American plate overriding an extension of the East Pacific Rise along a subduction zone paralleling the West Coast. Her synthesis, which accounts for seemingly disparate events (e.g., andesitic volcanism in northern California, strike-slip faulting along the San Andreas, compressional tectonics in the Transverse Ranges, rifting in the Gulf of California) was grounded in the kinematical principles of plate tectonics (*73*), and her paper did much to convince geologists that the new theory was a useful framework for understanding the complexities of continental tectonics.

Convergent plate boundaries in the oceans were observed to be broader than the other boundary types, with the zone of geologic activity on the surface encompassing the trench itself, the deformed sediments and basement rocks of the forearc sequence, the volcanic arc that overlies the subducting slab, and sometimes an extending back-arc basin (*74*). Nevertheless, the few-hundred-kilometer widths of the ocean-ocean convergence zones did not compare with the extensive orogenic terrains that mark major continental collisions. The controlling factors were recognized to be the density and strength of the silica-rich continental crust, which are significantly lower than those of the more iron- and magnesium-rich oceanic crust and upper mantle (*75*). When caught between two converging plates, the weak, buoyant continental crust resists subduction and piles up into arcuate mountain belts and thickened plateaus that erode into distinctive sequences of sedimentary rock. This distributed deformation also causes metamorphism and melting of the crust, generating siliceous magmas that intrude the crust's upper layers to form large granitic batholiths. In some instances, the redistribution of buoyancy-related stresses can lead to a reversal in the direction of subduction.

W. Hamilton used these consequences of plate tectonics to explain modern examples of mountain building, and J. Dewey and J. Bird used them to account for the geologic structures observed in ancient mountain belts (*76*).

Much of the early work on convergent plate boundaries interpreted mountain building in terms of two-dimensional models that consider deformations only in the vertical planes perpendicular to the strikes of the convergent zones. During a protracted continent-continent collision, however, crustal material is eventually squeezed sideways out of the collision zone along lateral systems of strike-slip faults. The best modern example is the Tethyian orogenic belt, which extends for 10,000 kilometers across the southern margin of Eurasia. At the eastern end of this belt, the convergence of the Indian subcontinent with Asia has uplifted the Himalaya, raised the great plateau of Tibet, re-elevated the Tien Shan Mountains to heights in excess of 5 kilometers, and caused deformations up to 2000 kilometers north of the Himalayan front. Earthquakes within these continental deformation zones have been frequent and dangerous.

In a series of studies, P. Molnar and P. Tapponnier explained the orientation of the major faults in southern Asia, their displacements, and the timing of key tectonic events as a consequence of the collision of the Indian continent with Asia (*77*). They investigated the active faulting in central Asia using photographs from the Earth Resources Technology Satellite, magnetic lineations on the ocean floor, and teleseismically determined focal mechanisms of recent earthquakes. By combining these remote-sensing observations with the plate-tectonic information, they demonstrated that strike-slip faulting has played a dominant role in the mature phase of the Himalayan collision (*78*).

The more diffuse nature of continental seismicity and deformation was consistent with the notion that the continental lithosphere is somehow weaker than the oceanic lithosphere, but a detailed picture required a better understanding of the mechanical properties of rocks. When subjected to differential compression at moderate temperatures and pressures, most rocks fail by brittle fracture according to the Coulomb criterion (Equation 2.1). Extensive laboratory experiments on carbonates and silicates showed that for all modes of brittle failure, the coefficient of friction μ usually lies in the range 0.6 to 0.8, with only a weak dependence on the rock type, pressure, temperature, and properties of the fault surface. This behavior has come to be known as *Byerlee's law* (*79*), and it implies that the frictional strength of continental and oceanic lithospheres should be about the same, at least at shallow depths.

Rocks deform by ductile flow, not brittle failure, when the temperature and pressure get high enough, however, and the onset of this ductility depends on composition. Investigations of ductile flow began in 1911 with Theodore von Kármán's triaxial tests on jacketed samples of marble.

It was found that the strength of ductile rocks decreases rapidly with increasing temperature and that their rheology approaches that of a viscous fluid. The brittle-ductile transition thus explained the plate-like behavior of the oceanic lithosphere and the fluid-like behavior of its subjacent, convecting mantle. Rock mechanics experiments further revealed that ductility sets in at lower temperatures in quartz-rich rocks than in olivine-rich rocks, typically at midcrustal depths in the continents. The ductile behavior of the lower continental crust inferred from laboratory data, which was consistent with the lack of earthquakes at these depths, thus explained the less plate-like behavior of the continents (*80*).

2.5 EARTHQUAKE MECHANICS

Gilbert and Reid recognized the distinction between fracture strength and frictional strength (*81*), and they portrayed earthquakes as frictional instabilities on two-dimensional faults in a three-dimensional elastic crust, driven to failure by slowly accumulating tectonic stresses—a view entirely consistent with plate tectonics. Although earthquakes surely involve some nonelastic, volumetric effects such as fluid flow, cracking of new rock, and the expansion of gouge zones, Gilbert and Reid's idealization still forms the conceptual framework for much of earthquake science, both basic and applied. Nevertheless, because the friction mechanism was not obviously compatible with deep earthquakes, as described below, their view that earthquakes are frictional instabilities on faults had, by the time Wilson wrote his 1965 paper on plate tectonics, been considered and rejected by some scientists.

The Instability Problem

Deep-focus earthquakes presented a major puzzle. Seismologists had found that the deepest events, 600 to 700 kilometers below the surface, are shear failures just like shallow-focus earthquakes and that the decrease in apparent shear stress during these events is on the order of 10 megapascals, about the same size as the stress drops estimated for shallow shocks. According to a Coulomb criterion (Equation 2.1), the shear stress needed to induce frictional failure on a fault should be comparable to the lithostatic pressure, which reaches 2500 megapascals in zones of deep seismicity. Shear stresses of this magnitude are impossibly high, and if the stress drop approximates the absolute stress, as most seismologists believe, they would conflict with the observations (*82*).

Furthermore, if earthquakes result from a frictional instability, the motion across a fault must at some point be accelerated by a drop in the frictional resistance. A spontaneous rupture like an earthquake thus requires some type of *strain weakening*, but the rock deformations observed in the laboratory at high pressure and temperature tended to display strain hardening during ductile creep. In a classic 1960 treatise *Rock Deformation*, D. Griggs and J. Handin (*83*) concluded that the old theory of earthquakes' originating by ordinary fracture with sudden loss of cohesion was invalid for deep earthquakes, although they did note that extremely high fluid pressures at depth could validate that same mechanism they presumed to hold for shallow events.

A renewed impetus was given to the frictional explanation in 1966, when W.F. Brace and Byerlee demonstrated that the well-known engineering phenomenon of stick-slip also occurs in geologic materials (*84*). Experimenting on samples with preexisting fault surfaces, they observed that the stress drops in the laboratory slip events were only a small fraction of the total stress. This implies that the stress drops during crustal earthquakes could be much smaller than the rock strength, eliminating the major seismological discrepancy. Subsequent experiments at the Massachusetts Institute of Technology found a transition from stick-slip behavior to ductile creep at about 350°C (*85*). Stick-slip instabilities thus matched the properties of earthquakes in the upper continental crust, which were usually confined above this brittle-ductile transition, although this could not explain the deeper shocks in subduction zones. In addition, Brace and Byerlee's work focused theoretical attention on how frictional instabilities depend on the elastic properties of the testing machine or fault system (*86*).

During the next decade, the servo-controlled testing machine was developed, in which the load levels and strain rates were precisely regulated, so that the postfailure part of the load-deformation curve in brittle materials could be followed without the stick-slip instabilities encountered with less stiff machines (*87*). Several new aspects of rock friction were investigated, including memory effects and dilatancy (*88*). The subsequent development of high-precision double-direct-shear and rotary-shear devices (*89*) allowed detailed measurements of friction for a wide range of materials under variable sliding conditions. This work documented three interrelated phenomena:

1. Static friction μ_s depends on the history of sliding and increases logarithmically with the time two surfaces are held in stationary contact (*90*).

2. Under steady-state sliding, the dynamic friction μ_d depends logarithmically on the slip rate V, with a coefficient that can be either positive (velocity strengthening) or negative (velocity weakening) (*91*).

3. When a slipping interface is subjected to a sudden change in the loading velocity, the frictional properties evolve to new values over a

characteristic slipping distance D_c, measured in microns and interpreted as the slip necessary to renew the microscopic contacts between the two rough surfaces (92).

During 1979 to 1983, J.H. Dieterich and A.L. Ruina (93) integrated these experimental results into a unified constitutive theory in which the slip rate V appears explicitly in the friction equation and the frictional strength evolves with a characteristic time set by the mean lifetime D_c/V of the surface contacts. The behavioral transition of Brace and Byerlee around 350 degrees, from stick-slip to creep, was interpreted by Tse and Rice (94) as a transition from rate weakening to rate strengthening in the crust and was shown to allow models of earthquake sequences in a crustal strike-slip fault to reproduce primary features inferred for natural events, such as the depth range of seismic slip and rapid after-slip below.

Scaling Relations

According to the dislocation model of earthquakes, slip on a small planar fault is equivalent to a double-couple force system, where the total moment M_0 of each couple is proportional to the product of the fault's area A and its average slip $\bar{u}$:

$$M_0 = GA\bar{u}. \tag{2.6}$$

The constant of proportionality G is the elastic shear modulus, a measure of the resistance to shear deformation of the rock mass containing the fault, which can be estimated from the shear-wave velocity. For waves that are large compared with the dislocation, the amplitude of radiation increases in proportion with M_0, so that this *static seismic moment* can be measured directly from seismograms. K. Aki made the first determination of seismic moment from the long-period surface waves of the 1964 Niigata earthquake (95). Many subsequent studies have demonstrated a consistent relationship between seismic moment and the various magnitude scales developed from the Richter standard; the results can be expressed as a general *moment magnitude* M_W of the form

$$M_W = \frac{2}{3}(\log M_0 - 9.05) \quad (M_0 \text{ in newton-meters}). \tag{2.7}$$

Equation 2.7 defines a *unified magnitude scale* (96) based on a physical measure of earthquake size. Calculating magnitude from seismic moment avoids the saturation effects of other magnitude estimates, and this procedure became the seismological standard for determining earthquake size. The 1960 Chile earthquake had the largest moment of any known seismic event, 2×10^{23} newton-meters, corresponding to $M_w = 9.5$ (Table 2.1).

TABLE 2.1 Size Measures of Some Important Earthquakes

Date	Location	M_S	M_W	M_0 (10^{18} N-m)
April 18, 1906	San Francisco	8.25	8.0	1,000
Sept. 1, 1923	Kanto, Japan	8.2	7.9	850
Nov. 4, 1952	Kamchatka	8.25	9.0	35,000
March 9, 1957	Aleutian Islands	8.25	9.1	58,500
May 22, 1960	Chile	8.3	9.5	200,000
March 25, 1964	Alaska	8.4	9.2	82,000
June 16, 1964	Niigata, Japan	7.5	7.6	300
Feb. 4, 1965	Aleutian Islands	7.75	8.7	12,500
May 31, 1970	Peru	7.4	8.0	1,000
Feb. 4, 1975	Haicheng, China	7.4	6.9	31
July 28, 1976	Tangshan, China	7.9	7.6	280
Aug. 19, 1977	Sumba	7.9	8.3	3,590
Oct. 28, 1983	Borah Peak	7.3	6.9	31
Sept. 19, 1985	Mexico	8.1	8.0	1,100
Oct. 18, 1989	Loma Prieta	7.1	6.9	27
June 28, 1992	Landers	7.5	7.3	110
Jan. 17, 1994	Northridge	6.6	6.7	12
June 9, 1994	Bolivia	7.0[a]	8.2	2,630
Jan. 16, 1995	Hyogo-ken Nanbu, Japan	6.8	6.9	24
Aug. 17, 1999	Izmit, Turkey	7.8	7.4	242
Sept. 20, 1999	Chi-Chi, Taiwan	7.7	7.6	340
Oct. 16, 1999	Hector Mine	7.4	7.1	60
Jan. 13, 2001	El Salvador	7.8	7.7	460
Jan. 26, 2001	Bhuj, India	8.0	7.6	340

NOTE: All events are shallow except Bolivia, which had a focal depth of 657 km. Moment magnitude M_W computed from seismic moment M_0 via Equation 2.7.
[a]Body-wave magnitude.
SOURCES: U.S. Geological Survey and Harvard University.

Unless otherwise noted, all magnitudes given throughout the remainder of this report are moment magnitudes.

Beginning in the 1950s, arrays of temporary seismic stations were deployed to study the aftershocks of large earthquakes. Aftershocks are caused by subsidiary faulting from stress concentrations produced by the main shock, owing to inhomogeneities in fault slippage and heterogeneities in the properties of the nearby rocks. Omori's work on the 1891 Nobi earthquake had demonstrated that the frequency of aftershocks decayed inversely with the time following the main shock (*97*). In its modern form, "Omori's law" states that the aftershock frequency obeys a power law of the form

$$n(t) = A(t + c)^{-p}, \tag{2.8}$$

where t is the time following the main shock and c and p are parameters of the aftershock sequence. Aftershock surveys confirmed that p is near unity (usually slightly greater) for most sequences. They also showed that the aftershock zone approximated the area of faulting inferred from geologic and geodetic measurements (*98*).

With independent information about rupture area A from aftershock, geologic, or geodetic information, Equation 2.6 can be solved for the average fault displacement $\overline{u}$. Aki obtained a value of about 4 meters for the 1964 Niigata earthquake by this method, consistent with echo-sounding surveys of the submarine fault scarp. A second method derived fault dimensions from the "corner frequency" of the seismic radiation spectrum, an observable value inversely proportional to the rupture duration (*99*). Corner frequencies were easily measurable from regional and teleseismic data and could be converted to fault lengths by assuming an average rupture velocity (*100*). Using this procedure, seismologists estimated the source dimensions for a much larger set of events, paving the way for global studies of the stress changes during earthquakes.

For an equidimensional rupture surface, the ratio $\overline{u} / \sqrt{A}$ measures the decrease in strain, or *strain drop*, during the faulting, and $\Delta\sigma \approx G \overline{u} \sqrt{A}$ is the *static stress drop*, the average difference between the initial and final stresses (*101*). Substituting this relationship into Equation 2.6 yields $M_0 \approx \Delta\sigma \approx A^{3/2}$. A logarithmic plot of seismic moment M_0 versus fault area A for a representative sample of crustal earthquakes on plate boundaries shows scatter about a linear relationship with a slope of about 1.5, implying that the stress drop is approximately constant across a large range of earthquake sizes, with an average value close to 3 megapascals (Figure 2.11) (*102*). The lack of any systematic variation in stress drop with event size was a fundamental observation that formed the basis for a series of earthquake scaling relations (*103*). Together with the Gutenberg-Richter and Omori power-law relations (Equations 2.5 and 2.8), near-constant stress drop suggested that many aspects of the earthquake process are scale invariant and that the underlying physics is not sensitive to the tectonic details.

Seismic Source Studies

Seismic moment measures the static difference between initial and final states of a fault, not what happens during the rupture. To investigate the dynamics of rupture process, seismologists had to tackle the difficult problem of determining the space-time distribution of faulting during an earthquake from its radiated seismic energy. In the 1960s, a simple kinematic dislocation model with uniform slip and rupture speed was devel-

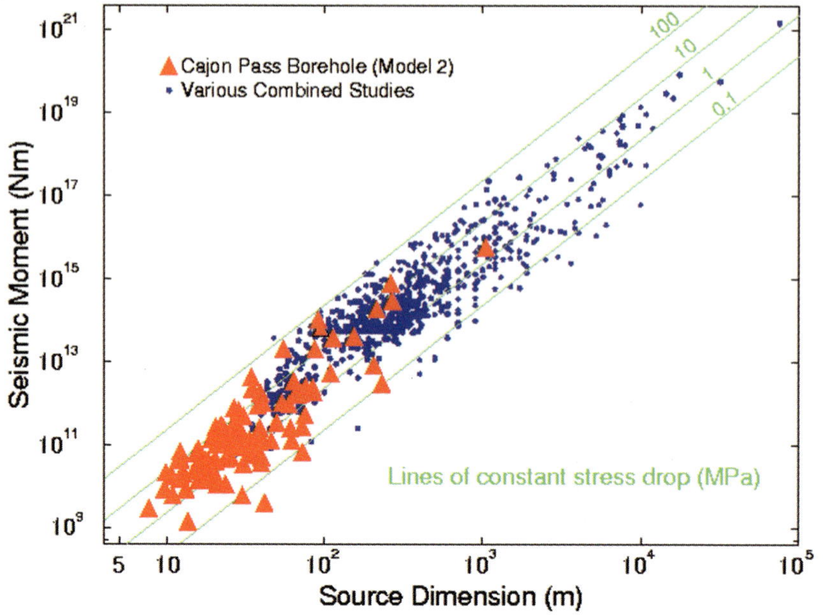

FIGURE 2.11 Seismic moment as a function of source dimension. When plotted on a log-log scale, the diagonal lines (slope = 3) indicate $M_0 \sim r^3$ and denote constant stress drop. Note that measurements of the static stress drop vary as the cube of the corner frequency, a sensitivity that contributes to substantial scatter in $\Delta\sigma$ for individual events. The value cited is based on a numerical study of rupture dynamics (R. Madariaga, Dynamics of an expanding circular fault, *Bull. Seis. Soc. Am.*, **66**, 639-666, 1976) and applies to shear waves radiated by a circular, cohesionless fault that stops suddenly around a circular periphery. Using a particular estimate for the source radius in terms of the corner frequency f_c yields $\Delta\sigma \approx 47M_0(f_c/\beta)^3$. SOURCE: R.E. Abercrombie, Earthquake source scaling relationships from –1 to 5 ML using seismograms recorded at 2.5-km depth, *J. Geophys. Res.*, **100,** 24,015-24,036, 1995. Copyright 1995 American Geophysical Union. Reproduced by permission of American Geophysical Union.

oped by N. Haskell to understand the energy radiation from an earthquake and the spectral structure of a seismic source (*104*). Haskell's model predicted that the frequency spectrum of an earthquake source is flat at low frequency and falls off as ω^{-2} at high frequency, where ω is the angular frequency. This simple model (generally called the omega-squared model) was extended to accommodate the much more complex kinematics of real seismic faulting, described stochastically (*105*), and it was found to approximate the spectral observations rather well, especially for small earthquakes.

The orientation of an elementary dislocation depends on two directions, the normal direction to the fault plane and the slip direction within this plane, so that the double-couple for a dislocation source is described by a three-dimensional, second-order *moment tensor* **M** proportional to M_0 (*106*). By 1970, it was recognized that the seismic moment tensor can be generalized to include an ideal (spherically symmetrical) explosion and another type of seismic source called a *compensated linear vector dipole* (CLVD). A CLVD mechanism was invoked as a plausible model for seismic sources with cylindrical symmetry, such as magma-injection events, ring-dike faulting, and some intermediate- and deep-focus events (Figure 2.8) (*107*).

The Stress Paradox

Plate tectonics accounted for the orientation of the stress field on simple plate boundaries, which could be classified according to Anderson's three principal types of faulting: divergent boundaries (normal faults), transform boundaries (strike-slip faults), and convergent boundaries (reverse faults). The stress orientations mapped on plate interiors using a variety of indicators—wellbore breakouts, volcanic alignments, and earthquake focal mechanisms—were generally found to be coherent over distances of 400 to 4000 kilometers and to match the predictions of intraplate stress from dynamic models of plate motions (*108*). This behavior implies that the spatial localization of intraplate seismicity primarily reflects the concentration of strain in zones of crustal weakness (*109*). Explaining the orientation of crustal stresses was a major success for the new field of *geodynamics*.

About 1970, a major debate erupted over the magnitude of the stress responsible for crustal earthquakes. Byerlee's law implies that the shear stress required to initiate frictional slip should be at least 100 megapascals, an order of magnitude greater than most seismic stress drops (*110*). The stresses measured during deep drilling generally agree with these predictions. If the average stresses were this large, however, the heat generated by earthquakes along major plate boundaries would greatly exceed the radiated seismic energy and the heat flowing out of the crust along active fault zones should be very high. Attempts to measure a heat flow anomaly on the San Andreas fault found no evidence of a peak (*111*). The puzzle of fault stress levels was further complicated as data became available in the middle to late 1980s on principal stress orientations in the crust near the San Andreas (*112*); the maximum stress direction was found to be steeply inclined to the fault trace and to resolve more stress onto faults at angles to the trace of the San Andreas fault than onto the San Andreas fault itself. These results, as well as data on subduction interfaces and oceanic transform faults, suggest that most plate-bounding faults operate at low overall driving stress, on the order of 20 megapascals or less. Various explanations have been put forward (*113*)—intrinsically weak materials in the fault zones, high fluid pore pressures, or dynamical processes that lower frictional resistance such as wave-generated decreases in normal stress during rupture—but the stress paradox remains a major unsolved problem.

2.6 EARTHQUAKE PREDICTION

Earthquake prediction is commonly defined as specifying the location, magnitude, and time of an impending earthquake within specified ranges. Earthquake predictions are customarily classified into long term (decades to centuries), intermediate term (months to decades), and short term (seconds to weeks). The following discussion is divided the same way, but the classification is not definitive because many proposed methods span the time boundaries. Because some predictions might be satisfied by chance, seismologists almost inevitably invoke probabilities to evaluate the success of an earthquake prediction. Many seismologists distinguish forecasts, which may involve relatively low probabilities, from predictions, which involve high enough probabilities to justify exceptional policy or scientific responses. This distinction, which is adopted here, implies that predictions refer to times when the earthquake probability is temporarily much higher than normal for a given region and magnitude range. Forecasts might or might not involve temporal variations. Even if they involve only estimates of the "normal" probability, long-term forecasts can be extremely useful for input to seismic hazard calculations and for decisions about building, retrofitting, insuring, and so forth. A clear statement of the target magnitude is crucial to evaluating a prediction because small earthquakes are so much more frequent than large ones. A prediction of a moment magnitude (M) 6 earthquake for a given region and time might be very bold, while a prediction of an M 5 event could easily be satisfied by chance.

Long-Term Forecasts

G.K. Gilbert issued what may have been the first scientifically based, long-term earthquake forecast in his 1883 letter to the *Salt Lake City Tribune* (*114*), in which he articulated the practical consequences of his field work along the seismically active Wasatch Front:

> Any locality on the fault line of a large mountain range, which has been exempt from earthquake for a long time, is by so much nearer to the date of recurrence. . . . Continuous as are the fault-scarps at the base of the Wasatch, there is one place where they are conspicuously absent, and that place is close to [Salt Lake City]. . . . The rational explanation of their absence is that a very long time has elapsed since their last renewal. In this period the earth strain has slowly been increasing, and some day it will overcome the friction, lift the mountains a few feet, and reenact on a more fearful scale the [1872] catastrophe of Owens Valley.

So far, Gilbert's forecast for Salt Lake City has not been fulfilled (*115*). H.F. Reid developed Gilbert's "principle of alternation" into a quantitative theory of earthquake forecasting. In his 1910 report for the Lawson Commission, he wrote: "As strains always precede the rupture and as the strains are sufficiently great to be easily detected before rupture occurs (*116*), . . . it is merely necessary to devise a method of determining the existence of strains; and the rupture will in general occur . . . where the strains are the greatest." He suggested that the time of the next major earthquake along that segment of the San Andreas fault could be estimated by establishing a line of piers at 1-kilometer spacing perpendicular to the fault and observing their positions "from time to time." When "the surface becomes strained through an angle of 1/2000, we should expect a strong shock." Reid noted that this prediction scheme relied on measurements commencing when the fault was in an "unstrained condition," which he presumed was the case following the 1906 earthquake (*117*).

The Gilbert-Reid forecast hypothesis—the idea that a large earthquake is due when the critical strain from the last large event has been recovered by steady tectonic motions—is the basis for the *seismic-gap method*. In its simplest form, this hypothesis asserts that a particular fault segment fails in a quasi-periodic series of earthquakes with a characteristic size and average recurrence interval. This interval can be estimated either from known dates of past characteristic earthquakes or from *D/V*, the ratio of the average slip in a characteristic quake to the long-term slip rate on the fault. A seismic gap is a fault segment that has not ruptured in a characteristic earthquake for a time longer than *T*. A. Imamura identified Sagami Bay, off Tokyo, as a seismic gap, and his prediction of an impeding rupture was satisfied by the disastrous Kanto earthquake of 1923 (*118*). Fedotov is generally credited with the first modern description of the seismic-gap method, publishing a map in 1965 showing where large earthquakes should be expected (*119*). His predictions were promptly satisfied by three major events (Tokachi-Oki, 1968; southern Kuriles, 1969; central Kamchatka, 1971).

Forecasting large earthquakes using the seismic-gap principle looked fairly straightforward in the early 1970s. Plate tectonics had established a precise kinematic framework for estimating the rates of geological deformation across plate boundaries, specifying a deformation budget that could be balanced against historic seismic activity. For example, Sykes divided the amount of co-seismic slip during the 1957, 1964, and 1965 Aleutian Trench earthquakes by the rate of relative motion between the North American and Pacific plates, obtaining recurrence intervals of a century or so for each of the three segments (*120*). Self-consistent models of the relative plate motions were derived from global data sets that included seafloor magnetic anomalies tied to the precise magnetic reversal time scale (*121*), allowing Sykes's calculation to be repeated for many of the major plate boundaries. Sykes and his colleagues produced maps in 1973 and 1979 showing plate boundary segments with high, medium, and low seismic potential based on the recent occurrence of large earthquakes (*122*) and published a more refined forecast in 1991 (*123*) (Figure 2.12).

While some form of the Gutenberg-Richter distribution is observed for almost all regions, Schwartz and Coppersmith (*124*) proposed that many individual faults, or segments of faults, behave quite differently. They proposed that most of the slip on a fault segment is released in large "characteristic" earthquakes having, for a given segment, similar magnitude, rupture area, and average displacement. It follows that characteristic earthquakes must be much more frequent, relative to smaller and larger earthquakes, than the Gutenberg-Richter relationship would predict. Wesnousky and colleagues (*125*) argue that earthquakes in a region obey the Gutenberg-Richter relationship because the fault segments there have a power-law distribution.

Characteristic earthquakes have profound implications for earthquake physics and hazards. For example, characteristic earthquakes can be counted confidently, and their average recurrence time would be an important measure of seismic hazard. The time of the last one would start a seismic clock, by which the probability of another such earthquake could be estimated. For Gutenberg-Richter earthquakes, the simple clock concept does not apply: for any magnitude of quake, there are many more earthquakes just slightly smaller but no different in character. The characteristic earthquake model has strong intuitive appeal, but the size of the characteristic earthquake and the excess frequencies of such events have been difficult to demonstrate experimentally (*126*).

The seismic-gap method met limited success as a basis for earthquake forecasting (*127*). Attempts to use it as a general tool were frustrated by the difficulty of specifying characteristic magnitudes and the lack of historical records needed to estimate the recurrence interval *T*. Moreover, the practical utility of the seismic-gap hypothesis was compromised by the intrinsic irregularity of the earthquake process and the tendency of earthquakes to cluster in space and time. The Gilbert-Reid idea that a

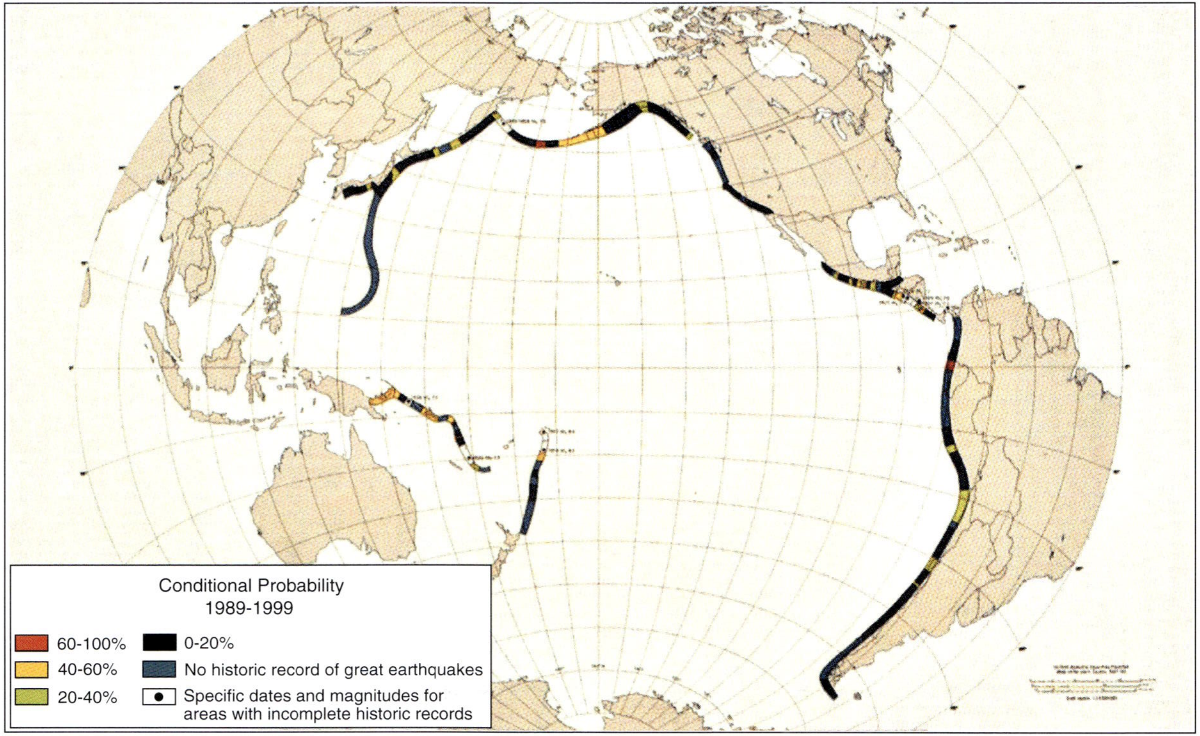

FIGURE 2.12 Circum-Pacific plate boundary segments with high, medium, and low seismic potential. Colors indicate time-dependent probability of the recurrence of either a large or a great shallow plate earthquake within a specified segment during the interval 1989 to 1999, conditional upon the event not having occurred prior to 1989. SOURCE: Modified from S.P. Nishenko, Circum-Pacific seismic potential: 1989-1999, *Pure Appl. Geophys.*, **135**, 169-259, 1991.

given fault segment will fail periodically assumes that the stress drop in successive earthquakes and the rate of stress accumulation between earthquakes are both constant. However, stick-slip experiments in well-controlled laboratory settings show variations in the time between slip events, which had incomplete and irregular stress drops, indicating variations in either the initial (rupture) stress or the final (postearthquake) stress, or both. Shimizaki and Nakata (*128*) discussed two special cases (Figure 2.13). In the "time-predictable" model, the initial stress is the same for successive large earthquakes, but the final stress varies. This implies that the time until the next earthquake is proportional to the stress drop, or average slip, in the previous event ($T_n = D_{n-1}/V$), while the size of the next quake D_n is not predictable. In the "slip-predictable" model, the initial stress varies from event to event, but the final stress is the same. This implies that the slip in the next earthquake is proportional to the time since the last one ($D_n = T_nV$), while the time T_n is not predictable. Shimazaki and Nakata found that the Holocene uplift data for several well-studied sites in Japan were consistent with a time-predictable model of the largest events.

Japanese seismologists and geologists have long been at the forefront of earthquake prediction studies, and their government has sponsored the world's largest and best-funded research programs on earthquake phenomena (*129*). One area of intense concern is the so-called Tokai seismic gap, southwest of Mt. Fuji (Box 2.4). This region is threatened by a

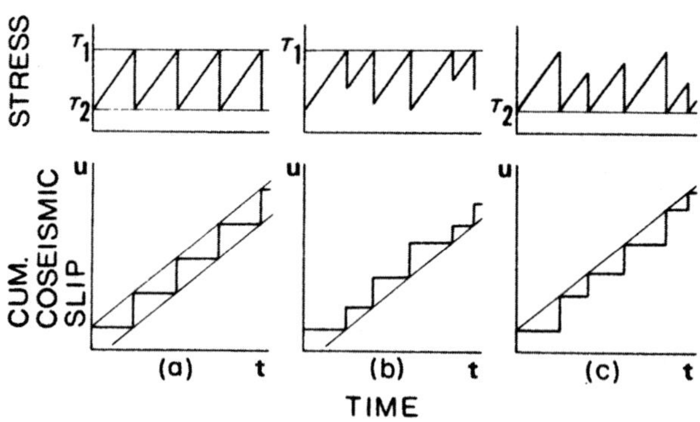

FIGURE 2.13 Schematic models of earthquake recurrence intervals for constant tectonic loading rates. (a) Constant stress drop, showing strictly periodic behavior (modified Reid model). (b) Variable stress drop, constant failure stress (time-predictable model). (c) Variable stress drop, constant final stress (slip-predictable model). SOURCE: K. Shimazaki and T. Nakata, Time-predictable recurrence model for large earthquakes, *Geophys. Res. Lett.*, **7**, 279-282, 1980. Copyright 1980 American Geophysical Union. Reproduced by permission of American Geophysical Union.

BOX 2.4 The Tokai Seismic Gap

Large earthquakes have repeatedly occurred in the Nankai Trough along the southwestern coast of Japan. The sequence during the past 500 years includes large (M ~ 8) earthquakes in 1498, 1605, 1707, 1854, and 1944-1946, with an approximate interval of about 120 years. In the early 1970s, several Japanese seismologists noticed that the 1944-1946 events were somewhat smaller than the 1854 and 1707 earthquakes, and they suggested that this rupture did not reach the northeastern part of the Nankai Trough, called the Suruga Trough. Given the historical evidence that the rupture of both the 1854 and the 1707 events extended all the way to the Suruga Trough, they concluded that this portion of the plate boundary, which became known as the "Tokai seismic gap," has the potential for a magnitude-8 earthquake in the near future.[1]

In 1978, the Japanese government introduced the Large-Scale Earthquake Countermeasures Act and embarked on an extensive project to monitor the Tokai gap. Many institutions deployed geophysical and other instrumentation, and very detailed plans for emergency relief efforts were made. This program specified the procedures for short-term prediction. When some anomaly is observed by the monitoring network, a special evaluation committee comprising technical experts is to decide whether it is a precursor for the predicted Tokai earthquake or not. If the anomaly is identified as a precursor, a large-scale emergency operation is to be initiated by the local and central governments. A detailed plan for this activity has been laid out as part of the prediction experiment.

After more than 23 years since the project began, no anomaly that requires initiation of the preplanned emergency operation has been detected. The chair of the Tokai evaluation committee, Professor K. Mogi, resigned in 1997, expressing doubts about the ability of the committee to perform its expected short-term prediction function, and the new chair, M. Mizoue, has voiced similar concerns. Also, a report released in 1997 by the Geodesy Council of Japan concluded that a technical basis for short-term prediction of the kind required by the Countermeasures Act does not currently exist in Japan, and that the time frame for establishing such a capability is not known.[2]

---

[1]K. Ishibashi, in *Earthquake Prediction—An International Review*, American Geophysical Union, Maurice Ewing Series 4, Washington, D.C., pp. 297-332, 1981; K. Mogi, *Earthquake Prediction*, Academic Press, New York, Chapter III-5, 1985.

[2]*State-of-the-Art Review of the National Programs for Earthquake Prediction*, Subcommittee for Review Drafting, Special Committee for Earthquake Prediction, Geodesy Council of the Ministry of Education, Science, and Culture, Tokyo, 137 pp., 1997.

potentially large earthquake on the thrust fault of the Suruga Trough, known to have ruptured in the great earthquakes of 1707 and 1854 and thought to be ripe for failure at any time. So far, the expected Tokai earthquake has not occurred. Many seismologists now agree that accurate forecasts are difficult even for plate boundaries such as this one that have seemingly regular historical sequences of earthquakes.

The Parkfield, California, earthquake prediction (*130*), arguably the boldest widely endorsed by the seismological community, was also based on the seismic gap theory. Moderate earthquakes of about M 6 on the San Andreas fault near Parkfield were recorded instrumentally in 1922, 1934, and 1966, and pre-instrumental data revealed that similar-size earthquakes occurred in 1857, 1881, and 1901. The regular recurrence of Parkfield events at an average interval of about 22-years and the similarity of the foreshock pattern in 1934 and 1966 led to the hypothesis that these events were characteristic earthquakes, breaking the same segment of the San Andreas with about the same slip. Estimates of the recurrence time from the ratio of earthquake displacement to fault slip agreed with the 22-year value above. Based on these and other data, the U.S. Geological Survey (USGS) issued an official prediction of an earthquake of about M 6 in about 1988, on an identified segment of the San Andreas, with 95 percent probability before the beginning of 1993. While the size and location of the predicted event were not precisely specified, no earthquake matching the description has occurred as of January 1, 2002 (*131*).

The seismic-gap model forms the basis of many other forecasts. Most involve low enough probabilities that they are not predictions by the usual definition, and they cannot yet be confirmed or rejected by available data. A notable example was the 1988 "Working Group Report" (*132*). The authors postulated specific segments of the San Andreas and other major strike-slip faults in California, and then tabulated characteristic magnitudes, average recurrence times, and 30-year probabilities for each segment. They estimated a 66 percent probability of at least one large characteristic earthquake on the four southern segments of the San Andreas fault before 2018, with a similar chance for northern California.

The 1989 Loma Prieta earthquake (M 6.9) occurred in an area where several seismologists (and the Working Group) had made long-term or intermediate-term forecasts of a large earthquake (*133*). It occurred near the southern end of the 1906 rupture, a segment of the San Andreas to which the Working Group assigned a 30-year probability of 30 percent. The earthquake was considered a successful forecast, especially as it happened just two years after the report was published. On the other hand success by chance cannot be ruled out, and the earthquake did not exactly match the forecasts (*134*).

Intermediate-Term Prediction

Intermediate-term prediction efforts are generally based on recognizing geophysical anomalies that might signal a state of near-critical stress approaching the breaking point. Apparent anomalies have been observed in small earthquake occurrence: accelerated strain or uplift; changes in the gravity field, magnetic field, electrical resistivity, water flow, groundwater chemistry, atmospheric chemistry; and many other parameters that might be sensitive to stress, cracks in rock, or changes in the frictional properties of rocks. The literature is extensive (*135*); only a few examples are discussed here.

A logical successor to the seismic-gap model is the hypothesis that earthquake occurrence is accelerated or decelerated by stress increments from previous earthquakes. One version of this hypothesis is the stress shadow model—that the occurrence of large earthquakes reduces the stress in certain neighborhoods about their rupture zones, thus decreasing the likelihood of both large and small earthquakes there until the stress recovers (*136*). The stress model differs from the seismic-gap model in that it applies not just to a fault segment, but to the region surrounding it. Furthermore, because stress is a tensor, it may encourage some faults and discourage others. In some regions near a ruptured fault segment, the stress is actually increased, offering an explanation for seismic clustering. At present, the model offers a good retrospective explanation for many earthquake sequences, but it has not been implemented as a testable prediction hypotheses because the stress pattern depends on details of the previous rupture, fault geometry, stress-strain properties of the crust, possible fluid flow in response to earthquake stress increments, and other properties that are very difficult to measure in sufficient detail.

Seismicity patterns are the basis of many prediction attempts, in part because reliable seismicity data are widely available. Mogi described a sequence of events that many feel can be used to identify stages in a repeatable seismic cycle involving large earthquakes (*137*). In this model a large earthquake may be followed by aftershocks of decreasing frequency, a lengthy period of quiescence, an increase of seismicity about the future rupture zone, a second intermediate-term quiescence, a period for foreshock activity, a third short-term quiescence, and finally the "big one." Any of the stages may be missing. This behavior formed the basis of an apparently successful prediction of the M 7.7 Oaxaca, Mexico, earthquake of 1978 (*138*). Unfortunately, there are no agreed-on definitions of the various phases that can be applied uniformly, nor has there been a comprehensive test of how Mogi's model works in general (*139*).

Computerized pattern recognition has been applied in several experiments to recognize the signs of readiness for large earthquakes. V. Keilis-Borok and Russian colleagues have developed an algorithm known as "M8" that scans a global catalog for changes in the earthquake rate, the ratio of large to small earthquakes, the vigor and duration of aftershock sequences, and other diagnostics within predefined circles in seismically active areas (*140*). They report significant success in predicting which circles are more likely to have large earthquakes (*141*). Since 1999, they have made six-month advance predictions for magnitude thresholds 7.5 and 8.0 accessible on their web page (*142*), and fully prospective statistical tests will be possible in the near future.

Short-Term Prediction

The "Holy Grail" of earthquake science has always been short-term prediction—anticipating the time, place, and size of a large earthquake in a window narrow and reliable enough to prepare for its effects (*143*). However, interest in the possibility of detecting earthquake precursors grew as new technologies were developed to monitor the crustal environment with increasing sensitivity. In the year following the destructive 1964 Alaskan earthquake, a select committee of the White House Office of Science and Technology issued a report called *Earthquake Prediction: A Proposal for a Ten Year Program of Research*, which called for a national program of research focused on this goal (*144*).

Optimism about the feasibility of short-term prediction was heightened in the mid-1970s by the apparent successes of empirical prediction schemes and the plausibility of physical process models, such as dilatancy diffusion. Laboratory studies had measured dilatant behavior in rocks prior to failure, caused by pervasive microcracking. Dilatancy creates measurable strain, changes the material properties, and increases the permeability of the samples (*145*). Field evidence for such effects came from the Garm region of the former U.S.S.R., where Soviet seismologists had identified changes in the ratio of shear and compressional velocities, V_S/V_P, as precursors to some moderate earthquakes (*146*). Positive results on V_S/V_P precursors were also reported in the United States (*147*). These observations prompted refinements of the dilatancy diffusion model and a wider search for related precursors.

A reported prediction of an M 7.3 earthquake in Haicheng, China, is widely regarded as the single most successful earthquake prediction. An international team that visited China shortly after the quake (*148*) reported that the region had already been subject to an intermediate-term earthquake forecast based on seismicity patterns, magnetic anomalies, and other geophysical data. Accelerating seismic activity (Figure 2.14) and rapid changes in the flow from local water wells prompted Chinese officials to issue a short-term prediction and to evacuate thousands of unsafe buildings. At 7:36 p.m. (local time) on February 4, 1975, less than 24 hours after the evacuation began, the main shock destroyed 90 percent of the city. Chinese officials stated that because of the evacuation the number of casualties was extremely low for such an earthquake. This reported success stimulated great optimism in the earthquake prediction community, but it did not signal a widespread breakthrough in predic-

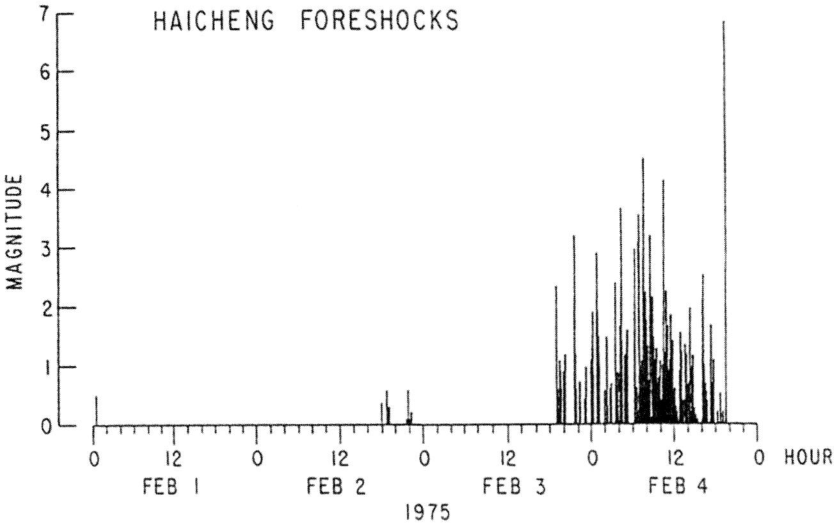

FIGURE 2.14 Plot of the magnitude versus time occurrence of the larger foreshocks of the Haicheng earthquake sequence, February 1-4, 1975. As indicated by the final large event, the earthquake occurred at 7:36 p.m. on February 4. SOURCE: P. Molnar, T. Hanks, A. Nur, B. Raleigh, F. Wu, J. Savage, C. Scholz, H. Craig, R. Turner, and G. Bennett, Prediction of the Haicheng earthquake, *Eos, Trans. Am. Geophys. Union*, **58**, 236-272, 1977. Copyright 1977 American Geophysical Union. Reproduced by permission of American Geophysical Union.

tion science. First, the foreshock series and hydrologic precursors were highly unusual, and similar phenomena have not been recognized before other large earthquakes. Second, the Chinese issued many false alarms, so the possibility of success by chance cannot confidently be rejected. Unfortunately, complete records of predictions and consequent actions are not accessible. The apparent triumph of the Haicheng prediction was soon overshadowed by disaster in July 1976, when a devastating (M 7.8) quake struck the Chinese city of Tangshan, resulting in the deaths of at least 240,000 people—one of the highest earthquake death tolls in recorded history. Although this area was also being monitored extensively, the disaster was not predicted.

Nevertheless, many prominent geophysicists were convinced that systematic short-term prediction was feasible and that the challenge remaining was to deploy adequate instrumentation to find and measure precursors of earthquake warnings (*149*). By 1976 a distinguished group of earthquake scientists convened by the National Research Council was willing to state (*150*):

> The Panel unanimously believes that reliable earthquake prediction is an achievable goal. We will probably predict an earthquake of at least magnitude 5 in California within the next five years in a scientifically sound way and with a sufficiently small space and time uncertainty to allow public acceptance and effective response.

In 1977, the U.S. government initiated the National Earthquake Hazards Reduction Program (Appendix A) to provide "data adequate for the design of an operational system that could predict accurately the time, place, magnitude, and physical effects of earthquakes." The USGS has the responsibility for issuing a prediction (statement that an earthquake will occur), whereas state and local officials have the responsibility for issuing a warning (recommendation or order to take defensive action).

The observational and theoretical basis for prediction soon began to unravel. Careful, repeated measurements showed that the purported V_S/V_P anomalies were not reproducible (*151*). At the same time, questions arose about the uniqueness of a posteriori reports of geodetic, geochemical, and electromagnetic precursors. Finally, theoretical models (*152*) incorporating laboratory rock dilatancy, microcracking, and fluid flow gave no support to the hypothesized V_S/V_P time history. By the end of the 1970s, most of the originally proposed precursors were recognized to be of limited value for short-term earthquake prediction (*153*).

Attention shifted in the 1980s to searching for transient slip precursors preceding large earthquakes. The hypothesis that such behavior might occur was based on the results of detailed laboratory sliding experiments and model simulations (*154*) and on qualitative field observations prior to an M 6 earthquake on the San Andreas fault near Parkfield, California (*155*). The preseismic slip observed under laboratory conditions was very subtle, but theoretical calculations suggested that under favorable conditions it might be observable in the field, provided that the critical slip distance D_c observed in the lab studies scaled to a larger size on natural faults.

To investigate these issues, the USGS launched a focused earthquake prediction experiment near Parkfield, in anticipation that an M 6 earthquake was imminent. Geodetic instrumentation, strainmeters, and tiltmeters were deployed to make continuous, precise measurements of crustal strains near the expected epicenter (Figure 2.15). The strain data were anticipated to place much stricter bounds on any premonitory slip. The predicted moderate earthquake has not occurred, so it is premature to evaluate the success of the search for short-term precursors. Nonetheless, the Parkfield experiment has contributed valuable data that improve our understanding of faults, deformation, and earthquakes.

After more than a century of intense research, no reliable method for short-term earthquake prediction has been demonstrated, and there is no

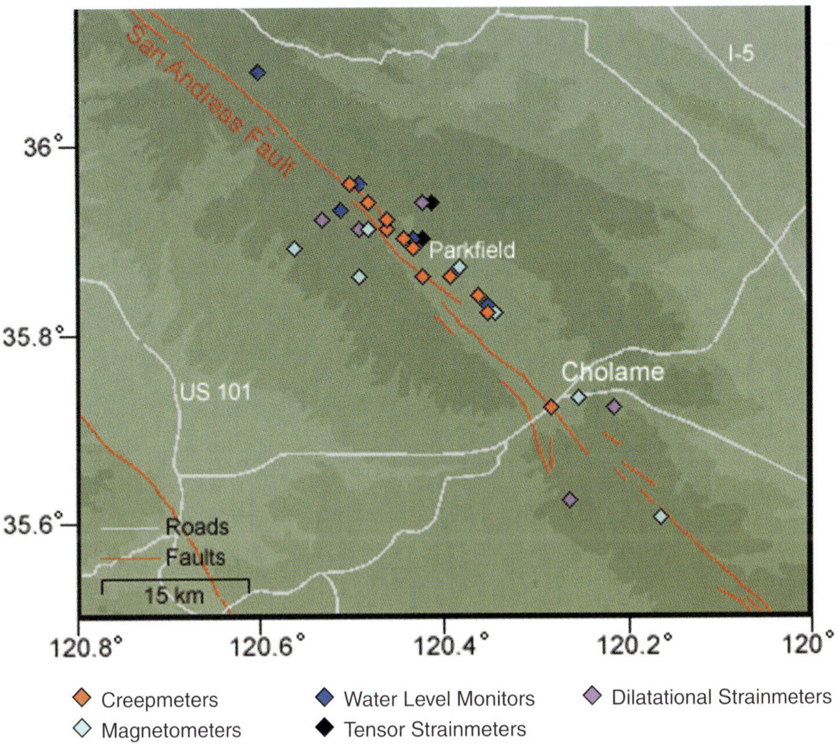

FIGURE 2.15 The network of crustal deformation instruments along the San Andreas fault maintained by the U.S. Geological Survey near Parkfield, California. SOURCE: U.S. Geological Survey.

guarantee that reliable short-term prediction will ever be feasible. At best, only a few earthquake "precursors" have been identified, and their applicability to other locations and earthquakes is questionable. Research continues on a broad range of proposed techniques for short-term prediction, as does vigorous debate on its promise (*156*). Most seismologists now agree that the difficulties of earthquake prediction were previously underestimated and that basic understanding of the earthquake process must precede prediction.

2.7 EARTHQUAKE ENGINEERING

The 1891 Nobi earthquake killed more than 7000 people and caused substantial damage to modern brick construction in the Nagoya region (*157*). Milne noted the extreme variability of ground shaking over short distances and reported that "buildings on soft ground . . . suffer more than those on hard ground." He laid the foundation for the development of codes regulating building construction by emphasizing that "we must construct, not simply to resist vertical stresses, but carefully consider effects due to movements applied more or less in horizontal directions" (*158*). Milne's conclusions were echoed in California following the 1906 San Francisco earthquake. J.C. Branner, a Stanford professor of geology on the Lawson Commission, supervised a detailed study of more than 1000 houses in San Mateo and Burlingame, and he noted that the local site response had a major influence on the level of damage: "The intensity of the shock was less on the hills than on the flat, in spite of the fact that the houses in the hills were nearer the fault line." Throughout California, the damage patterns were well correlated with the type of structure and building materials (*159*).

Early Building Codes

The first attempt to quantify the "earthquake design force" was made after the 1908 Messina-Reggio earthquake in southern Italy, which killed more than 83,000. In a report to the Italian government, M. Panetti, a professor of applied mechanics in Turin, recommended that new buildings be designed to withstand horizontal forces proportional to the vertical load (*160*). The Japanese engineer Toshikata Sano independently developed in 1915 the idea of a lateral design force V proportional to the building's weight W. This relationship can be written as $V = CW$, where C is a *lateral force coefficient*, expressed as some percentage of gravity (%g, where g = 9.8 m/s^2). The first official implementation of Sano's criterion was the specification C = 10 percent of gravity, issued as a part of the 1924 Japanese Urban Building Law Enforcement Regulations in response to the destruction caused by the great 1923 Kanto earthquake (*161*). In California, the Santa Barbara earthquake of 1925 motivated several communities to adopt codes with C as high as 20 percent of gravity. The first edition of the U.S. Uniform Building Code (UBC), published in 1927, also adopted Sano's criterion, allowing for variations in C depending on the region and foundation material (*162*). For building foundations on soft soil in earthquake-prone regions, the UBC's optional provisions corresponded to a lateral force coefficient equal to the Japanese value.

Measurement of Strong Ground Motions

By 1930, networks of permanent seismic observatories allowed the location and analysis of large earthquakes anywhere on the globe. However, the sensitive instruments could not register the strong (high-amplitude) ground motions close to large earthquakes, the primary cause of damage and loss of life, and were of little value to engineers. Consequently, engineers were forced to estimate the magnitude of the near-source ground accelerations from damage effects (e.g., overturned objects). The American engineer John Freeman voiced the frustration felt by many of his colleagues when he wrote in 1930 (*163*):

> The American structural engineer possesses no reliable accurate data about form, amplitude or acceleration of the motion of the earth during a great earthquake. . . . Notwithstanding there are upward of fifty seismograph stations in the country and an indefinitely large number of seismologists, professional and amateur; their measurements of earthquake motion have been all outside of the areas so strongly shaken as to wreck buildings.

Japanese seismologists were the first to attempt to obtain these data systematically. They began to record strong ground motions using long-period seismometers with little or no magnification, and by the 1930s, the development of broader-band, triggered devices allowed accurate measurement of the waves most destructive to buildings, those with shorter period and therefore higher acceleration. The Long Beach earthquake of 1933 was the first large event to be recorded by these improved strong-motion seismometers, several of which had been installed in the Los Angeles region just nine months before the earthquake. This new equipment recorded a peak acceleration of 29 percent of gravity on the vertical component and 20 percent of gravity on the horizontal component. The widespread damage caused by the 1933 Long Beach earthquake (Figure 2.16) spurred legislation for stricter building codes throughout California. One month after the event, the California Assembly passed the Field Act, which effectively prohibited masonry construction in public schools by instituting a lateral force requirement equivalent to 10 percent of the sum of the dead load (weight of the building) and the live load (weight of the contents). The Riley Act, also enacted in 1933, required all buildings in California to resist lateral forces of at least 2 percent of the total vertical design load. On September 6, 1933, the city of Los Angeles passed a law requiring a lateral force of 8 percent of the dead load plus 4 percent of the live load.

The success of the Long Beach recording can be credited to the Seismological Field Survey, which was established in California by the U.S. Department of Commerce at the urging of Freeman. A limited number of strong-motion instruments were deployed (*164*). One such instrument, located on the concrete block foundation of the Imperial Valley Irrigation District building in El Centro, recorded the next significant California event, the 1940 Imperial Valley earthquake (M 7.1). A peak horizontal

FIGURE 2.16 Jefferson Junior High School destroyed in the Long Beach earthquake. The Field Act, passed one month after the earthquake, prohibited this type of masonry construction for schools in California. SOURCE: Steinbrugge Collection, Earthquake Engineering Research Center, University of California, Berkeley. Photograph by Harold M. Engle.

acceleration of 33 percent of gravity was recorded at a distance of approximately 10 kilometers from the fault rupture. For the next 25 years, this was the largest measured ground acceleration, establishing the El Centro record as the de facto standard for earthquake engineering in the United States and Japan (Figure 2.17).

Response Spectra for Structural Analysis

Both the Long Beach and the El Centro data influenced the development of seismic safety provisions in building codes. However, the impact of seismometry on earthquake engineering was limited by the lack of data from a wider distribution of earthquakes, as well as by computational difficulties in performing a quantitative analysis of ground shaking and its effect on structures. Simplified techniques for structural analysis, such as H. Cross's moment-distribution method and K. Muto's *D*-value method, had been encoded in tables and figures by the early 1930s (*165*).

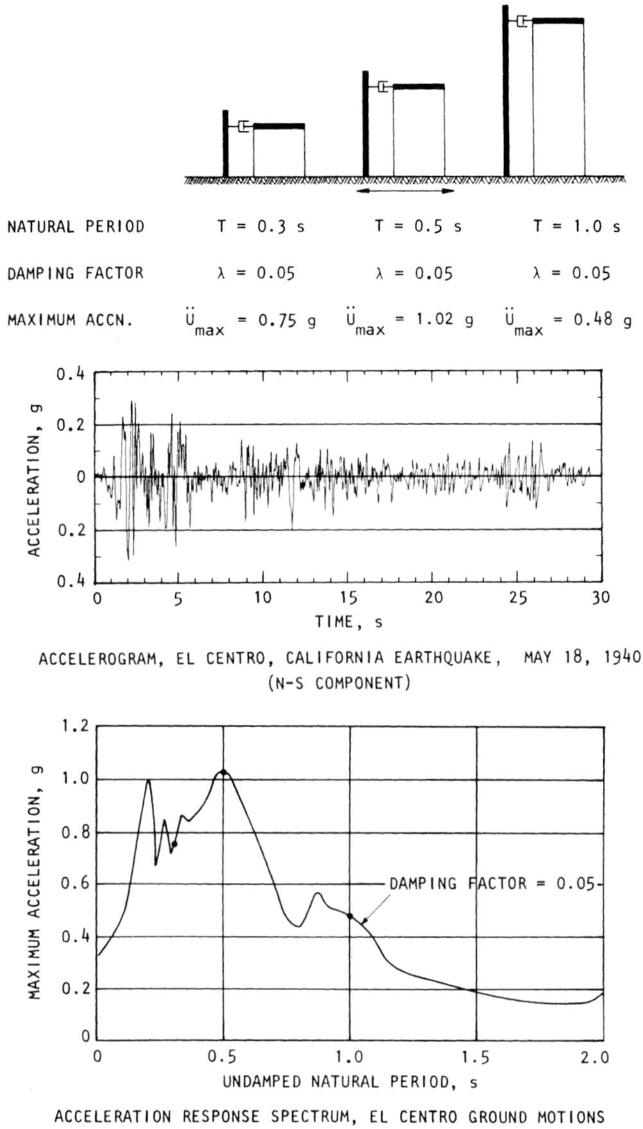

FIGURE 2.17 Accelerogram of the 1940 Imperial Valley earthquake (*center panel*), schematic representation of the response of a series of single degree-of-freedom oscillators having different natural periods (*top*), and their application to the construction of the response spectrum (*bottom panel*). SOURCE: H.B. Seed and I.M. Idriss, *Ground Motions and Soil Liquefaction During Earthquakes*, Earthquake Engineering Research Institute, Engineering Monograph on Earthquake Criteria, Structural Design, and Strong Motion Records 5, El Cerrito, Calif., 134 pp., 1982.

The advent of analog computers in the 1940s provided the first simulations of structural vibrations induced by the recorded ground motions (*166*) and allowed the automation of strong-motion spectral analysis (*167*). These early calculations showed that the spectra of earthquake accelerations are similar to "white noise" over a limited range of frequencies, a pivotal observation in the study of earthquake source processes. However, the immediate implication for earthquake engineering was the lack of a "dominant ground period" that might be destructive to particular structures (*168*). Without a characteristic frequency, earthquake engineering was recognized to be complex, requiring a comprehensive analysis of coupled vibrations between earthquakes and structures. George Housner outlined the issues in 1947:

> In engineering seismology, the response of structures to strong-motion earthquakes is of particular interest. . . . During an earthquake a structure is subjected to vibratory excitation by a ground motion which is to a high degree erratic and unpredictable. . . . Furthermore, the average structure, together with the ground upon which it stands, is an exceedingly complex system from the viewpoint of vibration theory. It is apparent the problem divides itself into two parts; first a determination of the characteristics of strong motion earthquakes, and second a determination of the characteristics of structures subjected to earthquakes.

Following an earlier suggestion by M.A. Biot, Housner put forward the concept of the response spectrum, the maximum response induced by ground motion in single degree-of-freedom oscillators ("buildings") with different natural periods but the same degree of internal damping (usually selected to be 5 percent) (*169*) (Figure 2.17). At shorter periods the maximum induced acceleration exceeds the recorded ground acceleration, whereas for longer periods it is less. When multiplied by the effective mass of a building, the response spectrum acceleration constrains the lateral force that a building must sustain during an earthquake. Computing response spectra over a wide range of frequencies using data from a wide range of earthquakes significantly improved understanding of the damage potential of strong motion.

Building Code Improvements Since 1950

The availability of strong-motion data began to transform earthquake engineering from a practice based on pseudostatic force criteria to a science grounded in an understanding of the complex coupling between ground motions and building vibrations. By the 1950s, strong-motion records were combined with response spectral analysis to demonstrate that structures can amplify the free-field accelerations (recorded on open ground). To approximate this dynamic behavior, a committee of the American Society of Civil Engineers and the Structural Engineers Association of Northern California proposed in 1952 that the lateral force requirement be revised to vary inversely with the building's fundamental period of vibration ($C \sim T^{-1}$). With only a handful of strong-motion recordings available at the time, the decrease in the response spectral accelerations with period remained uncertain. Particular attention was focused on the band from 0.5 to 5.0 seconds, which includes the fundamental periods of vibration for most midrise to high-rise buildings as well as many other large structures.

The lateral force coefficient was recast in the 1961 UBC with a weaker (inverse cubed root) dependence on the response period: $C \sim ZKT^{-1/3}$. This version introduced a seismic zone factor Z that represented the variability of the seismic hazard throughout the United States and a structural factor K that depended on building type and accounted for its dynamic response. The parameters were chosen to reproduce as well as possible the response spectral accelerations measured in previous earthquakes, which were still sparse. The uncertainties in the empirical coefficients remained high, but the form of the lateral force requirement did establish a firm connection between strong-motion measurements and the requirements of earthquake engineering.

The dearth of strong-motion data ended when the San Fernando earthquake (M 6.6) struck the Los Angeles region on February 9, 1971. It subjected a community of more than 400,000 people to ground accelerations greater than 20 percent of gravity and triggered in excess of 200 strong-motion recorders, more than doubling the size of the database. San Fernando provided the first well-resolved picture of the temporal and spatial variability of ground shaking during an earthquake (*170*). Short-period (0.1-second) accelerations varied widely, even among nearby sites with similar geologic conditions, while long-period (10-second) displacements were coherent over tens of kilometers (*171*). More important, this earthquake demonstrated that the ground motions could substantially exceed the maximum values observed in previous events. A strong-motion instrument on an abutment of the Pacoima Dam, 3 kilometers above the fault plane, recorded a sharp, high-amplitude (100-centimeter-per-second) velocity pulse in the first three seconds of the earthquake, as the rupture front passed under the dam (Figure 2.18). Four seconds later, after the rupture had broken the surface 5 kilometers away in the San Fernando Valley, the Pacoima instrument recorded an acceleration pulse exceeding 1.2 gravity in the horizontal plane. This value more than doubled the highest previously observed peak ground acceleration (PGA), measured during the 1966 Parkfield earthquake (M 5.5) on the San Andreas fault (*172*). The short acceleration pulse observed at Pacoima Dam

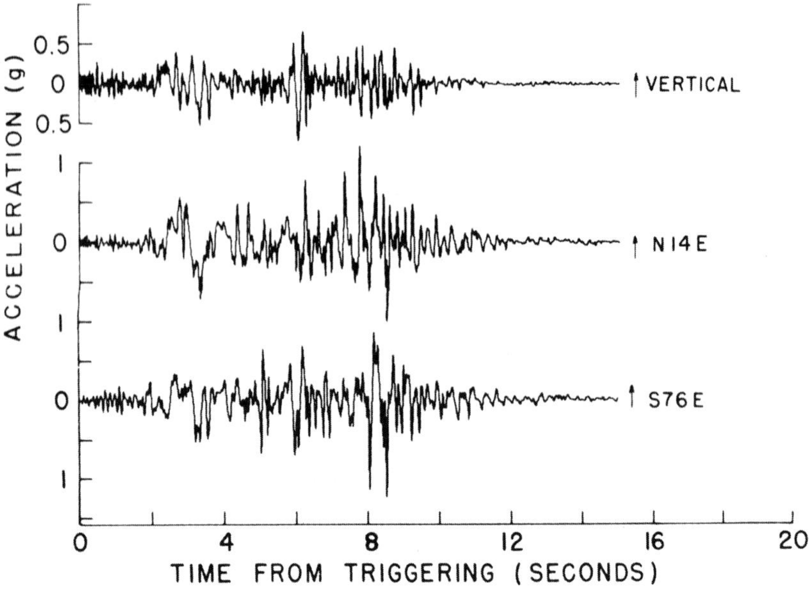

FIGURE 2.18 Three component records of acceleration, velocity, and displacement from the Pacoima Dam record of the San Fernando earthquake. The horizontal instruments are approximately parallel and perpendicular to the horizontal component of the rupture (N14E and S76E, respectively) and show accelerations exceeding 1 *g*. SOURCE: D.M. Boore and M.D. Zoback, Two dimensional kinematic fault modeling of the Pacoima Dam strong-motion recordings of February 9, 1971, San Fernando Earthquake, *Bull. Seis. Soc. Am.*, **64**, 555-570, 1974. Copyright Seismological Society of America.

engendered much discussion regarding the utility of PGA as a measure of seismic hazard. This pulse did not make a significant contribution to the overall response spectra values, except at the shortest periods (*173*), and when the data from all available earthquakes were considered, PGA was only weakly correlated with the size of the earthquake (*174*). From these and subsequent studies, it became clear that the PGA was not necessarily the best determinant of seismic hazard to structures; other characteristics—such as the response spectrum ordinates, anisotropic motions, and the occurrence of intense, low-frequency velocity pulses—were found to be more important.

After the 1971 San Fernando earthquake, policy makers tried to update building codes in light of the large amount of data on ground motion and building response collected from this urban event. The wealth of strong-motion data also prompted a 1976 revision to the UBC, which modified the period scaling in the lateral force equation from $T^{-1/3}$ to $T^{-1/2}$ and introduced a factor S based on local soil type. The newly formed Applied Technology Council (ATC) organized, with funding from the National Science Foundation (NSF) and the National Bureau of Standards, a national effort to develop a model seismic code. More than 100 professionals who volunteered for the work were organized into 22 committees. In a comprehensive report published in 1978 (*175*), the ATC proposed a more physically based lateral force coefficient of the form $C \sim A_v S R^{-1} T^{-2/3}$, where A_v is the effective peak ground velocity-related acceleration coefficient, S is a site-dependent soil factor, and R is a "response modification factor" dependent on the structure type. At shorter periods, this expression was replaced by a limiting value proportional to the effective peak acceleration coefficient A_a. The report also provides the first contoured maps of the ground-motion parameters A_a and A_v, derived from a probabilistic seismic hazard analysis conducted by the USGS.

Strong-motion data from a number of earthquakes, as well as laboratory test data and results of numerical site response models, demonstrated the need to modify the soil factor S to reflect nonlinear site response. The National Center for Earthquake Engineering Research, which NSF established in 1986, led the revision, recommending two sets of amplitude-dependent, site amplification factors derived for six site geology classifications. The factors were first incorporated in the 1994 National Earthquake Hazard Reduction Program (NEHRP) seismic provisions and then into the 1997 UBC.

Strong-motion data from the 1994 Northridge, California (M 6.7), and 1995 Kobe, Japan (M 6.9), earthquakes confirmed observations from several previous earthquakes that motions recorded close to the fault rupture had distinct pulse-like characteristics, which were not represented in the code's lateral force equation. The effect of these pulse motions was approximated by introducing near-fault factors into the lateral force equation. A single factor N was first introduced in the base isolation section of the 1994 UBC. This representation was replaced by two near-fault factors N_a and N_v in the lateral force provisions of the 1997 UBC lateral force requirement. The N_a factor was applied to the short-period, constant-acceleration portion of the design response spectrum, whereas the N_v factor was applied to the intermediate- and long-period constant-velocity portion, where the base shear is proportional to T^{-1}. Interestingly, after a 36-year absence, the T^{-1} proportionality was reintroduced in the 1997 UBC in part because it was judged to be a more accurate representation of the spectral character of earthquake ground motion (*176*).

Attenuation Relationships

Engineers wanted the scattered ground-motion observations reduced to simple empirical relationships that practitioners could apply, and the derivation of these relationships became a central focus of engineering seismology. A measure of shaking intensity was chosen (typically peak ground acceleration or velocity), and the observed variation of this intensity measure was factored into source, path, and site effects by identifying one or more independent control variables—typically, source magnitude, path distance, and site condition (e.g., soil or rock)—and fitting the observations with parameterized curves. The magnitude dependence or scaling and the fall-off of strong-motion amplitude with epicentral distance were together called the attenuation relation.

Lack of data precluded plotting PGA as a function of magnitude and epicentral distance until the 1960s. Figure 2.19 shows an attenuation relationship obtained from the strong-motion data for the 1979 Imperial Valley earthquake. The dispersion in the data resulted in a relative standard deviation of about 50 percent, which was typical. Other relationships described the site response in terms of the correlation between intensity measures and soil and rock conditions, including allowance for nonlinear soil behavior as a function of shaking intensity (Figure 2.20).

As the use of the response spectrum method increased, it became necessary to develop techniques to predict not only the PGA (equivalent to the response spectral value at zero period) but also the response spectra of earthquakes that might occur in the future. This was done initially by developing a library of response spectral shapes that varied with earthquake magnitude and soil conditions; the selected shape was anchored to a peak acceleration obtained from a set of attenuation relationships, each of which predicted the spectral acceleration at a specific period. Eventually, response spectra were computed directly from ground-motion attenuation relationships.

Seismic Hazard Analysis

By the 1960s, growing strong-motion databases and scientific understanding enabled site-specific seismic hazard assessments incorporating information about the length and distance of neighboring faults, the history of seismicity, and empirical predictions of ground-motion intensity for events of specified magnitude at specified distances. For major facilities in the western United States, in particular nuclear power plants such as San Onofre and Diablo Canyon (*177*), seismic hazard assessment focused on the maximum magnitude that each fault could produce, its closest distance to the site, and the PGA for these events. PGA was then the

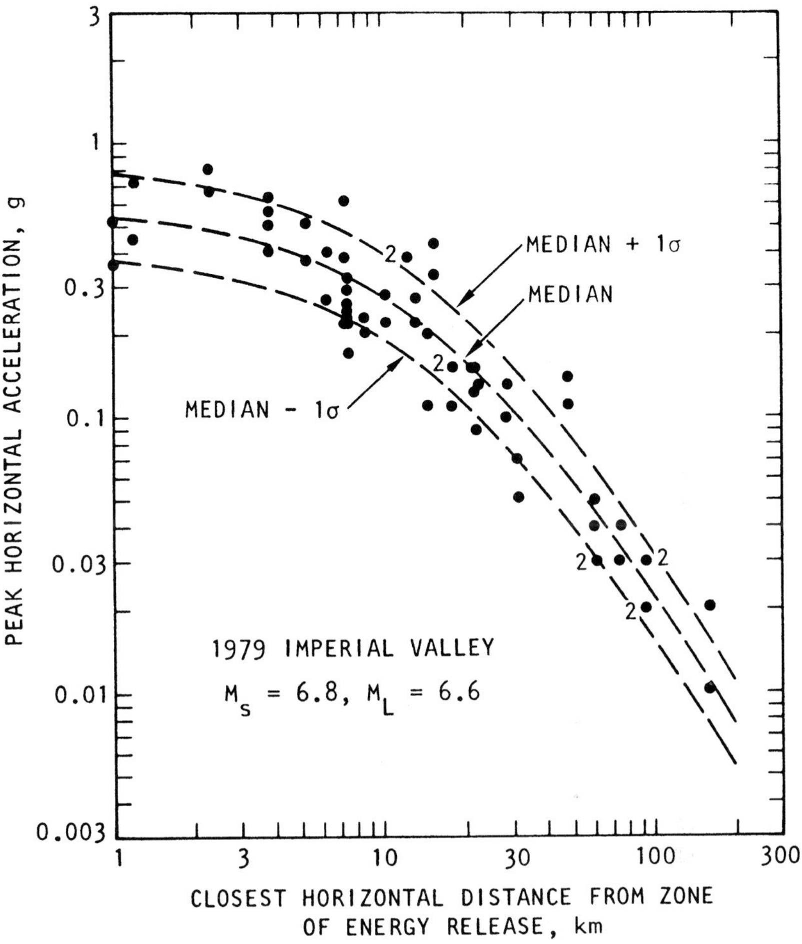

FIGURE 2.19 Recorded peak accelerations of the 1979 Imperial Valley earthquake and an attenuation relation that has been fit to the data. SOURCE: H.B. Seed and I.M. Idriss, *Ground Motions and Soil Liquefaction During Earthquakes*, Earthquake Engineering Research Institute, Engineering Monograph on Earthquake Criteria, Structural Design, and Strong Motion Records 5, El Cerrito, Calif., 134 pp., 1982.

primary scalar measure of ground-motion intensity for use in structural analysis and design. Typically PGA was used to scale a standard response spectral shape or, if the engineer requested more detailed ground-motion information, "standard" accelerograms, such as the El Centro record from the 1940 Imperial Valley earthquake. The basic motivation for these pro-

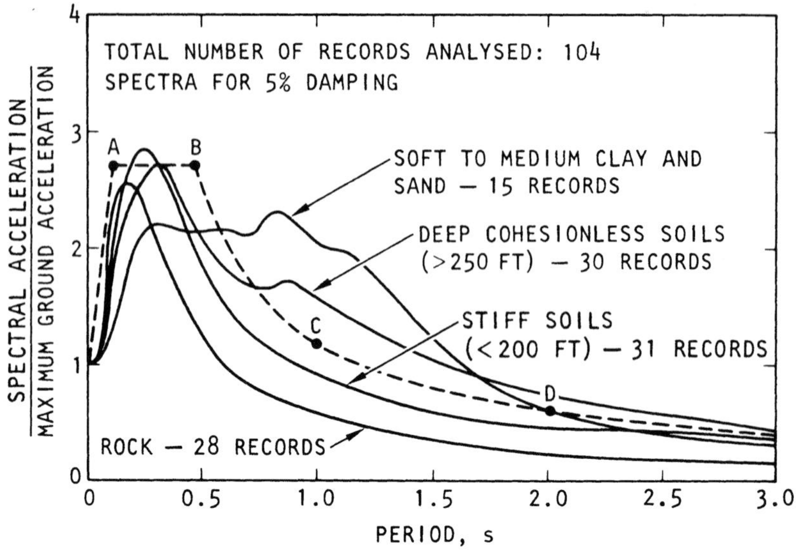

FIGURE 2.20 Average response spectral shapes for different soil categories derived from strong-motion recordings. The response spectral shapes are normalized to peak acceleration, which is equivalent to zero-period acceleration. The dashed line ABCD indicates a simplified response spectral shape for rock and stiff soil sites that was developed for use in building codes. SOURCE: H.B. Seed and I.M. Idriss, *Ground Motions and Soil Liquefaction During Earthquakes*, Earthquake Engineering Research Institute, Engineering Monograph on Earthquake Criteria, Structural Design, and Strong Motion Records 5, El Cerrito, Calif., 134 pp., 1982.

cedures was to identify a conservative or bounding value for the maximum potential threat at a specific site. These procedures are now referred to as deterministic to distinguish them from the probabilistic techniques that followed.

The deterministic methods of seismic hazard analysis developed for seismically active sites in California and other western states were unsuited to the tectonically stable environment of the eastern United States, where likely earthquake sources were largely unknown and strong-motion data had not yet been recorded. Therefore, a modified deterministic analysis had to be developed for the some 100 nuclear power plants east of the Rocky Mountains, based on "seismotectonic zones" developed from historic seismic activity and geologic trends. Application of these methods relies heavily on the judgment of scientists and engineers. Although the historical record for this analysis is comparatively long (about 300 years), estimates of past earthquake magnitude were limited to verbal accounts of earthquake effects translated into the 12 levels of the Modified Mercalli Intensity (MMI) scale (see endnote 27). The largest historic event in the zone became the basis for establishing the maximum event, typically the largest MMI or one-half intensity unit larger, depending on circumstances (e.g., nature of the facility, design rules, safety factors adopted by the engineers). The largest event in each zone was then presumed to occur as close to the site as the seismotectonic zone boundary permitted, except for events in the zone that contained the site, where a minimum separation was adopted to reflect the improbability of an event occurring very close to the site. Lacking sufficient ground-motion data, engineering seismologists used MMI data to develop attenuation relations, calibrating MMI to PGA with data from the western United States.

Probabilistic seismic hazard analysis (PSHA) was developed to characterize and integrate several effectively random elements of earthquake occurrence and ground-motion forecasting. The method uses probabilistic analysis to combine the seismic potential from several threatening faults, or spatially distributed across source zones characterized for each by an assumed frequency-magnitude distribution, to obtain an estimate of the total hazard, defined as the mean annual rate at which the chosen intensity measure, such as PGA, will exceed some specified threshold at a prescribed site (*178*). For each fault, the contribution to hazard was derived from a convolution of the mean annual rate of earthquakes with the probability that the shaking intensity will be exceeded for an event of specified magnitude. The method allows for assumed distributions of event location (e.g., randomly along the fault or within a region) and for variance about the predicted ground motions due to natural variability. The final result for a site is a hazard curve, a plot of mean annual frequency of exceedance at a specified intensity level.

By the late 1970s, the PSHA method had been tested and its application was growing throughout engineering seismology. As with the deterministic method, the practical application of PSHA at a specific site requires professional judgments based on local data and experience. In response to these difficulties, uncertainties in the model parameters associated with the limits of scientific information (e.g., in earthquake catalogs, fault locations, ground-motion prediction) are quantified and propagated through PSHA to produce quantitative confidence bounds on the resulting hazard curves. The objective of the hazard curve is to capture the randomness or "aleatory uncertainty" inherent in the forecasting of future events, while the confidence bounds reflect the current limits on professional knowledge, or "epistemic uncertainty," in such forecasts. Figure 2.21 presents an example of the analysis for a site in the San Francisco Bay area.

PSHA relies on a wider range of scientific information than deterministic analysis. It also satisfies modern engineering requirements for a

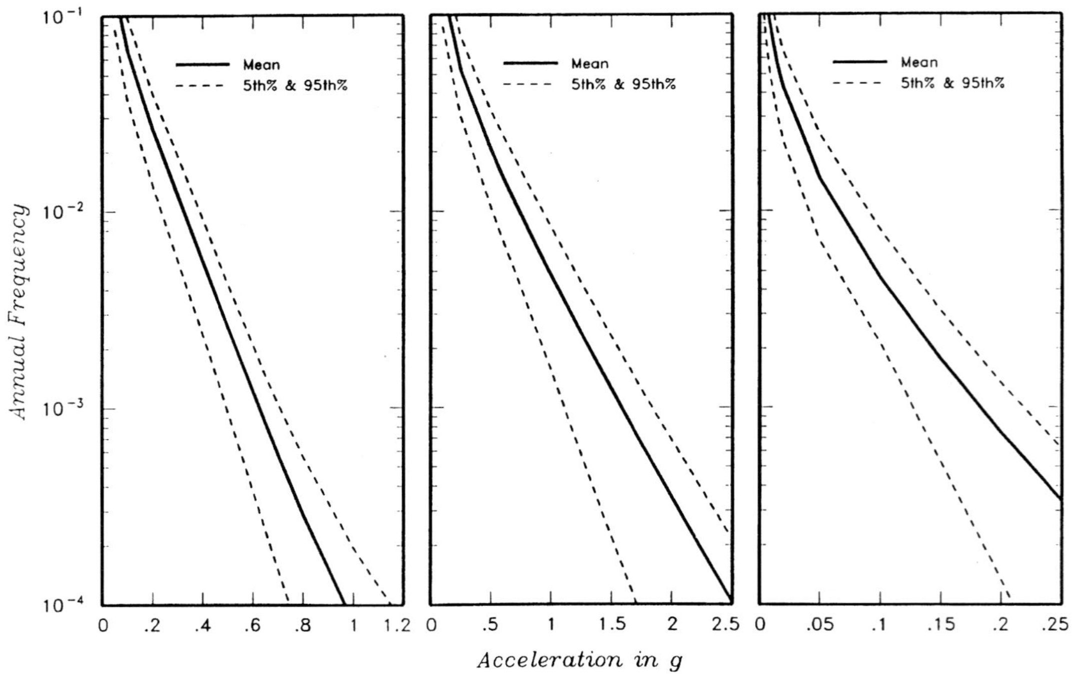

FIGURE 2.21 Mean estimate and 5/95 percent confidence bounds on the mean annual frequency hazard curves for the east end of the San Mateo-Hayward Bridge, San Francisco Bay area, California, for PGA and 5 percent damped spectral accelerations at oscillator periods of 0.3 and 3.0 seconds. SOURCE: Geomatrix Consultants, Inc., *Seismic Ground Motion Study for San Mateo-Hayward Bridge, Final Report*, prepared for CALTRANS, Division of Structures, Oakland, Calif,, 234 pp. + 6 appendixes, February 1993.

probabilistic definition of risk. Common engineering practice evolved to define a design ground-motion in terms of a specified frequency of exceedance. This value is lower (i.e., increases the design requirements) for facilities where structural failure involves more severe consequences.

Seismic Hazard Maps

Seismic hazard analysis for buildings, highway overpasses, and smaller structures has traditionally relied on design values mapped nationally or regionally. Early maps were quite crude owing to the typical building-code practice of using only four or five discrete zones with large relative differences (factor of 2) in ground-motion level. At first, the zones were drawn largely to reflect historic seismicity. For example, the first seismic probability map for the United States, distributed in 1948 by the U.S. Coast and Geodetic Survey (USCGS), simply used the locations of historic earthquakes and divided the country into four zones ranging from no expected damage to major damage (*179*). This basis led to understated earthquake hazards in the Pacific Northwest, the eastern Basin and Range Province, and other places with long recurrence intervals. The work was revised in 1958 and 1959 when Charles Richter published several maps based on the seismic regionalization technique that Soviet seismologists had developed in the 1940s (*180*). Richter also relied on historic seismicity and employed MMI as the intensity measure. In 1969, S.T. Algermission of the USCGS produced a national map with maximum MMI values from historic earthquakes contoured as zones, along with a table and map of earthquake recurrence rates. The maximum-intensity map was the basis for the UBC national zoning map published in 1970.

Several years later, Algermissen and coworkers at the USGS, using PSHA, repeated the national mapping (*181*). They produced a seismic hazard curve at each point on a grid; the PGA was calculated for a 10 percent probability of exceedance in 50 years; and these values were contoured to produce a national seismic hazard map. The maps provided quantitative estimates of the expected shaking (excluding site effects). They also furnished a compelling visual representation of the relative seismic hazard among different locations in the United States and were the basis for national building code zoning maps in 1979. The USGS updated the national seismic hazard maps in 1982, 1990, 1991, 1994, and 1996, incorporating new knowledge on earthquake sources and seismic-wave propagation. The 1991 maps were the first to display probabilistic values of response spectral ordinates and were published in the *NEHRP Recommended Provisions for Seismic Regulations for New Buildings*. The 1996 maps implemented a completely new PSHA methodology and provide the basis for the probabilistic portion of the seismic design guidelines in the 1997 and 2000 *NEHRP Provisions* and the 2000 International Building Code. These seismic hazard maps are also used in seismic provisions for highway bridge design, the International Residential Code, and many other applications.

Challenges Ahead

Establishing building codes, developing attenuation relationships and performing seismic hazard analysis are all examples of earthquake engineering activities that have helped quantify and reduce the threat posed by earthquakes; however, recent large earthquakes make it clear that significant challenges remain. For example, the 1995 Hyogo-ken Nanbu earthquake (Box 2.5, Figures 2.22 and 2.23) devastated the city of Kobe, in Japan—one of the most earthquake-prepared countries in the world. That this earthquake caused such tremendous damage and loss of life indicates

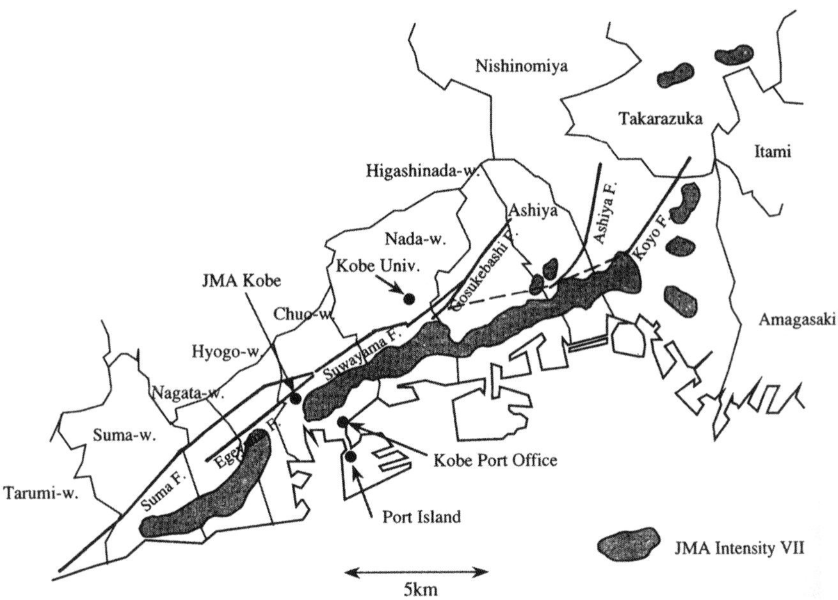

FIGURE 2.22 Kobe-Osaka region, annotated with major faults and geographic reference points, including inferred rupture during the 1995 earthquake and damage belt. SOURCE: H. Kawase, The cause of the damage belt in Kobe: "The basin-edge effect," constructive interference of the direct s-wave with the basin-induced diffracted/Rayleigh waves, *Seis. Res. Lett.*, **67**, 25-34, 1996. Copyright Seismological Society of America.

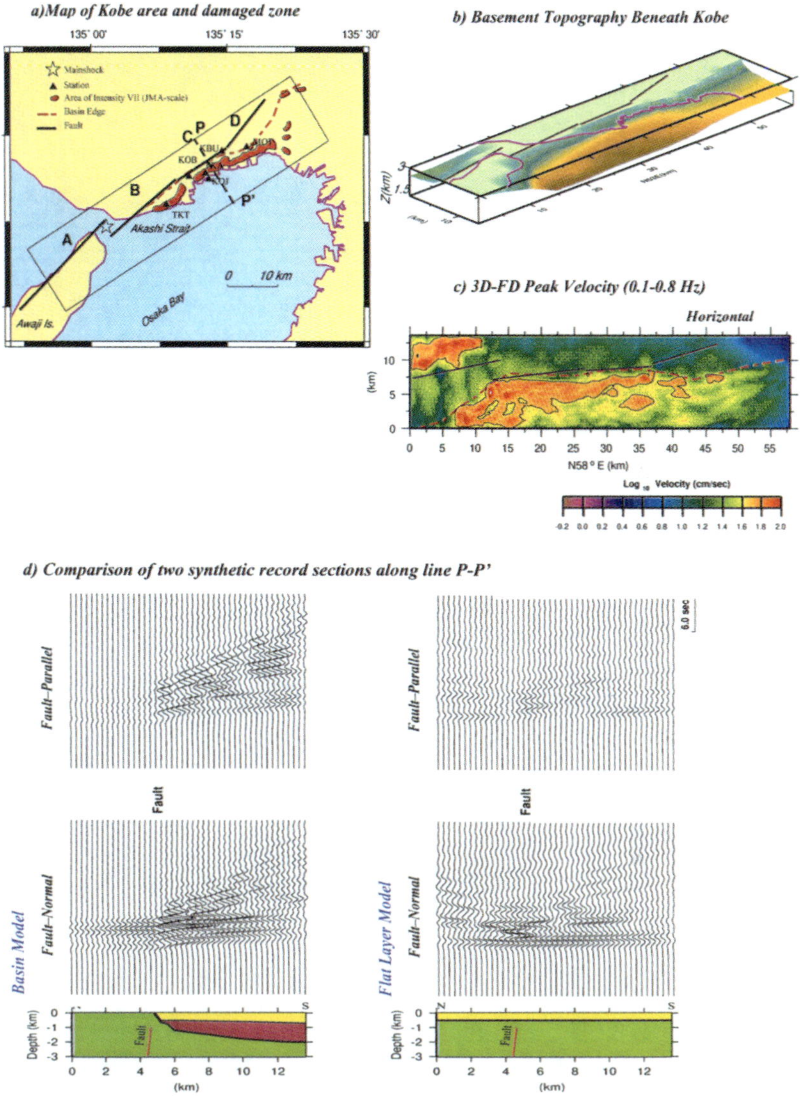

FIGURE 2.23 Numerical simulations of basin-edge effects in the 1995 Hyogo-ken Nanbu earthquake. SOURCE: A. Pitarka, K. Irikura, T. Iwata, and H. Sekiguchi, Three-dimensional simulation of the near-fault ground motion for the 1995 Hyogo-ken Nanbu (Kobe), Japan, earthquake, *Bull. Seis. Soc. Am.*, **88**, 428-440, 1998. Copyright Seismological Society of America.

BOX 2.5 Kobe, Japan, 1995

The official name of the M 6.9 earthquake that struck Kobe, Japan, on January 17, 1995 is Hyogo-ken Nanbu (Southern Hyogo Prefecture). It killed at least 5500 people, injured more than 26,000, and caused immense destruction throughout a metropolis of 1.5 million people. One-fifth of its inhabitants were left homeless, and more than 100,000 buildings were destroyed. The total direct economic loss has been estimated as high as $200 billion.[1] The Japanese call an earthquake with an epicenter directly under a city a *chokkagata*. History has demonstrated repeatedly that a direct hit on an urban center can be terribly destructive; for example, an earlier chokkagata wiped out the city of Tangshan, China, in 1976, killing at least 240,000. Nevertheless, given the rigorous Japanese building codes and disaster preparations, the extreme devastation to the city center was surprising. In contrast, the 1994 Northridge earthquake was of comparable size (M 6.7, only a factor of 2 smaller in seismic moment), and occurred in a densely populated region (the San Fernando Valley of California) but killed only 57 people and caused about $20 billion in damages.[2] The high losses in the Hyogo-ken Nanbu earthquake can be attributed to at least four independent factors:

1. *Rupture Directivity*: The hypocenter of the earthquake was on a nearly vertical fault at a depth of about 14 kilometers directly beneath the Kaikyo suspension bridge. The right-lateral rupture propagated southwestward toward Awaji Island, where surface displacement of 1 to 1.5 meters was mapped on the Nojima fault, and northeastward along the trend of the Rokko fault zone, straight into the city center.[3] This pattern of faulting radiated large-amplitude motions into the heart of downtown Kobe, as if it were at the end of a gun. In comparison, the seismic energy from the Northridge earthquake was preferentially radiated northward into the sparsely populated Santa Susanna Mountains.

2. *Basin-Edge Effect*: The surface outcrops of the Rokko fault—which mark the boundary between the hard granites in the hills northwest of the city and the soft sediments of the low, narrow coastal strip—were not displaced. Instead, in a narrow zone 500 to 1000 meters southeast of the fault, out in the sedimentary plain, building were heavily damaged and houses collapsed (see Figure 2.23). The very strong ground motions in this damage belt were caused by the constructive interference of two waves at the edge of the sedimentary basin bounded by the Rokko fault zone.[4] The basin-edge effect identified at Kobe may explain other peculiarities in the patterns of seismic damage, such as the localized regions of damage in Santa Monica at the northwestern edge of the Los Angeles basin during the 1994 Northridge earthquake.[5]

3. *Poor Construction and Maintenance of Buildings*: Differences in the vulnerability of buildings also contributed to the greater damage in Kobe than in Northridge. In California, engineers introduced ductility concepts into building design in 1971, a full decade before their adoption in Japan. While building codes are an effective way to reduce damage to newly erected structures, they do not address the seismic safety of older structures. Most of the buildings severely damaged in Kobe were built before 1981. Because of the progress in seismic engineering over the last several decades, exposure to seismic risk can be much higher for urban areas with many older buildings, such as the cities in the eastern United States.

(continued on next page)

4. *Poor Emergency Response*: The Japan Meteorological Agency (JMA) announced preliminary estimates of the magnitude and hypocenter of the earthquake only 18 minutes after the earthquake, but government officials were still accustomed to using seismic intensity, rather than magnitude and location, to characterize earthquake severity. Since the earthquake immediately knocked out telephone lines in the Kobe region, the highest intensities available to the JMA—5 on the Japanese scale—were from Kyoto and other locations away from the epicenter.[6] Emergency response officials did not comprehend the full extent of the disaster until four to six hours after the earthquake, when it became clear that the intensity in downtown Kobe had been as high as 7. This incident highlights the need for seismic information systems that can rapidly collect, analyze, and broadcast accurate information on peak accelerations and other measures of strong ground motions from a well-distributed set of instruments, as well as the event's magnitude and location, and they must be rugged enough to withstand large earthquakes. It also underscores the need for emergency response officials to be well educated in the proper interpretation of seismological data.

---

[1] The 1995 Hyogo-ken Nanbu earthquake also provided an example of the dire impact that earthquakes can have on global financial institutions. Nicholas Leeson of the Singapore office of Baring's Bank in London had been involved in unauthorized trading of Japanese government bonds, and the sudden drop in their value following the Hyogo-ken Nanbu earthquake caused the bank's collapse.

[2] USGS response to an urban earthquake <http://geohazards.cr.usgs.gov/northridge/norpub1.htm>.

[3] H. Kanamori, The Kobi (Hyogo-ken Nanbu), Japan, earthquake of January 16, 1995, *Seis. Res. Lett.*, **66**, 6-10, 1995.

[4] H. Kawase, The cause of the damage belt in Kobe: "The basin-edge effect," constructive interference of the direct s-wave with the basin-induced diffracted/Rayleigh waves, *Seis. Res. Lett.*, **67**, 25-34, 1996; A. Pitarka, K. Irikura, T. Iwata, and T. Kagawa, Basin structure effects in the Kobe area inferred from the modeling of ground motions from two aftershocks of the January 17, 1995 Hyogo-ken Nanbu earthquake, *J. Phys. Earth*, **44**, 563-576, 1996; A. Pitarka, K. Irikura, T. Iwata, and H. Sekiguchi, Three-dimensional simulation of the near-fault ground motion for the 1995 Hyogo-ken Nanbu (Kobe), Japan, earthquake, *Bull. Seis. Soc. Am.*, **88**, 428-440, 1998. According to these studies, the strongest ground motions were generated by the constructive interference of two simultaneous arrivals: the direct *S* (shear) wave propagating vertically through the soft sediments and a horizontally propagating surface wave diffracted into the basin. The latter was generated when the direct *S* wave traveling faster through the hard, granitic rocks of the sidewall encountered the sharp (fault-controlled) edge of the basin.

[5] R.W. Graves, A. Pitarka, and P. G. Somerville, Ground motion amplification in the Santa Monica area: Effects of shallow basin edge structure, *Bull. Seis. Soc. Am.*, **88**, 1224-1242, 1998; P.M. Davis, J.L. Rubinstein, K.H. Liu, S.S. Gao, and L. Knopoff, Northridge earthquake damage caused by geologic focusing of seismic waves, *Science*, **289**, 1746-1750, 2000; W.J. Stephenson, R.A. Williams, J.K. Odum, and D.M. Worley, High resolution seismic reflection surveys and modeling across an area of high damage from the 1994 Northridge earthquake, Sherman Oaks, California, *Bull. Seis. Soc. Am.*, **90**, 643-654, 2000.

[6] R. Geller, The role of seismology, *Nature*, **373**, 554, 1995; K. Yamakawa, The Prime Minister and the earthquake: Emergency management leadership of Prime Minister Murayama on the occasion of the great Hanshin-Awaji earthquake disaster, *Kansai Univ. Rev. Law and Politics*, **19**, 13-55, 1998.

that reducing, or even containing, the vulnerabilities to future earthquakes as urbanization of earthquake-prone regions increases, constitutes a major and continuing challenge for earthquake science and engineering.

NOTES

1. A general historical account is given by B.A. Bolt, *Earthquakes and Geological Discovery*, W.H. Freeman, New York, 229 pp., 1993. For a history of Japanese seismology, see T. Utsu, Seismological evidence for anomalous structure of island arcs, *Rev. Geophys. Space Phys.*, **9**, 839-890, 1971.

2. W. Whiston, translator, *The New Complete Works of Josephus*, Kregel Publications, Grand Rapids, 1143 pp., 1999.

3. The Aristotelian theory is the root of folkloric notions about "earthquake weather." In describing a series of earthquakes felt in London during 1750, Stephen Hales, a preacher, scientist, and follower of Isaac Newton, echoed Aristotle: "We find in the late earthquakes in London, that before they happen there is usually a calm air with a black sulfurous cloud which would probably be dispersed like a fog if there were a wind; which dispersion would prevent the earthquake which is probably caused by the explosive lightning of this sulfurous cloud; being both near the Earth and coming at a time when sulfurous vapors are rising from the Earth in greater quantity than usual which is often occasioned by a long period of hot and dry weather. Ascending sulfurous vapors in the Earth may probably take fire, and thereby cause Earth lightning which is first kindled at the surface and not at great depths as has been thought whose explosion is the immediate cause of an earthquake."

4. R. Mallet, *Neapolitan Earthquake of 1857. The First Principles of Observational Seismology*, Chapman and Hall, London, 2 vols., 831 pp., 1862. He also introduced the term *hypocenter* for the focus of the earthquake, which he presumed was a volcanic explosion, and deduced its location from the observed directions of ground motions, assumed to be excited by pure compressional waves. Despite the crudeness of his method, his estimate of the focal depth, about 10 kilometers, was probably not far off.

5. Much earlier than Lyell's text was the Book of Zachariah (14:4-5), which details a future scenario for a surface-faulting earthquake: "And his feet shall stand on that day upon the Mount of Olives, which is before Jerusalem on the east. And the Mount of Olives shall cleave in the midst thereof towards the east and towards the west. And there shall be a great valley and half of the mountain shall remove towards the north and half of it towards the south. And ye shall flee to the valley of the mountain as ye fled from before the earthquake in the days of Uzziah, King of Juda."

6. Darwin's observations were not actually new. In a report to the London Geographical Society (An account of some effects of the late earthquakes in Chili: Extracted from a letter to Henry Warburton, *Trans. Geol. Soc. London, Ser. 2*, **1**, 413-415, 1824), Maria Graham, an English travel writer, documented coastal uplift during an earlier earthquake near Valparaiso, Chile, in 1822: "I found the ancient bed of the sea laid bare and dry, with beds of oysters, mussels, and other shells adhering to the rocks on which they grew, the fish all being dead, and exhaling the most offensive effluvia."

7. Although Gilbert emphasized the normal component of faulting, it is now recognized that the 1872 event included a significant component of strike-slip.

8. G.K. Gilbert, *Lake Bonneville*, U.S. Geological Survey Monograph 1, U.S. Government Printing Office, Washington, D.C., 340 pp., 1890.

9. R.D. Oldham, *Report on the Great Earthquake of the 12th June, 1897*, Geological Survey of India, Memoir 29, Calcutta, 379 pp., 1899; C.S. Middlemiss, *The Kangra Earthquake of 4th April, 1905*, Geological Survey of India, Memoir 37, Calcutta, 409 pp., 1910. Middlemiss'

data set was good enough that it could be used eight decades later to model this event as a blind thrust (R. Chander, Interpretation of observed ground level changes due to the 1905 Kangra earthquake, northern Himalaya, *Tectonophysics*, **149**, 289-298, 1988).

10. During the first half of the nineteenth century, most geologists viewed vertical uplift by magmatic processes as the main cause of mountain building. The importance of horizontal compression was recognized in the context of Appalachian tectonics by W.B. Rogers and H.D. Rogers (On the physical structure of the Appalachian chain, as exemplifying the laws which have regulated the elevation of great mountain chains, generally, *Assoc. Am. Geol. Rep.*, **1**, 474-531, 1843) and championed by the supporters of Élie de Beaumont's theory (1829) that the Earth was cooling and therefore contracting. The latter included the great Austrian geologist, Eduard Suess, whose five-volume treatise *Das Antlitz der Erde* (*The Face of the Earth*) (Freytag, Leipzig, 158 pp., 1909) synthesized global tectonics in terms of the contraction hypothesis.

11. E.M. Anderson, Dynamics of faulting, *Trans. Geol. Soc. Edinburgh,* **8**, 387-402, 1905. He further developed his ideas in a monograph *The Dynamics of Faulting and Dyke Formation with Application to Britain* (2nd ed., Oliver & Boyd, Edinburgh, 206 pp., 1951).

12. M.K. Hubbert and W.W. Rubey, Mechanics of fluid-filled porous solids and its application to overthrust faulting, 1: Role of fluid pressure in mechanics of overthrust faulting, *Geol. Soc. Am. Bull.*, **70**, 115-166, 1959. In soil mechanics, the use of effective normal stress in the Coulomb criterion is sometimes called Terzaghi's principle, after the engineer who first articulated the concept (K. Terzaghi, Stress conditions for the failure of saturated concrete and rock, *Proc. Am. Soc. Test. Mat.*, **45**, 777-792, 1945). The historical development of the mechanical theory of faulting has been summarized by M.K. Hubbert in *Mechanical Behavior of Crustal Rocks: the Handin Volume* (N.L. Carter, M. Friedman, J.M. Logan, and D.W. Sterns, eds., Geophys. Mono. 24, American Geophysical Union, Washington, D.C., pp. 1-9, 1981).

13. The hydrostatic pressure at depth *h* is the pressure of a water column that deep, whereas the lithostatic pressure is the full weight of the overlying rocks; the latter is greater than the former by the rock-to-water density ratio, a factor of about 2.7.

14. State Earthquake Investigation Commission, *The California Earthquake of April 18, 1906*, Publication 87, vol. I, Carnegie Institution of Washington, 451 pp., 1908, and vol. II, with Atlas (by H.F. Reid), 192 pp., 1910; reprinted 1969. The Lawson Commission submitted a preliminary report almost immediately, on May 31, 1906, but no state or federal funds were available to continue the investigation, so that most of the research following the event had to be underwritten by a private organization, the Carnegie Institution of Washington.

15. The correlation between earthquake damage and "made ground" was noted 38 years before the 1906 earthquake when San Francisco's financial district was badly damaged in the 1868 Hayward earthquake. The 1906 quake caused extensive damage to the same area.

16. The State Earthquake Investigation Commission reports on the 1906 earthquake have been the principal source of data for the study of strong ground motions by D.M. Boore (Strong-motion recordings of the California earthquake of April 18, 1906, *Bull. Seis. Soc. Am.*, **67**, 561-577, 1977), the reconstruction of the space-time sequence of rupture by D.J. Wald, H. Kanamori, D.V. Helmberger, and T.H. Heaton (Source study of the 1906 San Francisco earthquake, *Bull. Seis. Soc. Am.*, **83**, 981-1019, 1993), and the recent reinterpretation of the geodetic measurements by W. Thatcher, G. Marshall, and M. Lisowski (Resolution of fault slip along the 470-kilometer-long rupture of the great 1906 San Francisco earthquake and its implications, *J. Geophys. Res.*, **102**, 5353-5367, 1997). These studies, which applied state-of-the-art techniques to old data, form the basis for the reconstruction of the faulting events outlined in Box 2.2.

17. H.F. Reid, The elastic-rebound theory of earthquakes, *Univ. Calif. Pub. Bull. Dept. Geol. Sci.*, **6**, 413-444, 1911. This paper was the transcription of the first of the Hitchcock Lectures delivered at the University of California, Berkeley, in the spring of 1911.

18. For example, the word "strength" in the first proposition would be modified to "frictional strength." Reid clearly understood that the strength of faults was governed by the friction across fault surfaces, rather than the strength of intact rocks. In his 1911 paper, he discussed the role of friction in experiments on elastic rebound using jelly sheets (p. 422), recognizing that in nature "the surface rocks on opposite sides of the fault are not identical as is the jelly" (p. 430); he considered the role of fault friction in the generation of slip irregularities that cause strong ground motions (pp. 435-436), and in discussing slickensides on the limbs of folds, he states, "It seems quite certain that, as the rocks were being folded by horizontal pressure, the friction would at first prevent any such slipping of the strata; but as the elastic forces become stronger, slipping would occur suddenly with an elastic rebound of the adjacent strata, which would constitute an earthquake" (p. 437).

19. Although Reid's formulation of the elastic rebound hypothesis preceded plate tectonics by more than 50 years, notions about the horizontal mobility of the Earth's crust were already in the air. Reid sought to provide some geological mechanism for the "slow displacements" required by his theory in a footnote on p. 28 of his report: "Mr. Baily Willis, on account of the forms of the mountain ranges bordering the Pacific Ocean, has concluded that the bed of the ocean is spreading and crowding against the land. He thinks in particular that there is a general sub-surface flow towards the north which would produce strains and earthquakes along the western coast of North America."

20. The story of the 1906 San Francisco earthquake presented here is based on modern reconstructions that have been updated using geophysical analysis techniques and geological knowledge that were unavailable at the time. An example is the estimation of the event's nucleation point (hypocenter) and origin time. H.F. Reid's original determination, given in his 1910 report, used local, imprecise estimates of the beginning of shaking to fix the origin time at 13:12:28 GMT and place the hypocenter at a depth of about 20 kilometers between the town of Olema and the southern end of Tomales Bay (i.e., north and west of San Francisco). Fortunately, this event was one of the first large earthquakes to be recorded by a global network of continuously recording seismometers. The network was sparse and poorly distributed by today's standards, but the State Earthquake Investigation Commission was able to collect copies of seismograms or arrival times of seismic waves from 96 observatories, 83 of which were outside the conterminous United States. In 1906, the structure of the Earth's interior was still too poorly known to predict the travel times of distant seismic waves, which is why Reid did not use them in his estimates. However, from an analysis of the teleseismic and local records reproduced in Reid's report, B.A. Bolt (The focus of the 1906 California earthquake, *Bull. Seis. Soc. Am.*, **58**, 457-471, 1968) found that the epicenter was similar to that of the small March 22, 1957, earthquake (37.67°N, 122.48°W), from which he obtained an origin time of 13:12:21 GMT. This position, near the point on the San Francisco peninsula where the San Andreas goes out to sea, is consistent with further studies of the archived seismograms.

21. The Milne seismograph grew out of a collaboration of British scientists (Ewing, Gray, and Milne) in Japan in the early 1880s. They solved the problem of how to obtain a long-period response from a physically short pendulum by inclining it so that the gravitational restoring force is reduced. The Milne seismographs used a horizontal bracket pendulum to attain a period of about 12 seconds, and they were recorded photographically. The early history of seismometry is discussed by J. Dewey and P. Byerly (The early history of seismometry—Up to 1900, *Bull. Seis. Soc. Am.*, **59**, 183-227, 1969).

22. The Jesuits established seismographic stations at their educational institutions in Europe, North and South America, Asia, Africa, and Australia, instrumenting them first

with seismoscopes (the earliest dates back to 1868 in Manila) and later with successively improved types of seismographs. In particular, the first standardized network of seismographic stations in North America was deployed in 1908-1911 by the newly formed Jesuit Seismological Service. Each of the fifteen U.S. and one Canadian stations was equipped with a Wiechert inverted-pendulum seismograph with an 80-kilogram mass, stabilized by springs and free to oscillate in any horizontal direction. The history of Jesuit seismology is outlined by A. Udías and W. Stauder (The Jesuit contribution to seismology, *Seis. Res. Lett.*, **67**, 10-19, 1996).

23. R.D. Oldham, The constitution of the Earth, *Quaternary. J. Geol. Soc. London*, **62**, 456-475, 1906. Prior to Oldham's analysis, there was considerable confusion over the identification of shear and surface waves. It was later established that there are two basic types of surface waves, those with retrograde-elliptical particle motions in the vertical plane of the source and receiver (*LR* or Rayleigh waves) and those with motions transverse to this plane (*LQ* or Love waves).

24. Although Milne's simple graphical method for locating earthquake epicenters was routinely employed by seismological observatories for many years, numerical methods were also formulated. In 1912, L. Geiger (Probability method for the determination of earthquake epicenters from the arrival time only, *Bull. St. Louis Univ.*, **8**, 60-71) applied the Gauss-Newton method to the iterative, least-squares solution of the nonlinear equations relating the space-time location parameters to the arrival times of seismic waves, and in the 1930s, Jeffreys showed how the least-squares normal equations could be solved efficiently by successive approximations. The first implementations of Geiger's method on electronic computers was in 1960 (B. Bolt, The revision of earthquake epicentres, focal depths and origin-times using a high-speed computer, *Geophys. J. R. Astron. Soc.*, **3**, 433-440, 1960; E.A. Flinn, Local earthquake location with an electronic computer, *Bull. Seis. Soc. Am.*, **50**, 467-470, 1960), and the International Seismological Summary adapted Bolt's code for routine location in 1961.

25. The collection and analysis of earthquake arrival times was facilitated by the establishment of the International Association of Seismology in 1905. Data reporting from the early networks was centralized at the International Seismological Centre, established in Strasbourg in 1906. The distribution of seismological data was standardized through Turner's publication of the International Seismological Summary, which began in 1923.

26. H. Jeffreys and K.E. Bullen, *Seismological Tables*, British Association for the Advancement of Science, London, 50 pp., 1940. These tables were derived from a model describing the variation of the seismic compressional-wave and shear-wave velocities as a function of radius. Although superseded by more precise estimates of radial structure, the Jeffreys-Bullen (J-B) model is still employed to locate earthquakes by the International Seismological Centre, a testament to its remarkable success.

27. The Rossi-Forel scale assigned intensity values from I to X, based on commonly observable effects of the shaking at a particular point. These range from "barely perceptible to an experienced observer" (I) through "felt generally by everyone" (V) to "great disaster" (X). The Modified Mercalli Scale of 1931, developed for building and social conditions of California by H.O. Wood and F. Neumann (Modified Mercalli Intensity Scale of 1931, *Bull. Seis. Soc. Am.*, **21**, 277-283, 1931) and close to the one still in use today, comprises 12 grades; the following are abbreviated descriptions of the higher levels: X. "Some well-built wooden structures destroyed; most masonry and frame structures destroyed with foundations; ground badly cracked. Rails bent. Landslides considerable from river banks and steep slopes. Shifted sand and mud. Water splashed, slopped over banks." XI. "Few, if any masonry structures remain standing. Bridges destroyed. Broad fissures in ground. Underground pipelines completely out of service. Earth slumps and land slips in soft ground. Rails bent greatly." XII. "Damage total. Waves seen on ground surface. Lines of sight and

level distorted. Objects thrown in air." Regions of constant intensity mapped by averaging over a geographic distribution of local values are separated by contours called *isoseismals.*

28. C.F. Richter, An instrumental earthquake magnitude scale, *Bull. Seis. Soc. Am.*, **21**, 28-46, 1935. At the time of this study, Caltech operated a southern California network that included seven stations with standardized Wood-Anderson seismometers. The active element in the standard instrument was a copper mass suspended vertically on a torsion spring with a free period of 0.8 second whose rotation was magnetically attenuated (damping constant of 0.8) and photographically recorded by a light beam reflected from a mirror mounted on the mass; the apparatus was sensitive only to horizontal ground motions, which it recorded with a nominal static magnification of 2800.

29. To calibrate how A_0, the amplitude of the magnitude-zero reference earthquake, decreased with distance from the epicenter, Richter used the data for 11 earthquakes recorded by these stations in the month of January 1932 (see Figure 22-2 of his textbook, *Elementary Seismology*, W.H. Freeman, San Francisco, 768 pp., 1958). He then tested the results on a set of 21 well-located local earthquakes during 1929-1931. Correcting the observations for the instrument-specific average residuals did not improve his original calibration curve, which remains the standard for southern California.

30. The magnitude scale is open ended; it places no limits on the minimum or maximum sizes of earthquakes. Richter's choice of A_0, though arbitrary, ensured that essentially all shocks recorded by the southern California network would have positive magnitudes and that most events locatable by this network would have magnitudes greater than 3. Beginning with T. Asada's work in Japan in the late 1940s, networks of instruments with higher sensitivity have been set up in some seismically active regions to locate *microearthquakes,* defined to be those with Richter magnitudes less than 3 (W.H.K. Lee and S.W. Stewart, *Principles and Applications of Microearthquake Networks*, Academic Press, New York, 293 pp., 1981).

31. The relationships given in Chapter 22 of Richter's textbook (op. cit.) are $m_b = 2.5 + 0.63M_S = 1.7 + 0.8M_L - 0.01M_L^2$. The values of m_b and M_S agree at magnitude 6.75; above this value, $m_b < M_S$, and below it, $m_b > M_S$. The former inequality is because the m_b scale uses shorter-period waves than M_S (~5 seconds versus 20 seconds) and thus saturates at a smaller magnitude.

32. Trinity was recorded on Caltech's permanent array of seismographs, as reported by Gutenberg in the open literature (Interpretation of records obtained from the New Mexico atomic bomb test, July 16, 1945, *Bull. Seis. Soc. Am.*, **36**, 327-330, 1946). The equipment to measure the time at the shot point failed, so that his seismologically determined origin time (12:29:12 GMT, July 16, 1945) was adopted by the popular press as the official beginning of the Atomic Age.

33. B. Gutenberg and C.F. Richter, Earthquake magnitude, intensity energy and acceleration, *Bull. Seis. Soc. Am.*, **46**, 105-145, 1956. They had initially estimated that the energy transmitted as seismic waves from the Baker test was about 10^{15} joules; however, this value exceeded its total explosive yield and became among the data that forced a revision of their original magnitude-energy formula. They used body-wave rather than surface-wave magnitude to measure explosion size, because explosions are relatively inefficient in exciting surface waves. This empirical observation was later shown to provide one of the better methods for discriminating underground nuclear explosions from earthquakes.

34. Multinational discussions to ban the testing of nuclear weapons were begun in 1958 in Geneva, where a Conference of Experts attempted to outline the requirements for treaty verification by seismological and other means. It was agreed that existing techniques were adequate for identifying nuclear explosions on the surface and in the atmosphere, but the prospects for detecting small underground nuclear explosions and discriminating them from earthquakes proved controversial, with the U.S. and U.S.S.R. delegations split on the

reliability of existing techniques. Following these negotiations, the United States convened a Panel on Seismic Improvement under the chairmanship of Lloyd V. Berkner. Noting that "the annual budget in the United States from all sources of seismological research amounts to roughly several hundred thousand dollars," the panel recommended a research program of $53 million for the first two years, including the establishment of a worldwide seismic detection system (*The Need for Fundamental Research in Seismology*, Panel on Seismic Improvement, Department of State, Washington, D.C., 212 pp., 1959). The recommendations of the Berkner Panel led to Project Vela, begun by the Department of Defense Advanced Research Projects Agency (ARPA) in 1959, and to the establishment of the World Wide Standardized Seismographic Network (WWSSN) in 1961. Between 1960 and 1971, about $245 million were expended by Vela Uniform, the underground explosion component of Project Vela. The history of this period is detailed by B. Bolt in *Nuclear Explosions and Earthquakes: The Parted Veil* (W.H. Freeman, San Francisco, 309 pp., 1976).

35. B. Gutenberg and C.F. Richter, *Seismicity of the Earth*, 2nd Edition, Princeton, N.J., 310 pp., 1954. Their first paper was published under the same title as Special Paper 34 of the Geological Society of America (1941).

36. B. Gutenberg and C.F. Richter, Magnitude and energy of earthquakes, *Ann. Geofisica*, **9**, 1-15, 1956. Prior to Gutenberg and Richter's publication of this equation, M. Ishimoto and K. Iida (Seismological observation by tremometer, 1. Magnitude and distribution pattern, *Bull. Earthquake Res. Inst. Univ. Tokyo Univ.*, **17**, 443-478, 1939) had discovered that the maximum trace amplitude A for Japanese earthquakes at approximately equal focal distances is related to their frequency of occurrence n by $A^m n = k$, which is equivalent to the Gutenberg-Richter relation for $m = b + 1$, which they estimated to be about 1.7.

37. A pure power-law distribution contains no information about characteristic values of lengths or energies and is consistent with the notion that earthquake dynamics is scale invariant. The power-law distribution of frequency versus size must obviously break down for very large earthquakes, because fault dimensions are finite and all realistic friction laws imply a minimum scale for dynamic nucleation, but the role of these inner and outer scales in earthquake mechanics remains controversial.

38. In their 1954 book, Gutenberg and Richter applied an energy-magnitude formula that gave a total seismic energy rate of about 10^{27} ergs per year, which was only about a factor of six less than Bullard's (Thermal history of the Earth, *Nature*, **156**, 35-36, 1945) estimate of total terrestrial heat flow. Two years later Gutenberg used the improved formula (Equation 2.3) to revise this number downward by two orders of magnitude.

39. The static theory of "distorsioni" in elastic media was worked out by V. Volterra and C. Somigliana about the same time as Reid formulated his elastic-rebound hypothesis, and the modern English usage of *dislocations* to describe these planar discontinuities began with A.E.H. Love in the second edition of his *Treatise on the Mathematical Theory of Elasticity*, Cambridge University Press, Cambridge, U.K., 643 pp., 1906. However, their connection to the Reid hypothesis was not formalized as a quantitative tool until a half-century later, when the static solution for a finite Volterra dislocation in a semi-infinite elastic medium became available (I.A. Steketee, Some geophysical applications of the elasticity theory of dislocations, *Canadian J. Phys.*, **36**, 192-205, 1958; I.A. Steketee, On Volterra's dislocations in a semi-infinite elastic medium, *Canadian J. Phys.*, **36**, 1168-1198, 1958; M.A. Chinnery, The deformation of the ground around surface faults, *Bull. Seis. Soc. Am.*, **51**, 355-372, 1961).

40. Fusakichi Omori observed consistent patterns to the first motions of earthquakes in the early part of the twentieth century, and T. Shida noted the division of compressions and dilatations into quadrants for the May 18, 1917, Shizuoka earthquake.

41. H. Nakano, Notes on the nature of the forces which give rise to the earthquake motions, *Seis. Bull. Centr. Met. Obs. Japan*, **1**, 92-120, 1923. Nakano's method is summarized

by P.A. Byerly, I. Mei, and C. Romney (Dependence on azimuth of the amplitudes of P and PP, *Bull. Seis. Soc. Am.*, **39**, 269-284, 1949).

42. An influential work in promoting this understanding was by R. Burridge and L. Knopoff, Body force equivalents of seismic dislocations, *Bull. Seis. Soc. Am.*, **54**, 1875-1888, 1964. They developed a general theory for earthquake radiation by proving that a time-dependent Volterra dislocation on an arbitrary surface in a heterogeneous, anisotropic medium is equivalent to a surface distribution of double couples. The equivalence between a pointwise dislocation and double couple had been realized by a number of previous authors, including F.R.N. Nabbaro (1951), V. Vvedenskaya (1956), and Knopoff and F. Gilbert (1960). Essentially the same results had been published by T. Marauyama (On force equivalents of dynamic elastic dislocations with reference to the earthquake mechanism, *Bull. Earthquake Res. Inst. Tokyo*, **41**, 467-486, 1963). Prior to the Burridge-Knopoff paper, however, many seismologists maintained the intuitive but incorrect notion that the single displacement couple of an elementary dislocation must correspond to a single force couple. Exceptions included H. Honda and his colleagues in Japan who plotted the initial motions and amplitudes of both *P* and *S* waves on the focal sphere; they showed that the *S*-wave patterns for deep Japanese earthquakes were not consistent with single-couple mechanisms, which required an *S*-wave node in the plane of faulting, but rather with double couples, which did not (H. Honda, Earthquake mechanism and seismic waves, *J. Phys. Earth*, **10**, 1-98, 1962). With hindsight, the solution to this controversy seems obvious; both a single couple and a double couple impart no net linear momentum, but only the latter conserves angular momentum, as required for an indigenous seismic source.

43. M.A. Chinnery, The deformation of the ground around surface faults, *Bull. Seis. Soc. Am.*, **51**, 355-372, 1961.

44. Wegener elaborated his continental-drift theory in his 1915 book *Die Entstehung der Kontinente und Ozeane*, written while recovering from war wounds. An expanded version was published in 1920 and a third edition in 1922, which was translated into English, French, Russian, and Spanish; the fourth edition was issued in 1929, the year before his death during an expedition on the Greenland icecap.

45. Wegener was optimistic that continental drift could be observed directly on trans-Atlantic profiles by geodetic methods: "Compared with all other theories of similarly wide scope, drift theory has the great advantage that it can be tested by accurate astronomical position-finding. If continental displacement was operative for so long a time, it is probable that the process is still continuing, and it is just a question of whether the rate of movement is enough to be revealed by our astronomical measurements in a reasonable period of time." He believed it was, because he surmised (incorrectly) that the youngest glacial moraines in Greenland and Europe had been connected, which implied that the North Atlantic had opened at rates of 10-30 meters per year. He persuaded his colleague, J.P. Koch, to compare the astronomical measurements of longitude from the expeditions they had made to Greenland in 1912-1913 and 1906-1908 with earlier determinations in 1823 and 1870; the shift in longitude was of the right order and in the right direction. He considered his theory confirmed when further measurements by the Danish Survey Organization in 1922 and 1927 yielded a drift rate of 36 ± 4 meters per year. Unfortunately for Wegener, the quoted uncertainties did not account for unappreciated sources of bias, and this value turned out to be about 1000 times too large. Continental drift was not measured by geodetic techniques until the development of ultraprecise (±1 centimeter) Very Long Baseline Interferometry more than 50 years after his death (T.A. Herring, I.I. Shapiro, T.A. Clark, C. Ma, J.W. Ryan, B.R. Schupler, C.A. Knight, G. Lundqvist, D.B. Shaffer, N.R. Vandenberg, B.E. Corey, H.F. Hinteregger, A.E.E. Rogers, J.C. Webber, A.R. Whitney, G. Elgered, B.O. Ronnang, and J.L.

Davis, Geodesy by radio interferometry: Evidence for contemporary plate motion, *J. Geophys. Res.*, **91**, 8341-8347, 1986).

46. W.A. van der Gracht and J.M. van Waterschoot, eds., *Theory of Continental Drift, A Symposium*, American Association of Petroleum Geologists, Tulsa, 240 pp., 1928.

47. M. Ewing and B. C. Heezen, Some problems of Antarctic submarine geology, in *Antarctic in the International Geophysical Year*, A. Carey, L.M. Gould, E.O. Hulburt, H. Odishaw, and W.E. Smith, eds., American Geophysical Union Monograph 1, 75-81, 1956. The first published study of mid-ocean seismicity was by E. Tams (Die seismichen verhaltnisse des offenen Atlantischen Ozeans, *Zeitschr. Geophys.*, **3**, 361-363, 1927).

48. H.H. Hess, History of ocean basins, in *Petrologic Studies: A Volume in Honor of A.F. Buddington*, A.E. Engle, H.L. James, and B.F. Leonard, eds., Geological Society of America, pp. 599-620, 1962; R.S. Dietz, Continent and ocean basin evolution by spreading of the sea floor, *Nature*, **190**, 854-857, 1961. An excellent insider's account of the postwar oceanographic expeditions that led to the discovery of seafloor spreading and plate tectonics can be found in H.W. Menard's historical memoir, *The Ocean of Truth* (Princeton University Press, Princeton, 353 pp., 1986).

49. Reprints of the original papers and historical commentaries on the confirmation of seafloor spreading by marine magnetic data can be found in *Plate Tectonics and Geomagnetic Reversals* by A. Cox (W.H. Freeman, San Francisco, 702 pp., 1973).

50. M.L. Hill and T.W. Dibblee, Jr., San Andreas, Garlock and Big Pine faults, California—A study of their character, history and tectonic significance of their displacements, *Geol. Soc. Am. Bull.*, **64**, 443-458, 1953; H.W. Wellman, Structural outline of New Zealand, *Bull. N.Z. Dept. Sci. Indust. Res.*, **121**, 35 pp. + map, 1956.

51. H.W. Menard and his colleagues at the Scripps Institution of Oceanography mapped a regularly spaced set of very long (>3000 kilometers), linear fracture zones in the eastern Pacific Ocean; the first discovered was the Mendocino Fracture Zone, which intersects the San Andreas system at Cape Mendocino, California (H.W. Menard and R.S. Dietz, Mendocino submarine escarpment, *J. Geol.*, **60**, 266-278, 1952). A decade later, V. Vacquier, A.D. Raff, and R.E. Warren (Horizontal displacements in the floor of the northeastern Pacific Ocean, *Geol. Soc. Am. Bull.*, **72**, 1251-1258, 1961) recognized a left-lateral offset of 1170 kilometers in the magnetic anomaly patterns across the Mendocino Fracture Zone.

52. J.T. Wilson, A new class of faults and their bearing on continental drift, *Nature*, **207**, 343-347, 1965. In the opening paragraphs of this paper, Wilson first used the term *plate* in its modern form: "Many geologists have maintained that movements of the Earth's crust are concentrated in mobile belts, which may take the form of mountains, mid-ocean ridges or major faults with large horizontal movements. These features and the seismic activity along them often appear to end abruptly, which is puzzling. . . . This article suggests that these features are not isolated, that few come to dead ends, but that they are connected into a continuous network of mobile belts about the Earth which divide the surface into several large rigid plates." He defined a *transform* as the juncture where one type of mobile belt (plate boundary) changes into another and a *transform fault* as a strike-slip fault terminated by transforms, which can be contrasted with a *transcurrent fault* that ends at points unconnected to other zones of deformation. He cataloged the types of transform faults that can connect two segments of oceanic ridges (spreading centers), two segments of island/mountain arcs (subduction zones), or one of each type.

53. Motivated by the Berkner panel, ARPA prepared a plan to upgrade and expand the global network of permanent seismic observatories in 1960. The execution of the plan was assigned to the U.S. Coast and Geodetic Survey (USCGS), which held the operational responsibility for earthquake monitoring within the federal government. The new instrumentation comprised three Benioff short-period seismometers (1-second free period), three Press-Ewing long-period seismometers (15- or 30-second free period, 90-second galvanom-

eter period), the photographic drum recording apparatus, and a radio-synchronized crystal clock. Installations began in 1961, reaching a total of about 120 stations in 60 countries by 1967; because some of the expenses were borne by the host countries, the total cost to Project Vela Uniform was only about $10 million.

54. The possibility that the more mobile, interior layers of the Earth are convecting was discussed as early as 1839 by W. Hopkins (Researches in physical geology, *Phil. Trans. Roy. Soc. Lond.*, **129**, 381-423, 1839). The convection hypothesis was proposed as the driving mechanism for mountain building by A. Holmes (Geophysics—The thermal history of the Earth, *J. Wash. Acad. Sci.*, **23**, 169-195, 1933) and F.A. Vening Meinesz (The mechanism of mountain formation in geosynclinal belts, *Proceedings of the Section of Sciences Koninklijke Nederlandse Akademie van Wetenschappen*, **36**, 372-377, 1933), and the theory of mantle convection was further developed in the 1930s by C. Pekeris, A. Hales, and D. Griggs.

55. D. Griggs, A theory of mountain building, *Am. J. Sci.*, **9**, 611-650, 1939.

56. The pre-plate-tectonics thinking among the mobilists was documented in the proceedings of a 1950 colloquium in Hershey, Pennsylvania, on "Plastic Flow and Deformation Within the Earth" (B. Gutenberg, H. Benioff, J. M. Burgers, and D. Griggs, *Eos, Trans. Am. Geophys. Union*, **32**, 497-543, 1951).

57. B.C. Heezen, The rift in the ocean floor, *Sci. Am.*, **203**, 98-110, 1960. Earth expansion was first proposed as a mechanism for continental drift in the 1920s (see A.A. Meyerhoff and A. Holmes, Originator of spreading ocean floor hypothesis, *J. Geophys. Res.*, **73**, 6563-6565, 1968, for references), and the hypothesis was revitalized in the geological studies of L. Egyed (The change of the Earth's dimensions determined from paleogeographical data, *Geof. Pura Appl.*, **33**, 42-48, 1956) and S.W. Carey (The tectonic approach to continental drift, in *Continental Drift, A Symposium*, University of Tasmania, Hobart, pp. 177-355, 1958). Its popularity was broadened considerably by the recognition that Earth expansion might be the observable consequence of a secular decrease in Newton's universal gravitational constant, as predicted by a class of cosmological theories that attempted to explain the Hubble redshift without a Big Bang expansion of the universe (R.H. Dicke, Principle of equivalence and the weak interactions, *Rev. Mod. Phys.*, **29**, 355-362, 1957).

58. Subduction, an old term employed by Alpine geologists, was not widely used to describe the sinking of the lithospheric slabs beneath volcanic arcs until it was reintroduced by D. Roeder at a Penrose conference on "The Meaning of the New Global Tectonics for Magmatism, Sedimentation, and Metamorphism in Orogenic Belts" in December 1969 (W.R. Dickinson, Global tectonics, *Science*, **168**, 1250-1259, 1970).

59. Benioff's major papers on the subject were published in the same years as the two editions of Gutenberg and Richter's *Seismicity of the Earth* (1949, 1954), and he drew heavily from their catalogs to support his reverse-faulting hypothesis.

60. R.R. Coats, in *Crust of the Pacific Basin*, G.A. Macdonald and H. Kuno, eds., American Geophysical Union Monograph 6, Washington, D.C., pp. 92-109, 1962.

61. When fault-plane solutions became more common in the 1950s, the results were often in poor agreement with other information about fault orientations and slip directions. J.H. Hodgson of the Dominion Observatory in Canada concluded, for example, that much of the circum-Pacific faulting was strike-slip, in direct conflict with Benioff's hypothesis. In his 1954 paper, Benioff discussed the directions of earthquake slip vectors inferred from first-motion radiation patterns, citing studies by Dutch and Japanese seismologists in support of his reverse-faulting hypothesis, but he was clearly skeptical of their reliability. He noted, for example, that Hodgson's strike-slip solution for the Peru earthquake of November 10, 1946, was contradicted by the field observations of E. Silgado, which indicated dip-slip faulting. The purported dominance of strike-slip faulting in the circum-Pacific region was an incorrect inference by North American seismologists, who employed graphical methods based on Perry Byerly's "extended station distances" that were inferior to the

focal-sphere projections developed by the Dutch and Japanese. The confusion of this period is evident in the proceedings of two symposia convened by Hodgson on focal-mechanism studies (*The Mechanics of Faulting, With Special Reference to the Fault-Plane Work*, Pub. Dominion Obs. Ottawa, 20, 215-418, 1957; *A Symposium on Earthquake Mechanism*, Pub. Dominion Obs. Ottawa, 24, 299-397, 1960).

62. Early free-oscillation studies of this earthquake were done using strainmeters and specialized ultralow-frequency seismometers by A.A. Nowroozi (Eigenvibrations of the earth after the Alaskan earthquake, *J. Geophys. Res.*, **70**, 5145-5156, 1965) and S.W. Smith (Free oscillations excited by the Alaskan earthquake, *J. Geophys. Res.*, **71**, 1183-1193, 1968). A major advance in this subject came from the laborious digitization of the analog, long-period records of the 1964 event from more than 100 WWSSN stations, which allowed A.M. Dziewonski and F. Gilbert (Observations of normal modes from recordings of the Alaskan earthquake 1964, *Geophys. J. R. Astr. Soc.*, **27**, 393-446, 1972; ibid., Observations of normal modes from recordings of the Alaskan earthquake 1964, 2, *Geophys. J. R. Astron. Soc.*, **35**, 401-437, 1973) to apply stacking techniques to measure a large set of free-oscillation eigenfrequencies and use these data to derive improved models of Earth structure.

63. F. Press, Displacements, strains, and tilts at teleseismic distances, *J. Geophys. Res.*, **70**, 2395-2412, 1965. Press used a strainmeter developed by Benioff in the 1930s, which measured the change in distance between two piers anchored to the ground about 10 meters apart. A rigid quartz rod attached to one pier was extended almost to the second pier, and a sensitive capacitance transducer monitored small changes in the gap caused by either the passage of long-period seismic waves or permanent deformation of the crust.

64. The first arrival times of seismic waves at stations around the world demonstrated that the focus was shallow, but the closest recording seismograph was at College, Alaska—440 kilometers from the epicenter—so that seismologists were unable to fix the depth to the nucleation point precisely enough to help the geologists. The rupture process for the 1964 Alaska earthquake turned out to be very complex. From an analysis of the short-period *P* waves, M. Wyss and J.N. Brune (The Alaska earthquake of 28 March 1964: A complex multiple rupture, *Bull. Seis. Soc. Am.*, **57**, 1017-1023, 1967) identified nine subevents during the first 72 seconds of rupture. Recently, D.H. Christensen and S.L. Beck (The rupture process and tectonic implications for the great 1964 Prince William Sound earthquake, *Pure Appl. Geophys.*, **142**, 29-53, 1994) used long-period *P* waves from the 20 stations with low-gain, on-scale recordings (including those that had been diminished by diffraction around the Earth's core) to model the rupture process; their solution shows two major episodes of fault slippage, one near the epicenter during the first 100 seconds of rupture and a second near Kodiak Island, starting at about 160 seconds and lasting for 40 seconds.

65. The quote is from G. Plafker, Tectonic deformation associated with the Alaska earthquake 1964, USA, *Science*, **148**, 1675-1687, 1965. Plafker's detailed field observations were synthesized in *Tectonics of the March 27, 1964, Alaska Earthquake* (U.S. Geological Survey Professional Paper 543-I, U.S. Government Printing Office, Washington, D.C., 74 pp., 1969).

66. W. Stauder and G.A. Bollinger, Focal mechanism of Alaska earthquake 1964 March 28 and of its aftershocks, *J. Geophys. Res.*, **71**, 5283-5296, 1966; The s-wave project for focal mechanism studies, earthquakes of 1963, *Bull. Seis. Soc. Am.*, **56**, 1363-1371, 1966).

67. D.P. McKenzie and R.L. Parker, The north Pacific. An example of tectonics on a sphere, *Nature*, **216**, 1276-1280, 1967. These authors were students at Cambridge University when Bullard first employed finite Euler rotations to quantify continental drift (E.C. Bullard, J.E. Everett, and A.G. Smith, Symposium on continental drift. The fit of continents around the Atlantic, *Phil. Trans. Roy. Soc. Lond.*, **A258**, 41-75, 1965), and they adapted this concept to describe the instantaneous relative motion between two plates as an angular velocity vector. They were the first to appreciate that slip vectors inferred from double-couple fault-plane solutions could be used to constrain the relative angular velocity vector.

68. W.J. Morgan, Rises, trenches, great faults, and crustal blocks, *J. Geophys. Res.*, **73**, 1959-1982, 1968. In this paper, Morgan established plate tectonics as a quantitative, global theory by showing that the present-day pole positions estimated from transform-fault azimuths were consistent with the gradients in seafloor-spreading rates observed from marine magnetic anomalies. His study of instantaneous block rotations was followed in the same year by X. Le Pichon's (Sea-floor spreading and continental drift, *J. Geophys. Res.*, **73**, 3661-3697, 1968) reconstruction of the Mesozoic and Cenozoic history of seafloor spreading and continental drift in terms of finite rotations constrained by magnetic anomaly data.

69. Walter Elsasser, a colleague of Hess and Morgan at Princeton, circulated his ideas about plates as stress guides in a 1967 preprint that was not published until two years later (in *The Application of Modern Physics to the Earth and Planetary Interiors*, S.K. Runcorn, ed., Wiley-Interscience, New York, pp. 223-246, 1969). The thermal model of plates as the cold, outer boundary layer of a convecting mantle was discussed by D. Turcotte and R. Oxburgh (Finite amplitude convection cells and continental drift, *J. Fluid Mech.* **28**, 29-42, 1967) and modified to include plates of finite thickness by D. McKenzie (Some remarks on heat flow and gravity anomalies, *J. Geophys. Res.*, **72**, 6261-6273, 1967).

70. B.L. Isacks, J. Oliver, and L.R. Sykes, Seismology and the new global tectonics, *J. Geophys. Res.*, **73**, 5855-5899, 1968.

71. Japanese seismologists recognized the anomalous velocity and attenuation structure beneath their home islands before the first structural study by J. Oliver and B.L. Isacks (Deep earthquake zones, anomalous structures in upper mantle and lithosphere, *J. Geophys. Res.*, **72**, 4259-4275, 1967), but they lacked a satisfactory geodynamic explanation. See T. Utsu (Seismological evidence for anomalous structure of island arcs, *Rev. Geophys. Space Phys.*, **9**, 839-890, 1971) for a discussion of this early work.

72. T. Atwater, Implications of plate tectonics for the Cenozoic tectonic evolution of western Northern America, *Geol. Soc. Am. Bull.*, **81**, 3513-3536, 1970.

73. In particular, she showed that the distribution of many geological events in space and time could be understood by the migration of the triple junction between the Pacific, North American, and (now extinct) Farallon plates. The kinematics of such triple junctions had first been discussed by D.P. McKenzie and J.W. Morgan (Evolution of triple junctions, *Nature*, **224**, 125-133, 1969).

74. T. Matsuda and S. Uyeda (On the Pacific-type orogeny in its model-extension of paired belts concept and possible origin of marginal seas, *Tectonophysics*, **11**, 5-27, 1971) demonstrated that this paired structure of "Pacific-type" orogenic belts, first recognized by A. Miyashiro (Evolution of metamorphic belts, *J. Petrol.*, **2**, 277-311, 1961), could be explained by plate tectonics.

75. These facts were first noted in a plate-tectonic context by D.P. McKenzie (Speculations on the consequences and causes of plate motions. Tectonics, *Geophys. J. Roy. Astron. Soc.*, **18**, 1-32, 1969).

76. W. Hamilton, in *Proceedings of Andesite Conference*, A.R. McBirney, ed., *Oregon Dept. Geol. Mineral Industries Bull., No. 65*, 175-184, 1969; J.F. Dewey and J.M. Bird, Lithospheric plate-continental margin tectonics and the evolution of the Appalachian orogen, *J. Geophys. Res.*, **75**, 2625-2647, 1970.

77. P. Molnar and P. Tapponnier, Cenozoic tectonics of Asia. Effects of a continental collision, *Science*, **189**, 419-426, 1975; Active tectonics of Tibet, *J. Geophys. Res.*, **83**, 5361-5375, 1978.

78. The satellite photos and seismic evidence available to Molnar and Tapponnier did not show evidence for a large amount of crustal shortening across Tibet or for the existence of a shallow-dipping fault that would allow the Indian crust to underplate the Tibetan crust. The reason for the high elevation of Tibet thus remained problematic, with these authors preferring some thermal or magmatic source in the mantle for the uplift. Later

seismological studies would confirm the great thickness of the Tibetan crust and lead others to infer that the Tibetan crust was thickened by ductile flow in the lower crust.

79. J.D. Byerlee, Friction of rocks, *Pure Appl. Geophys.*, **116**, 615-626, 1978. Byerlee obtained a best fit to a diverse data set with $\tau_0 = 0$, $\mu = 0.85$ at low normal stresses ($\sigma_n < 200$ megapascals) and $\tau_0 = 50$ megapascals, $\mu = 0.6$ for normal stresses in the range of 200 megapascals to 20 gigapascals. The latter corresponds to pressures in a depth range of 6-60 kilometers. J.C. Jaeger (The frictional properties of joints in rocks, *Geofisica Pura Appl.*, **43**, 148-159, 1959) had earlier reported roughly comparable values of friction for various rocks. Exceptions to Byerlee's law included some of the clay minerals, which show lower values of internal friction, $0.2 < \mu < 0.4$ (C. Morrow, B. Radney, and J. Byerlee, Frictional strength and the effective pressure law of montmorillonite and illite clays, in *Fault Mechanics and Transport Properties of Rocks*, B. Evans and T.-F. Wong, eds., Academic Press, London, pp. 69-88, 1992).

80. D. McKenzie, Speculations on the causes and consequences of plate motions, *Geophys. J. Roy. Astr. Soc.*, **18**, 1-32, 1969; W.F. Brace and D. Kohlstedt, Limits of lithospheric stress imposed by laboratory experiments, *J. Geophys. Res.*, **85**, 6348-6252, 1980.

81. In discussing his theory of earthquakes in the Great Basin, Gilbert (A theory of the earthquakes of the Great Basin, with a practical application, *Am. J. Sci., 3rd Ser.*, **27**, 49-53, 1884) commented explicitly on the role of fault friction: "The upthrust produces a local strain in the crust, involving a certain amount of compression and distortion, and this strain increases until it is sufficient to overcome the starting friction along the fractured surface. Suddenly, and almost instantaneously, there is an amount of motion sufficient to relieve the strain, and this is followed by a long period of quiet, during which the strain is gradually reimposed." In his 1910 volume for the Lawson Commission (*The California Earthquake of April 18, 1906*, Publication 87, vol. II, Carnegie Institution of Washington, 192 pp., 1910; reprinted 1969), Reid recognized the San Andreas as a persistent plane of weakness: "We must therefore conclude that former ruptures of the fault-plane were by no means entirely healed, but that this plane was somewhat less strong than the surrounding rock and yielded to a smaller force than would have been necessary to break the solid rock" (p. 21). He also objected to the stress-free crack hypothesis on dynamical grounds: "If the break had been sharp, with no friction at the fault-plane, . . . the rock at the fault-plane would have made rapid but short vibrations back and forth during the 2.2 seconds necessary for it to reach the equilibrium position. This, however, is not what actually occurred; small slips took place at different parts of the fault-plane, and as the results of these successive slips and great friction, some 30 to 60 seconds were required before the rock came to rest; and even then certain parts of the rock were apparently still held in a strained condition by strong friction, and from time to time gave way, producing the aftershocks . . ." (p. 39).

82. M. Orowan, Mechanism of seismic faulting, in *Rock Deformation*, D. Griggs and J. Handin, eds., Geological Society of America Memoir 79, New York, pp. 323-345, 1960; M.A. Chinnery, The strength of the earth's crust under horizontal shear stress, *J.* Geophys. *Res.*, **69**, 2085-2089, 1964. Jeffreys (Note on fracture, *R. Soc. Edinburgh Proc.*, **56**, 158-163, 1936) was the first to raise objections to frictional instabilities as the mechanism for deep earthquakes. H. Benioff (Earthquake source mechanisms, *Science*, **143**, 1399-1406, 1964) attributed deep-focus seismicity to sudden phase changes, arguing that a long-period waveform of a deep (600-kilometer) Peruvian earthquake recorded with his strain seismometer close to the epicenter differed radically from the waveform produced by a faulting source. Recent studies have demonstrated that almost all deep-focus earthquakes can be explained with a simple planar fault model with stress drops ranging from 10 bar to 1 kilobar (e.g., H. Kawakatsu, Insignificant isotropic component in the moment tensor of deep earthquakes, *Nature*, **351**, 50-53, 1991).

83. D. Griggs and J. Handin, *Rock Deformation*, Geological Society of America Memoir 79, Boulder, Colo., 382 pp., 1960.

84. W.F. Brace and J.D. Byerlee, Stick slip as a mechanism for earthquakes, *Science*, **153**, 990-992, 1966. In an early series of experiments at Harvard University, P.W. Bridgman (Shearing phenomena at high pressure of possible importance for geology, *J. Geol.*, **44**, 653-669, 1936) had observed strike-slip between thin layers of materials subjected to torsional stress. The term *stick-slip* was introduced by F.P. Bowden and L. Leben (The nature of sliding and the analysis of friction, *Proc. R. Soc. Lond.*, **A169**, 371-391, 1939), and the theory was developed with a focus on machine engineering issues (E. Rabinowicz, *Friction and Wear of Materials*, John Wiley, New York, 50 pp., 1965). The phenomenon was notoriously difficult to study in the laboratory, however, primarily because most testing machines could not be adequately regulated to obtain reproducible results. Brace and Byerlee's success depended on their use of a new, "stiffer" testing machine.

85. W.F. Brace and J.D. Byerlee, California earthquakes: Why only shallow focus, *Science*, **168**, 1575, 1968.

86. Although Brace and Byerlee recognized the importance of the dynamical characteristics of the testing machine in the generation of stick-slip instabilities, a quantitative understanding of the dynamics of stable and unstable sliding was not achieved for another 10 years, when the concept of a critical stiffness k_c was precisely formulated for strain-weakening materials (J.W. Rudnicki, The inception of faulting in a rock mass with a weakened zone, *J. Geophys. Res.*, **82**, 844-854, 1977). Instabilities associated with velocity-weakening "rate-state" constitutive laws were investigated later by J.R. Rice and A.L. Ruina (Stability of steady frictional slipping, *J. Appl. Mech.*, **50**, 343-349, 1983).

87. F. Rummel and C. Fairhurst, Determination of the post-failure behavior of brittle rock using a servo-controlled testing machine, *Rock Mech.*, **2**, 189-204, 1970; R. Houpert, The uniaxial compressive strength of rocks, *Proceedings of the 2nd International Society of Rock Mechanics*, vol. II, Jaroslav Cerni Institute for Development of Water Resources, Belgrade, pp. 49-55, 1970; H.R. Hardy, R. Stefanko, and E.J. Kimble, An automated test facility for rock mechanics research, *Int. J. Rock Mech. Min. Sci.*, **8**, 17-28, 1971.

88. Dilatancy is the volume expansion due to the application of a shear stress, a well-known property of granular materials. Brace and his colleagues observed this phenomenon during rock-shearing experiments in the laboratory, which they interpreted as being due to pervasive microcracking in their specimens with a concomitant increase in void space (W.R. Brace, B.W. Paulding, and C.H. Scholz, Dilatancy of the fracture of crystalline rocks, *J. Geophys. Res.*, **71**, 3939-3953, 1966).

89. T.E. Tullis and J.D. Weeks, Constitutive behavior and stability of frictional sliding of granite, *Pure Appl. Geophys.*, **124**, 10-42, 1986; N.M. Beeler, T.E. Tullis, M.L. Blanpied, and J.D. Weeks, Frictional behavior of large displacement experimental faults, *J. Geophys. Res.*, **101**, 8697-8715, 1996.

90. J.H. Dieterich, Time-dependent friction in rocks, *J. Geophys. Res.*, **77**, 3690-3697, 1972.

91. C. Scholz, P. Molnar, and T. Johnson, Frictional sliding of granite and earthquake mechanism implications, *J. Geophys. Res.*, **77**, 6392-6406, 1972.

92. J.H. Dieterich, Time-dependent friction and the mechanics of stick-slip, *Pure Appl. Geophys.*, **116**, 790-806, 1978. The concept of a critical slip distance as a parameter characterizing frictional changes was first introduced by E. Rabinowicz (The nature of static and kinetic coefficients of friction, *J. Appl. Phys.*, **22**, 1373-1379, 1951; The intrinsic variables affecting the stick-slip process, *Proc. R. Soc. Lond.*, **A71**, 668-675, 1958), who interpreted it as the typical dimension of surface contact junctions.

93. J.H. Dieterich, Modelling of rock friction 1. Experimental results and constitutive equations, *J. Geophys. Res.*, **84**, 2161-2168, 1979; J.H. Dieterich, Constitutive properties of faults with simulated gouge, in *Mechanical Behavior of Crustal Rocks*, N.L. Carter, M. Friedman, J.M.

Logan, and D.W. Sterns, eds., American Geophysical Union Monograph 24, Washington, D.C., pp. 103-120, 1981; A. Ruina, Slip instability and state variable friction laws, *J. Geophys. Res.*, **88**, 10,359-10,370, 1983 (see also J.R. Rice, Constitutive relations for fault slip and earthquake instabilities, *Pure Appl. Geophys.*, **121**, 443-475, 1983). The historical development of the rate-state theory is summarized in the review papers by C. Marone (Laboratory-derived friction laws and their application to seismic faulting, *Ann. Rev. Earth Planet. Sci.*, **26**, 643-696, 1998) and C.H. Scholz (Earthquakes and friction laws, *Nature*, **391**, 37-42, 1998).

94. S.T. Tse and J.R. Rice, Crustal earthquake instability in relation to the depth variation of frictional slip properties, *J. Geophys. Res.*, **91**, 9452-9472, 1986. Subsequent depth-variable crustal models were based on friction data for saturated granite gouge studied over the range of crustal temperatures by M.L. Blanpied, D.A. Lockner, and J.D. Byerlee (Fault stability inferred from granite sliding experiments at hydrothermal conditions, *Geophys. Res. Letters*, **18**, 609-612, 1991; Frictional slip of granite at hydrothermal conditions, *J. Geophys. Res.*, **100**, 13,045-13,064, 1995); a recent example is N. Lapusta, J.R. Rice, Y. Ben-Zion, and G. Zheng, Elastodynamic analysis for slow tectonic loading with spontaneous rupture episodes on faults with rate- and state-dependent friction, *J. Geophys. Res.*, **105**, 23,765-23,789, 2000.

95. K. Aki, Generation, propagation of G waves, Niigata earthquake 1964 June 16. Pt 1, *Bull. Earthquake Res. Inst. Tokyo Univ.*, **44**, 23-88, 1966. Formally speaking, the right-hand side of Equation 2.6 is the *static* moment, so that an accurate estimate of M_0 for an earthquake of finite size requires the measurement of waves whose periods are long compared to the duration of the faulting. For the 1964 Niigata earthquake, Aki used surface waves with periods up to 200 seconds, which satisfied this criterion. He found $M_0 \approx 3 \times 10^{20}$ newton-meters.

96. Moment magnitude M_W was introduced by H. Kanamori (The energy release in great earthquakes, *J. Geophys. Res.*, **82**, 2981-2987, 1977; Quantification of earthquakes, *Nature*, **271**, 411-414, 1978), and its agreement with other magnitude scales in their unsaturated ranges was discussed by T.C. Hanks and H. Kanamori (A moment magnitude scale, *J. Geophys. Res.*, **84**, 2348-2350, 1979).

97. Omori established his empirical hyperbolic law of aftershock frequency, $n(t) \sim t^{-1}$, in his studies of the 1891 Nobi earthquake (F. Omori, On aftershocks, *Report by the Earthquake Investigation Committee*, **2**, 103-139, 1894; ibid., *Report by the Earthquake Investigation Committee*, **30**, 4-29, 1900). This relationship was extended first by R. Hirano and later by T. Utsu (A statistical study on the occurrence of aftershocks, *Geophys. Mag.*, **30**, 521-605, 1961) to a power law in the form of Equation 2.8, which is now called the *modified Omori law*.

98. The first detailed study of aftershocks using portable seismometers was conducted by Caltech seismologists following the 1952 Kern County earthquake in central California (H. Benioff, *Earthquakes in Kern County, California, 1952*, California Division of Mines, State of California, Bulletin 171, 283 pp., 1955). Based on this and other data, Benioff (Seismic evidence for crustal structure and tectonic activity, in *Crust of the Earth. A Symposium*, A Poldervaart, ed., Geological Society of America, New York, pp. 61-75, 1955) hypothesized that the spatial distribution of aftershocks defines the segment of the fault that had ruptured during the mainshock. Subsequent work has shown this to be approximately true for the initial sequence of aftershocks, although the aftershock zone typically expands to a larger area after a week or so (F. Tajima and H. Kanamori, Global survey of aftershock area expansion patterns, *Phys. Earth Planet. Int.*, **40**, 77-134, 1985).

99. K. Aki used a spectral representation to establish that earthquakes of varying size had spectra of similar shape, differing primarily in the low-frequency amplitude, proportional to seismic moment, and the location of the "characteristic frequency," which he related to the characteristic length scale of an earthquake (Scaling law of seismic spectrum, *J. Geophys. Res.*, **72**, 1217-1231, 1967). Subsequent studies by J.N. Brune (Tectonic stress and

the spectra of seismic shear waves from earthquakes, *J. Geophys. Res.*, **75**, 4997-5009, 1970) and J.C. Savage (Relation of corner frequency to fault dimensions, *J. Geophys. Res.*, **77**, 3788-3795, 1972) related the corner frequency to the dimensions of the fault plane.

100. Theoretical studies of propagating fractures by the engineer G.R. Irwin and his associates (see L.B. Freund, *Dynamic Fracture Mechanics*, Cambridge University Press, Cambridge, U.K., 563 pp., 1990) had shown that the speed limit for an *antiplane* crack (slip direction parallel to the dislocation line) was the shear-wave velocity v_S, while that for an *in-plane* crack (slip direction perpendicular to the dislocation line) was the Rayleigh-wave velocity. Seismological observations indicated that the actual rupture velocities for earthquakes are 10-20 percent lower than these theoretical limits.

101. This equation for the stress drop omits a nondimensional scaling constant c_S of order unity, which depends on the rupture geometry. V. Keilis Borok (On the estimation of the displacement in an earthquake source and of source dimensions, *Ann. Geofisica*, **12**, 205-214, 1959) derived $c_S = 7\pi/16 \approx 1.37$ for a circular crack, and he applied this formula to obtain a stress drop for the 1906 San Francisco earthquake.

102. The observation that the static stress drop is approximately constant for earthquakes of different sizes in similar tectonic environments was established in publications by K. Aki (e.g., Earthquake mechanism, *Tectonophysics*, **13**, 423-446, 1972) and various other authors in the early 1970s; see the summary by T. Hanks (Earthquake stress drops, ambient tectonic stresses and stresses that drive plate motions, *Pure Appl. Geophys.*, **115**, 441-458, 1977).

103. H. Kanamori and D.L. Anderson, Theoretical basis of some empirical relations in seismology, *Bull. Seis. Soc. Am.*, **65**, 1073-1095, 1975.

104. N. Haskell, Total energy and energy spectral density of elastic wave radiation from propagating faults, *Bull. Seis. Soc. Am.*, **54**, 1811-1842, 1964.

105. N. Haskell, Total energy and energy spectral density of elastic wave radiation from propagating faults, 2, A statistical source model, *Bull. Seis. Soc. Am.*, **56**, 125-140, 1966; K. Aki, Scaling law of seismic spectrum, *J. Geophys. Res.*, **72**, 1217-1231, 1967.

106. B.V. Kostrov, Teoriya ochagov tektonicheskikh zemletryaseniy. Tectonic earthquake focal theory, *Izv. Akad. Nauk. S.S.R.*, Earth Physics, **4**, 84-101, 1970; M.J. Randall, Elastic multiple theory and seismic moment, *Bull. Seis. Soc. Am.*, **61**, 1321-1326, 1971; F. Gilbert, Excitation of the normal modes of the earth by earthquake sources, *Geophys. J. R. Astr. Soc.*, **22**, 223-226, 1971. The moment tensor can be represented a 3×3 matrix, which must be symmetric to conserve angular momentum; the most general moment tensor is therefore specified by six real numbers.

107. L. Knopoff and M.J. Randall, Compensated linear-vector dipole-possible mechanism of deep earthquakes, *J. Geophys. Res.*, **75**, 4957-4963, 1970. A CLVD represents the strain associated with a volume-preserving extension (or compression) of an infinitesimal cylinder along its axis of symmetry. The evidence and interpretation of earthquake focal mechanisms that do not conform to simple double couples have recently been reviewed by B.R. Julian, A.D. Miller, and G.R. Foulger, Non-double-couple earthquakes: 1. Theory, *Rev. Geophys.*, **36**, 525-549, 1998.

108. D. Forsyth and S. Uyeda (On the relative importance of the driving forces of plate motion, *Geophys. J. Roy. Astr. Soc.*, **43**, 163-200, 1975) showed that the primary driving forces are "ridge push" (compression due to gravitational sliding of newly formed lithosphere away from mid-ocean ridge highs) and "slab pull" (tension due to the gravitational sinking of the old subducting slabs). Subsequent analyses attempted to explain plate motions using the lateral density variations inferred from seismic tomography as the buoyancy forces in self-consistent models of mantle convection (B.H. Hager and R.J. O'Connell, A simple global model of plate dynamics and mantle convection, *J. Geophys. Res.*, **86**, 4843-4867, 1981; Y. Ricard and C. Vigny, Mantle dynamics with induced plate tectonics, *J. Geophys. Res.*, **94**,

17,543-17,559, 1989; C.W. Gable, R.J. O'Connell, and B.J. Travis, Convection in three dimensions with surface plates: Generation of toroidal flow, *J. Geophys. Res.*, **96**, 8391-8405, 1991).

109. Some intraplate earthquakes occur along preexisting zones of weakness in the continental crust associated with the landward extensions of oceanic fracture zones (L.R. Sykes, Intraplate seismicity reactivation of preexisting zones of weakness, alkaline magmatism, and other tectonism postdating continental fragmentation, *Rev. Geophys. Space Phys.*, **16**, 621-688, 1978), and some occur along reactivated faults that were first formed during continental rifting (A.C. Johnston, K.H. Coppersmith, L.R. Kanter, and C.A. Cornell, *The Earthquakes of Stable Continental Regions: Assessment of Large Earthquake Potential*, J.F. Schneider, ed., Electric Power Research Institute, Technical Report 102261, Palo Alto, Calif., 4 vols., 2985 pp., 1994).

110. At 10-kilometer depth, the effective normal stress σ_n^{eff} is about 180 megapascals (assuming hydrostatic pore pressure P_f), so that reasonable values of the coefficient of friction (μ = 0.6-0.8) imply that the shear stress to initiate frictional slip should be 110-140 megapascals.

111. The lack of detectable heat flow anomaly was first reported by J.N. Brune, T.L. Henyey, and R.F. Roy (Heat flow, stress and rate of slip along San Andreas fault, California, *J. Geophys. Res.*, **74**, 3821-3827, 1969) and confirmed by A.H. Lachenbruch and J.H. Sass (Thermo-mechanical aspects of the San Andreas fault system, in *Proceedings of the Conference on Tectonic Problems of the San Andreas Fault System*, R.L. Kovach and A. Nur, eds., Stanford University Publications in Geological Science 13, Stanford, Calif., pp. 192-205, 1973). Both of these studies quoted an upper bound on the average stress of 20 megapascals.

112. The direction of maximum principal compressive stress can be constrained from the orientations of borehole breakouts, in some cases by direct stress measurements, and by the orientations of known faults and the inferred focal mechanism orientation for smaller earthquakes in a region (assuming that these involve slip in the maximally shear-stressed direction on the causative fault). Such data showed that the maximum principal stress direction near the San Andreas fault was steeply inclined to the fault trace, in places approaching 80 degrees, much different than the 30- to 45-degree inclination expected if the fault was the most highly stressed feature in the region and was close to frictional failure (V. Mount and J. Suppe, State of stress near the San Andreas Fault—Implications of wrench tectonics, *Geology*, **15**, 1143-1146, 1987; M.D. Zoback, M.L. Zoback, V. Mount, J. Eaton, J. Healy, D. Oppenheimer, P. Reasenberg, L. Jones, B. Raleigh, I. Wong, O. Scotti, and C. Wentworth, New evidence on the state of stress of the San Andreas fault system, *Science*, **238**, 1105-1111, 1988).

113. S.H. Hickman, R.H. Sibson, and R.L. Bruhn, Introduction to special session: Mechanical involvement of fluids in faulting, *J. Geophys. Res.*, **100**, 12,831-12,840, 1995.

114. Gilbert's article "A Theory of the Earthquakes of the Great Basin, with a Practical Application" appeared in the *Salt Lake City Tribune* on September 30, 1883 (reprinted in *Am. J. Sci., 3rd Ser.*, **27**, 49-54, 1884).

115. Recent paleoseismic studies have shown that during the last 6000 years, the Wasatch fault has not ruptured all at once along its entire 340-kilometer length, but rather in smaller independent segments (M.N. Machette, S.F. Personius, A.R. Nelson, D.P. Schwartz, and W.R. Lund, The Wasatch fault zone, *J. Struct. Geol.*, **13**, 137-149, 1991). About 10 such segments have been identified, ranging in length from about 11 to 70 kilometers. During the last 6000 years, one of these segments has ruptured every 350 years on average, producing earthquakes with magnitudes ranging from 6.5 to 7.5. Individual segments appear to show irregular patterns of recurrence, however, with interseismic intervals lasting as long as 4000 years.

116. Reid's comment about precursory strain is based on the geodetic measurements of displacement of the Farallon Islands in the half-century before the 1906 earthquake. This

motion, which was about half that of the co-seismic displacement, was a key observation in the development of his hypothesis.

117. By "strong earthquake," Reid clearly meant something like the 1906 San Francisco earthquake, and he was implicitly assuming a characteristic earthquake model. Modern geodetic measurements show a rate of angle change of a few hundred parts per billion per year, with very modest changes over time, in the region of the 1906 quake (W.H. Prescott and M. Lisowski, Strain accumulation along the San Andreas fault system east of San Francisco Bay, California, *Tectonophysics*, **97**, 41-56, 1983). Thus, it would take a few thousand years to accumulate Reid's critical strain of 1/2000. Recent paleoseismic investigations indicate that surface-rupturing earthquakes on the 1906 rupture zone have an average interval of a few hundred years (D.P. Schwartz, D. Pantosti, K. Okumura, T.J. Powers, J.C. Hamilton, Paleoseismic investigations in the Santa Cruz Mountains, California; Implications for recurrence of large-magnitude earthquakes on the San Andreas fault, *J. Geophys. Res.*, **103**, 17,985-18,001, 1998). Reid may have overestimated the critical strain, the paleo-earthquakes may not have all been as large as the 1906 event, or the strain may not have been reset to zero in 1906.

118. K. Aki, Possibilities of seismology in the 1980s (Presidential address to the Seattle meeting), *Bull. Seis. Soc. Am.*, **70**, 1969-1976, 1980. Imamura is also reported to have forecast the occurrence of the great Nankaido earthquakes of 1944 and 1946 (On the seismic activity of central Japan, *Jap. Jour. Astron. Geophys.*, **6**, 119-137, 1928, *Jap. Jour. Astron. Geophys.*, **6**, 119-137, 1928; see S.P. Nishenko, Earthquakes, hazards and predictions, in *The Encyclopeida of Solid-Earth Geophysics*, D.E. James, ed., Van Nostrand Reinhold, New York, pp. 260-268, 1989).

119. S.A. Fedotov, Regularities of distribution of strong earthquakes in Kamchatka, the Kuril Islands and northern Japan (in Russian), *Akad. Nauk. SSSR Inst. Fiziki Zemli Trudi*, **36**, 66-93, 1965. Fedotov's map of seismic gaps include earthquakes of M 7.75 and larger; it was reproduced by K. Mogi in *Earthquake Prediction*, Academic Press, London, p. 82, 1985.

120. L.R. Sykes, Aftershock zones of great earthquakes, seismicity gaps and prediction, *J. Geophys. Res.*, **76**, 8021-8041, 1971. G. Plafker and M. Rubin (Uplift history and earthquake recurrence as deduced from marine terraces on Middleton Island, in *Proceedings of Conference VI: Methodology for Identifying Seismic Gaps and Soon-to-Break Gaps*, U.S. Geological Survey Open File Report 78-943, Reston, Va., pp. 687-722, 1978) later argued from direct geological evidence that this estimate of repeat time was short by a factor of two to four.

121. C.G. Chase, The n plate problem of plate tectonics, *Geophys. J. R. Astron. Soc.*, **29**, 117-122, 1972; J.B. Minster, T.H. Jordan, P. Molnar, and E. Haines, Numerical modelling of instantaneous plate tectonics, *Geophys. J. Roy. Astron. Soc.*, **36**, 541-576, 1974; J.B. Minster and T.H. Jordan, Present-day plate motions, *J. Geophys. Res.*, **83**, 5331-5354, 1978.

122. J.A. Kelleher, L.R. Sykes, and J. Oliver, Criteria for prediction of earthquake locations, Pacific and Caribbean, *J. Geophys. Res.*, **78**, 2547-2585, 1973; W.R. McCann, S.P. Nishenko, L.R. Sykes, and J. Krause, Seismic gaps and plate tectonics: Seismic potential for major boundaries, *Pure App. Geophys.*, **117**, 1082-1147, 1979. McCann et al. defined seismic gap as follows: "The term seismic gap is taken to refer to any region along an active plate boundary that has not experienced a large thrust or strike slip earthquake for more than 30 years . . . Segments of plate boundaries that have not been the site of large earthquakes for tens to hundreds of years (i.e., have been seismic gaps for large shocks) are more likely to be the sites of future large shocks than segments that experience rupture during, say, the last 30 years."

123. S.P. Nishenko, Circum-Pacific seismic potential: 1989-1999, *Pure Appl. Geophys.*, **135**, 169-259, 1991. Nishenko defined for each of 98 plate boundary segments a characteristic earthquake with a magnitude sufficient to rupture the entire segment, and a probability that such an earthquake would occur within 5, 10, and 20 years beginning in 1989 was

specified. The probability was based on the assumption that characteristic earthquakes are quasi periodic, with an average recurrence time that can be estimated from historic earthquakes or from the rates of relative plate motion. In the latter method, the mean recurrence time is estimated from the ratio of the average displacement in a characteristic earthquake to the relative plate velocity.

124. D.P. Schwatz and K.J. Coppersmith, Fault behavior and characteristic earthquakes: Examples from the Wasatch and San Andreas faults, *J. Geophys. Res.*, **89**, 5681-5698, 1984.

125. T. Hirata, Fractal dimension of fault systems in Japan: Fractal structure in rock fracture geometry at various scales, *Pure Appl. Geophys.*, **131**, 157-170, 1989; P. Segall and D. Pollard, Joint formation in granitic rock of the Sierra Nevada, *Geol. Soc. Am. Bull.*, **94**, 563-575, 1983; S. Wesnousky, C. Scholz, and K. Shimazaki, Earthquake frequency distribution and the mechanics of faulting, *J. Geophys. Res.*, **88**, 9331-9340, 1983.

126. In principle, the characteristic magnitude can be determined from the magnitudes of previous earthquakes on a fault segment or estimated from the length or area of the segment. Both methods present practical difficulties. Because the characteristic magnitude is thought to vary by segment, only events on a specific segment pertain. Uncertainties in magnitude and location of early quakes make it difficult to identify truly characteristic earthquakes in the seismic or geologic record. Fault geometry is not very definitive because there is insufficient understanding of the fault features that would prevent earthquake rupture from propagating further. S. Wesnousky (The Gutenberg-Richter or characteristic earthquake distribution: Which is it?, *Bull. Seis. Soc. Am.*, **84**, 1940-1959, 1994) examined the magnitude distribution on large sections of the San Andreas and other faults. He compared the rate of large earthquakes required to match the observed fault slip with that inferred by extrapolating the rate of smaller events with the Gutenberg-Richter relationship. For several sections, he found that the rate based on fault slip exceeded the Gutenberg-Richter rate, suggesting that the characteristic model was more appropriate there. However, Kagan disputed these findings (Y. Kagan and S. Wesnousky, The Gutenberg-Richter or characteristic earthquake distribution, Which is it? Discussion and reply, *Bull. Seis. Soc. Am.*, **86**, 274-291, 1996.). He argued that the result was biased because the regions were chosen around past earthquakes. Furthermore, the fault slip could be explained by a Gutenberg-Richter distribution if the maximum magnitude chosen was large enough.

127. S.P. Nishenko (Earthquakes, hazards and predictions, in *The Encyclopeida of Solid-Earth Geophysics*, D.E. James, ed., Van Nostrand Reinhold, New York, pp. 260-268, 1989) listed 15 large earthquakes that occurred in previously identified seismic gaps, including 8 earthquakes of M 8 and larger. Y.Y. Kagan and D.D. Jackson (Seismic gap hypothesis, *J. Geophys. Res.*, **96**, 21,419-21,431, 1991) reevaluated Nishenko's conclusions for the 10 events that occurred after Kelleher, Sykes, and Oliver (Possible criteria for predicting earthquake locations and their application to major plate boundaries of the Pacific and the Caribbean, *J. Geophys. Res.*, **78**, 2547-2585, 1973) provided geographically specific gap definitions. Kagan and Jackson found five unqualified successes, two mixed successes (earthquake on gap boundary), two events where gaps were not defined), and one earthquake in a "filled" gap. The track record of the gap hypothesis has been quite controversial, in part because early definitions were subjective. For further discussion, see S.P. Nishenko and L.R. Sykes, Comment on "Seismic gap hypothesis: Ten years after" by Y.Y. Kagan and D.D. Jackson, *J. Geophys. Res.*, **98**, 9909-9916, 1993; D.D. Jackson and Y.Y. Kagan, "Comment on 'Seismic gap hypothesis: Ten years after", Reply to S.P. Nishenko and L.R. Sykes, *J. Geophys. Res.*, **98**, 9917-9920, 1993; Y.Y. Kagan and D.D. Jackson, New seismic gap hypothesis: Five years after, *J. Geophys. Res.*, **100**, 3943-3959, 1995.

128. K. Shimazaki and T. Nakata, Time-predictable recurrence model for large earthquakes, *Geophys. Res. Lett.*, **7**, 279-282, 1980. The time-predictable model had been proposed

earlier by C.G Bufe, P.W. Harsh, and R.O. Burford (Steady-state seismic slip—A precise recurrence model, *Geophys. Res. Lett.*, **4**, 91-94, 1977).

129. The Japanese program for earthquake research, like its U.S. counterpart, is a broadly based, multidisciplinary effort directed at mitigating seismic hazards, but unlike the National Earthquake Hazards Reduction Program (NEHRP), its principal focus has been event-specific, "practical earthquake prediction." The program involves six agencies, with the scientific lead given to the Japan Meteorological Agency. The total expenditures over the first 30 years of the program's existence (1964-1993) were nearly 2000×10^8 yen, or about \$1.7 billion (in as-spent dollars). In units of 10^8 yen (approximately \$1 million), recent budgets for the program were 743 for 1995, 177 for 1996, and 214 for 1997. The 1995 number is high because it includes two supplements of 370 and 265 that were allocated following the destructive Hyogo-ken Nambu (Kobe) earthquake. The NEHRP expenditures for the same years, in millions of dollars, were 103, 106, and 98, respectively. The differences in expenditures between the two programs are even larger because the Japanese budgets do not include salaries, which are funded separately, whereas the U.S. budgets do.

130. W.H. Bakun and A.G. Lindh, The Parkfield, California earthquake prediction experiment, *Science*, **229**, 619-624, 1985.

131. Postmortem explanations cover a wide range. The 1983 Coalinga earthquake may have reduced the stress (R.W. Simpson, S.S. Schulz, L.D. Dietz, and R.O. Burford, The response of creeping parts of the San Andreas fault to earthquakes on nearby faults: Two examples, *Pure Appl. Geophys.*, **126**, 665-685, 1988). The rate of Parkfield earthquakes may be slowing because of postseismic relaxation from 1906 (Y. Ben-Zion, J.R. Rice, and R. Dmowska, Interaction of the San Andreas fault creeping segment with adjacent great rupture zones, and earthquake recurrence at Parkfield, *J. Geophys. Res.*, **98**, 2135-2144, 1993), or the Parkfield sequence may be a chance occurrence rather than a characteristic earthquake sequence (Y.Y. Kagan, Statistical aspects of Parkfield earthquake sequence and Parkfield prediction, *Tectonophysics*, **270**, 207-219, 1997).

132. Working Group on California Earthquake Probabilities, *Probabilities of Large Earthquakes Occurring in California on the San Andreas Fault*, U.S. Geological Survey Open-File Report 88-398, Reston, Va., 62 pp., 1988.

133. R.A. Harris summarized more than 20 published or broadcast statements that could be interpreted as scientific forecasts of the Loma Prieta earthquakes (Forecasts of the 1989 Loma Prieta, California, earthquake, *Bull. Seis. Soc. Am.*, **88**, 898-916, 1998). The 1906 surface displacement was smaller than it was to the north, suggesting incomplete release of the accumulated strain. Small earthquakes were notably absent on this segment, a pattern that was thought to appear before large earthquakes. From these observations, several earthquake forecasts were generated expressing the rupture length, magnitude, and approximate timing in probabilistic terms. For example, in 1983 Alan Lindh forecast an M 6.5 earthquake with a probability of 0.30 in 20 years (A.G. Lindh, *Preliminary Assessment of Long-Term Probabilities for Large Earthquakes Along Selected Fault Segments of the San Andreas Fault System in California*, U.S. Geological Survey Open-File Report 83-63, Menlo Park, Calif., 15 pp., 1983). In 1984 Sykes and Nishenko forecast an M 7.0 earthquake with a probability of 0.19 to 0.95 in 20 years (S.P. Nishenko, Probabilities of occurrence of large plate rupturing earthquakes for the San Andreas, San Jacinto, and Imperial faults, California, 1983-2003, *J. Geophys. Res.*, **89**, 5905-5927, 1984).

134. The earthquake was not on the San Andreas fault in a strict sense, but rather on a subsidiary, the Sargent fault, that dips about 70 degrees to the southwest and does not intersect the San Andreas. The fault slip had a large vertical component, which was different from expected San Andreas motion. Moreover, historical strain data suggest that significant slip occurred at depth on the San Andreas fault during the 1906 earthquake, so there may have been little or no slip deficit to warrant the initial forecasts.

135. For a recent compilation, including long-, intermediate-, and short-term prediction, see the conference proceedings introduced by L. Knopoff, Earthquake prediction: The scientific challenge, *Proc. Natl. Acad. Sci.*, **93**, 3719-3720, 1996. Articles by many other authors follow in sequence. For a brief, cautiously skeptical review see D.L. Turcotte, Earthquake prediction, *Ann. Rev. Earth Planet. Sci.*, **19**, 263-281, 1991. For a detailed, negative assessment of the history of earthquake prediction research, see R.J. Geller, Earthquake prediction: A critical review, *Geophys. J. Int.*, **131**, 425-450, 1997.

136. J. Deng and L. Sykes, Evolution of the stress field in southern California and triggering of moderate-size earthquakes; A 200-year perspective, *J. Geophys. Res.*, **102**, 9859-9886, 1997; R.A. Harris and R.W. Simpson, Stress relaxation shadows and the suppression of earthquakes; Some examples from California and their possible uses for earthquake hazard estimates, *Seis. Res. Lett.*, **67**, 40, 1996; R.A. Harris and R.W. Simpson, Suppression of large earthquakes by stress shadows; A comparison of Coulomb and rate-and-state failure, *J. Geophys. Res.*, **103**, 24,439-24,451, 1998.

137. K. Mogi, *Earthquake Prediction*, Academic Press, Tokyo, 355 pp., 1985. Mogi's "donut" hypothesis is summarized succinctly in C. Scholz, *The Mechanics of Earthquakes and Faulting*, Cambridge University Press, New York, pp. 340-343, 1990.

138. M. Ohtake, T. Matumoto, and G. Latham, Seismicity gap near Oaxaca, southern Mexico, as a probable precursor to a large earthquake, *Pure Appl. Geophys.*, **113**, 375-385, 1977. Further details are given in M. Ohtake, T. Matumoto, and G. Latham, Evaluation of the forecast of the 1978 Oaxaca, southern Mexico earthquake based on a precursory seismic quiescence, in *Earthquake Prediction—An International Review*, D. Simpson and P. Richards, eds., American Geophysical Union, Maurice Ewing Series 4, Washington, D.C., pp. 53-62, 1981. Interpretation of the success of the prediction and the reality of the precursor is complicated by a global change in earthquake recording because some large seismic networks were closed in 1967. For more details, see R.E. Habermann, Precursory seismic quiescence: Past, present, and future, *Pure Appl. Geophys.*, **126**, 277-318, 1988.

139. A comprehensive test requires a complete record of successes and failures for predictions made using well-defined and consistent methods. Otherwise, the likelihood of success by chance cannot be evaluated.

140. V.I. Kieilis Borok, and V.G. Kossobokov, Premonitory activation of seismic flow: Algorithm M8, *Phys. Earth Planet. Int.*, **61**, 73-83, 1990.

141. J.H. Healy, V.G. Kossobokov, and J.W. Dewey, *A Test to Evaluate the Earthquake Prediction Algorithm M8*, U.S. Geological Survey Open-File Report 92-401, Denver, Colo., 23 pp. + 6 appendixes, 1992; V.G. Kossobokov, L.L. Romashkova, V.I. Keilis-Borok, and J.H. Healy, Testing earthquake prediction algorithms: Statistically significant advance prediction of the largest earthquakes in the circum-Pacific, 1992-1997, *Phys. Earth Planet. Int.*, **111**, 187-196, 1999.

142. See <http://www.mitp.ru/predictions.html>. A password is needed to access predictions for the current six-month time interval.

143. John Milne noted this quest in his treatise *Earthquakes and Other Earth Movements* (D. Appelton and Company, New York, pp. 301 and 310, 1899): "Ever since seismology has been studied, one of the chief aims of its students has been to discover some means which could enable them to foretell the coming of an earthquake, and the attempts which have been made by workers in various countries to correlate these occurrences with other well-marked phenomena may be regarded as attempts in this direction." Milne himself proposed short-term prediction schemes based on measurements of ground deformation and associated phenomena, such as disturbances in the local electromagnetic field. "As our knowledge of earth movements, and their attendant phenomena, increases there is little doubt that laws will be gradually formulated and in the future, as telluric disturbances increase, a large black ball gradually ascending a staff

may warn the inhabitants on land of a coming earthquake, with as much certainty as the ball upon a pole at many seaports warns the mariner of coming storms." Milne's optimism was not shared by all seismologists, notably Charles Richter, who in his textbook (C.F. Richter, *Elementary Seismology*, W.H. Freeman, San Francisco, pp. 385-386, 1958) dismissed short-term earthquake prediction as a "will-o'-the wisp," questioning whether "any such prediction will be possible in the foreseeable future; the conditions of the problem are highly complex. One may compare it to the situation of a man who is bending a board across his knee and attempts to determine in advance just where and when the cracks will appear."

144. Ad Hoc Panel on Earthquake Prediction, *Earthquake Prediction: A Proposal for a Ten Year Program of Research*, White House Office of Science and Technology, Washington, D.C., 134 pp., 1965.

145. Dilatancy is the volume expansion due to the application of a shear stress, a well-known property of granular materials. Brace and his colleagues observed this phenomenon during rock-shearing experiments in the laboratory, which they interpreted as being due to pervasive microcracking in their specimens with a concomitant increase in void space (W.R. Brace, B.W. Paulding, and C.H. Scholz, Dilatancy of the fracture of crystalline rocks, *J. Geophys. Res.*, **71**, 3939-3953, 1966).

146. I.L. Nersesov, A.N. Semenov, and I.G. Simbireva, Space time distribution of travel time ratios of transverse and longitudinal waves in the Garm Area, in the physical basis of foreshocks, *Akad. Nauk USSR*, pp. 88-89, 1969.

147. See C.H. Scholz, L.R. Sykes, and Y.P. Aggarwal, Earthquake prediction: A physical basis, *Science*, **181**, 803-810, 1973, for a description for the laboratory and field observations that lead to the development of the dilatancy-diffusion model.

148. For a comprehensive description of the Haicheng prediction, see P. Molnar, T. Hanks, A. Nur, B. Raleigh, F. Wu, J. Savage, C. Scholz, H. Craig, R. Turner, and G. Bennett, Prediction of the Haicheng earthquake, *Eos, Trans. Am. Geophys. Union*, **58**, 236-272, 1977.

149. At the time there were many articles in the popular and scientific press about the impending breakthroughs in prediction capabilities. For example, see F. Press, Earthquake prediction, *Sci. Am.*, **232**, 14-23, 1975.

150. National Research Council, *Predicting Earthquakes: A Scientific and Technical Evaluation—With Implications for Society*, National Academy Press, Washington, D.C., 62 pp., 1976.

151. C.R. Allen and D.V. Helmberger, Search for temporal changes in seismic velocities using large explosions in southern California, in *Proceedings of the Conference on Tectonic Problems of the San Andreas Fault System*, R.L. Kovach and A. Nur, eds., Stanford University Publications in Geological Science 13, Stanford, Calif., pp. 436-452, 1973.

152. J.R. Rice and J.W. Rudnicki, Earthquake precursory effects due to pore fluid stabilization of a weakening fault zone, *J. Geophys. Res.*, **84**, 2177-2193, 1979.

153. In the reauthorization of the NEHRP program in 1990, references to earthquake prediction as a goal of the program were specifically removed. See Office of Technology Assessment, *Reducing Earthquake Losses*, OTA-ETI-623, U.S. Government Printing Office, Washington, D.C., 162 pp., 1995.

154. J.H. Dieterich, Preseismic fault slip and earthquake prediction, *J. Geophys. Res.*, **83**, 3940-3948, 1978; J.R. Rice, Theory of precursory processes in the inception of earthquake rupture, *Gerlands Beitr. Geophys.*, **88**, 91-127, 1979.

155. J.J. Lienkaemper and W.H. Prescott, Historic surface slip along the San Andreas fault near Parkfield, California, *J. Geophys. Res.*, **94**, 17,647-17,670, 1989; T. Donalee, Historical vignettes of the 1881, 1901, 1922, 1934, and 1966 Parkfield earthquakes, in Parkfield; the Prediction . . . and the Promise, *Earthquakes and Volcanoes*, **20**, 52-55, 1988.

156. Examples of current research on earthquake prediction techniques include the application of pattern recognition techniques (e.g., V.I. Keilis-Borok, L. Knopoff, I.M.

Rotwain, and C.R. Allen, Intermediate term prediction of occurrence time of strong earthquakes, *Nature*, **335**, 690-694, 1988; M. Eneva, and Y. Ben-Zion, Techniques and parameters to analyze seismicity patterns associated with large earthquakes, *J. Geophys. Res.*, **102**, 17,785-17,795, 1997), investigations of geochemical precursors (e.g., H. Wakita, Geochemical challenge to earthquake prediction, *Proc. Nat Acad. Sci*, **93**, 3781-3786, 1996), and the electromagnetic techniques associated with the VAN hypothesis. This last work has generated considerable controversy (e.g., R.J. Geller, ed., Debate on "VAN," *Geophys. Res. Lett.*, **23**, 1291-1452, 1996). For a negative assessment of the history of earthquake prediction research, see R. Geller, Earthquake prediction: A critical review, *Geophys. J. Int.*, **131**, 425-450, 1997.

157. Western-style masonry construction had been introduced in Japan to reduce the fire hazard associated with traditional wooden structures.

158. J. Milne and W.K. Burton, *The Great Earthquake in Japan, 1891*, 2nd ed., Lane, Crawford & Co., Yokohama, Japan, 69 pp + 30 plates, 1891.

159. State Earthquake Investigation Commission, *The California Earthquake of April 18, 1906*, Publication 87, vol. I, Carnegie Institution of Washington, pp. 365-366, 1908.

160. G.W. Housner, Historical view of earthquake engineering, *Proceedings of the Eight World Conference on Earthquake Engineering*, San Francisco, pp. 25-39, 1984. Panetti specifically proposed that the first story should be built to withstand one-twelfth of its superposed weight and the second and third stories one-eighth of their superposed weights.

161. A. Whittaker, J Moehle, and M Higashino, Evolution of seismic design practice in Japan, *Struct. Design Tall Build.*, **7**, 93-111, 1998.

162. The January 1927 edition of the UBC was published by the Pacific Coast Building Officials Conference. The provisions in the 1927 UBC to increase the design forces for structures on poor soils were removed in 1949, only to be reintroduced in the 1974 UBC.

163. J.R. Freeman, Engineering data needed on earthquake motion for use in the design of earthquake-resisting structures, *Bull. Seis. Soc. Am.*, **20**, 67-87, 1930.

164. For a discussion of the early strong-motion program, see G.W. Housner, *Connections, The EERI Oral History Series*, Earthquake Engineering Research Institute, Oakland, Calif., pp. 67-88, 1997; and D.E. Hudson, ed., *Proc. Golden Anniversary Workshop on Strong Motion Seismometry*, University of Southern California, Los Angeles, March 30-31, 1983.

165. H. Cross, Analysis of continuous frames by distributing fixed end moments, *Proc. Am. Soc. Civil Engineers*, **56**, 919-928, 1930; AIJ Standard for Structural Calculation of Reinforced Concrete Structures, *Kenchiku Zassh I (Architectural Institute of Japan)*, **47**, 62 pp., 1933.

166. M.A. Biot, A mechanical analyzer for the prediction of earthquake stresses, *Bull. Seis. Soc. Am.*, **31**, 151-171, 1940.

167. G.W. Housner and G.D. McCann, The analysis of strong-motion earthquake records with the electric analog computer, *Bull. Seis. Soc. Am.*, **39**, 47-56, 1949.

168. G.W. Housner, Characteristics of strong-motion earthquakes, *Bull. Seis. Soc. Am.*, **37**, 19-31, 1947.

169. G.W. Housner, R. Martel, and J.L. Alford, Spectrum analysis of strong motion earthquakes, *Bull. Seis. Soc. Am.*, **42**, 97-120, 1953.

170. National Research Council, *The San Fernando Earthquake of February 9, 1971: Lessons from a Moderate Earthquake on the Fringe of a Densely Populated Region*, National Academy Press, Washington, D.C., 24 pp., 1971.

171. High-frequency ground motions were analyzed by D.E. Hudson (Local distribution of strong earthquake ground motions, *Bull Seis. Soc. Am.*, **62**, 1765-1786, 1972), and low-frequency motions by T.C. Hanks (Strong ground motion of the San Fernando, California earthquake: Ground displacements, *Bull. Seis. Soc. Am.*, **65**, 193-225, 1975).

172. G.W. Housner and M.D. Trifunac, Analysis of Accelerograms—Parkfield Earthquake, *Bull. Seis. Soc. Am.*, **57**, 1193-1220, 1967.

173. M.D. Trifunac and D.E. Hudson, Analysis of the Pacoima Dam Accelerogram—San Fernando, California Earthquake of 1971, *Bull. Seis. Soc. Am.*, **61**, 1393-1141, 1971.

174. T.C. Hanks and D.A. Johnson, Geophysical assessment of peak accelerations, *Bull. Seis. Soc. Am.*, **66**, 959-968, 1976.

175. Applied Technology Council, *Tentative Provisions for the Development of Seismic Regulations for Buildings*, Applied Technology Council Publications ATC-3-06, Palo Alto, Calif., 505 pp., 1978. These provisions served as the basis for the seismic provisions of the 1988 Uniform Building Code and the Federal Emergency Management Agency publication, *NEHRP Recommended Provisions for Seismic Regulation for New Buildings, 1994 Edition*, Building Seismic Safety Council, Federal Emergency Management Agency Report FEMA-222A (Provisions, 290 pp. 15 maps, 1990) and FEMA-223A (Commentary, 335 pp., 1995), Washington, D.C.

176. See Section 5.3.2 of Part 2—Commentary, *NEHRP Recommended Provisions for Seismic Regulations for New Buildings and Other Structures, 1997 Edition*, Building Seismic Safety Council, Federal Emergency Management Agency Report FEMA 303, Washington, D.C., February 1998.

177. L. Reiter, *Earthquake Hazard Analysis; Issues and Insights*, Columbia University Press, New York, 233 pp., 1990.

178. C.A. Cornell, Engineering seismic risk analysis, *Bull. Seis. Soc. Am.*, **58**, 1583-1606, 1968.

179. E.B. Roberts and F.P. Ulrich, Seismological activities of the U.S. Coast and Geodetic Survey in 1949, *Bull. Seis. Soc. Am.*, **41**, 205-220, 1949.

180. C.F. Richter, Seismic regionalization, *Bull. Seis. Soc. Am.*, **49**, 123-162, 1959.

181. S.T. Algermissen and D.M. Perkins, *A Probabilistic Estimate of the Maximum Ground Acceleration in Rock in the Contiguous United States*, U.S. Geological Survey Open File Report 76-416, Denver, Colo., 45 pp., 1976.

3

Facing the Earthquake Threat

Earthquakes rival all other natural disasters in the threat they pose to human life and habitat. Unlike floods, hurricanes, and volcanic eruptions, specific earthquakes cannot be predicted with the short-term accuracy required for effective emergency management. The science is now capable of identifying where earthquakes will happen and how big they might be, but such forecasts are valid only for intervals measured in decades or even centuries. Once an event has occurred, there is very little time for warning and action; the fast-moving seismic ground waves do most of their damage in a macroseismic zone within the first minute or so after the rupture nucleates (*1*). Preparation and rapid emergency response are therefore the bulwarks of a good seismic defense. This chapter describes the context of current efforts to improve seismic safety and performance by summarizing what is known about the principal types of earthquake hazards, their distribution across the nation and the world, and the knowledge-based approaches to reducing earthquake risk. It concludes by addressing the issue of how scientists can help to implement the knowledge gained through research by stimulating civic actions that actually reduce risk.

3.1 TYPES OF SEISMIC HAZARDS

Earthquakes pose several types of threats that often proceed as chain reactions. The *primary hazards* are the breaks in the ground surface caused when faults rupture, the seismic shaking radiated from the fault slip during rupture, and the permanent subsidence and uplift. Strong ground motion may, in turn, cause ground failure—slumps, landslides, liquefaction, and lateral spread—depending on shaking intensity (usually stronger nearer the source) and local site conditions. If it occurs offshore, fault displacement can generate tsunamis capable of inundating nearby and distant shorelines. Ground failure and tsunamis are examples of *secondary hazards* (2).

Fault Rupture

Tectonic earthquakes are spontaneous releases of tectonic stress that produce macroscopic, permanent displacements across fault surfaces (ruptures) and within the rock mass around faults (co-seismic deformations). Most fault ruptures are confined to buried regions of the crust where brittle behavior allows stick-slip instabilities to nucleate (e.g., between 2 and 20 kilometers deep in most continental deformation zones). Such ruptures propagate to the surface only in larger earthquakes. When this happens, however, almost any structure built across the rupture path will be deformed by the severe strains characteristic of primary ground failure (Figure 3.1). Predicting the magnitude and extent of fault rupture is therefore a major issue in seismic hazard analysis.

Ruptures tend to occur along faults that have produced large earthquakes in the past, so a map of active faults is a first-order representation of the rupture hazard. The average amount of co-seismic slip increases systematically with earthquake magnitude (3), and the maximum displacement tends to occur toward the middle of the rupturing segment. These behaviors can be used to quantify the hazard along well-defined active faults. For example, where the Hollywood subway crosses the Hollywood fault in Los Angeles, California, the maximum expected slip is estimated to be 1 to 2 meters. In anticipation, the Metropolitan Transportation Authority overbored the subway tunnel to allow the tracks to be realigned after such an earthquake.

Mapped faults are often categorized as active and inactive, but doing so is problematic because the maximum magnitude, frequency of rupture, and other measures of activity can be highly variable among faults in the same tectonic province. Even within a single zone, the distribution of recent faulting can be considerably more complex than the simple traces that represent active faults on small-scale geologic maps. Detailed mapping reveals a wide range of features, such as segmentation, stepovers, and faulting at conjugate angles, often with self-similar scaling (Figure 3.2). The faulting patterns observed in large earthquakes show similar complexity, which can vary rapidly along strike. In some places, the rupture may be a single, clean break, while elsewhere it may occupy a zone tens or hundreds

FIGURE 3.1 Damage due to fault rupture during the September 20, 1999, Chi-Chi, Taiwan, earthquake (magnitude [M] 7.6) was extensive along the Chelungpu fault. Structures that were built to withstand strong ground motion nevertheless did not survive severe dislocations along the fault. The left abutment of this bridge across the Ta-An River was constructed through the fault plane (*top*). The thrust fault slipped about 10 meters during the earthquake, severely deforming the pillar and, thus, destroying the bridge. After the earthquake, the abutment was reconstructed in the same location. Most reinforced multistory concrete structures survived the shaking, but those on the fault trace collapsed or suffered severe tilts and other distortions, which rendered them uninhabitable (*bottom*). SOURCE: Photographs courtesy of Kerry Sieh, Caltech.

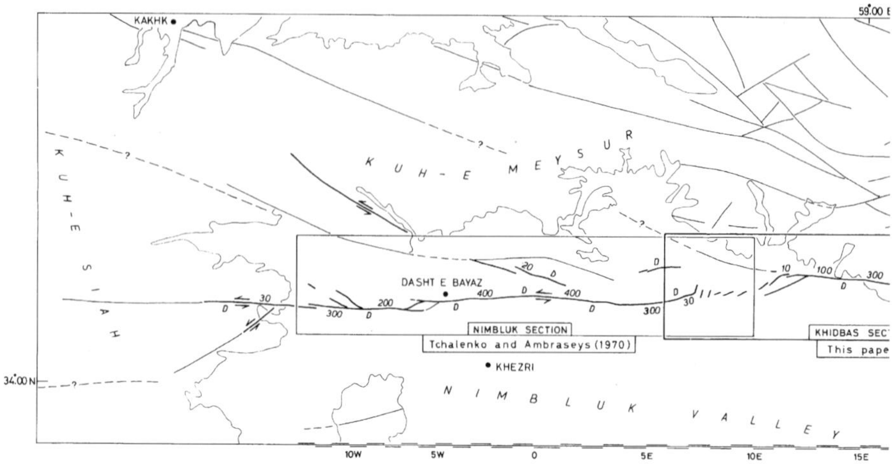

FIGURE 3.2 Map of the surface trace of the 1968 Dasht-e-Bayez, Iran, earthquake rupture (M 7.3). SOURCE: J.S. Tchalenko and M. Berberian, Dasht-e-Bayez fault, Iran: Earthquake and earlier related structures, *Geol. Soc. Am. Bull.*, **86**, 703-709, 1975.

of meters wide comprising en echelon offsets, anastomosing fractures, mole tracks, nonbrittle warping, and other types of co-seismic deformation (*4*). A more complete characterization of the rupture hazard will require a better understanding of how the distribution of surface breaks depends on the details of the fault slip at depth and how fault movements interact with a variety of structural factors, including topography, near-surface sedimentary layering, and fault-zone complexity.

Ground Shaking

Ground shaking is typically the primary cause of earthquake damage to the built environment. Shaking occurs during the passage of seismic waves as they propagate away from the rupturing fault. The most destructive shaking is usually the horizontal ground motion from *S* waves and surface waves, although the vertical component of motion can also excite a damaging structural response. The severity of the shaking is typically measured by the peak ground acceleration (PGA) or peak ground velocity (PGV), as recorded on strong-motion seismographs in the free field (i.e., on open ground away from buildings or other structures), or by the spectral response of a standard oscillator, either spectral acceleration Sa or spectral velocity Sv, calculated from the observed "time history" of

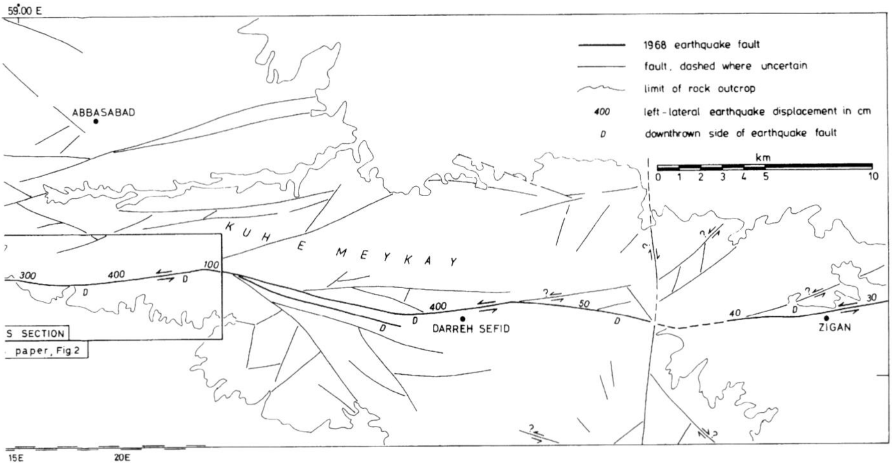

the shaking (5). Measurements by strong-motion instruments near large earthquakes have shown that the time histories can be complex and can vary rapidly from place to place, especially at high frequency where interference effects are typically strong.

Seismic waves are generated by fault slip during an earthquake. The distribution of the slip in space and time determines the radiation pattern (i.e., how the wave amplitudes vary with direction away from the fault). Small earthquakes can usually be approximated by the beachball-looking radiation patterns described in Section 2.3, but larger events show significant complications and asymmetries. For instance, the propagation of a rupture along a fault may produce a directivity pulse of coherent, high-amplitude shear motion at locations in the propagation direction. Rupture directivity effects, amplified by basin-edge effects, were the primary cause of the damage in the 1995 Hyogo-ken Nanbu earthquake. During the 1994 Northridge earthquake, the ground motion at frequencies below 2 hertz was observed to be highest at locations around the top edge of the fault to the north of the hypocenter, consistent with the directivity pulse expected from the Northridge rupture. At higher frequency, the radiation of waves from fault surfaces becomes less coherent, owing to small-scale fluctuations in fault slip and nearby material irregularities, causing the rupture directivity effect to become subdued and other effects such as proximity to the fault to dominate. In the Northridge earthquake, ground-motion amplitude greater than 2 hertz was observed to be highest above the hypocenter on the hanging wall side (*6*), not around the top edge of the fault, which experienced the longer-period directivity effects. Systematic differences in ground motion have been observed for different faulting types (*7*), but as yet no clear explanation of this exists. Elucidating how faults radiate seismic energy across the entire frequency band relevant to earthquake engineering (0.1-10 hertz) is a research challenge of major importance.

The amplitude of seismic waves generally decreases as the waves propagate away from the source (as required to conserve energy), but the measurement from large earthquakes always exhibits a high degree of scatter. In seismic hazard analysis, the decay of ground-motion intensity with distance is represented by an attenuation relation, usually derived by fitting smooth functions to the scattered data (see Section 2.7). An objective of current research is to explain the variations in shaking intensity through a more fundamental understanding of the wave propagation process (*8*). Important physical effects include refraction by variation in the seismic velocity, reflection from surfaces of material discontinuity, and damping by the anelastic response of the rock and soil media. Some of the strongest variations are associated with horizontal layering of the crust and upper mantle. In the 1989 Loma Prieta earthquake, shear waves, critically refracted from the M discontinuity at the base of the crust (*SmS* waves), were partially responsible for the shaking that damaged parts of San Francisco nearly 90 kilometers from the epicenter (*9*). Data from aftershocks of the 1994 Northridge earthquake demonstrated that reflections from midcrustal interfaces can increase the shaking from shallow sources at certain shorter distances.

Seismic waves can be amplified or attenuated by three-dimensional structures such as fault-bounded blocks and sedimentary basins. Earthquakes can excite resonance in the deep basins, shaking the soft sediments like jelly. A striking example was the massive destruction and loss of life during the 1985 Michoacan earthquake (moment magnitude [M] 8.0) in the parts of Mexico City underlain by soft, lake-bed clays. The source was in a subduction zone more than 350 kilometers away, which under normal circumstances would have caused little damage; however, sediment resonance was observed to amplify the spectral acceleration at low frequencies (about 0.5 hertz) by factors as large as 8 to 50 times relative to hard-rock sites (*10*). Other mechanisms for amplification include the focusing of waves by lens-like structures (*11*) and the generation of surface waves by the fault-bounded edges of sedimentary basins. Basin-edge effects of the latter type were partly responsible for the extreme damage to the Japanese city of Kobe in the 1995 earthquake (Box 2.4).

The response of soils at shallow depth to strong shaking is a complex phenomenon (*12*). Amplitude builds as the waves slow (another consequence of energy conservation), so seismic shaking is typically amplified in the soft soils and unconsolidated sediments near the ground surface, where the wave speed can be much lower than in hard rock. For this reason, the average shear velocity in the upper 30 meters or so has become the primary basis for the National Earthquake Hazard Reduction Program (NEHRP) site classification used in many building codes, including the 1997 Uniform Building Code (UBC), 2000 International Building Code (IBC), and 2000 American Society of Civil Engineers (ASCE) Standards 7-98 (*13*). The high amplitude predicted by the linear wave theory is thought to be reduced by the nonlinear response of the unsaturated near-surface layers. Laboratory tests clearly demonstrate nonlinear strain behavior in soils under dynamic loading, but the importance of nonlinearity during actual earthquakes continues to be debated. Using available ground-motion data to differentiate nonlinear strain behavior from other wave propagation effects has usually been difficult. For example, interpretations of the data collected in Mexico City from the 1985 Michoacan earthquake reached conflicting conclusions on the importance of the nonlinearity of the city's soft clay deposits (*14*). On the other hand, direct evidence of significant nonlinear soil response was clearly observed in the motions recorded by surface and subsurface (borehole) instruments at saturated sandy sites that liquefied during the 1987 Superstition Hills, California, and the 1995 Hyogo-ken Nanbu earthquakes (*15*). Aside from these extreme cases where the soil failed, indirect evidence of nonlinear site response on soils that remained stable during strong shaking is becoming more apparent with the greater number of seismograms being recorded in strong-motion arrays throughout the world (*16*). However, more of these data are clearly needed to better understand and predict this phenomenon.

Another interesting aspect of seismic shaking is that it can vary substantially from one tectonic setting to another. For example, the motion from similar-sized earthquakes is observed to be stronger in the central and eastern United States than west of the Rocky Mountains. Felt areas and areas of specific intensity (isoseismals) are also larger for earthquakes in the central and eastern United States compared to those of earthquakes with similar magnitudes in the western United States. Earthquakes in the older, stronger regions of the continent generally have greater stress drops and therefore radiate more high-frequency energy for a given amount of fault slip; moreover, their seismic waves propagate with less attenuation compared to earthquakes in plate boundary deformation zones. The attenuation difference is probably attributable to lower temperature, reduced scattering, and more continuous waveguide for crustal shear energy (*Lg* waves) in the more stable crust of the central and eastern United States. Regional studies that deploy seismometers more densely will be needed to clarify these explanations and to understand how the vertical and lateral structure of the crust controls ground motions.

Subsidence and Uplift

Large thrust earthquakes in subduction zones can cause sudden, permanent elevation changes with damaging effects to coastal areas. Uplift and subsidence related to fault slippage on shallow thrusts have been documented in New Zealand, Japan, Chile, and southeast Alaska (*17*). During the great 1964 Alaska earthquake (Box 2.3), the shorelines of Prince William Sound rose in some places by several meters, draining small-craft harbors, while they dropped in others, causing the streets of coastal towns to flood at high tide. Submerged marshlands in several estuaries along the coasts of Washington, Oregon, and northern California indicate that similar episodes of sudden subsidence have resulted from large thrust events in the Cascadia subduction zone (described in Section 3.2). The pattern of uplift and subsidence during an earthquake can be predicted from elastic dislocation models if the area of the fault plane and the distribution of slip within that plane are known (*18*). Anticipating the damage from elevation changes in future events can thus be approached by combining theoretical studies with seismic, paleoseismic, and geodetic observations.

Secondary Ground Failures

The secondary hazards caused by seismic shaking include forms of mass wasting—such as landslides, rockfalls, and slumps—as well as soil failures associated with compaction, liquefaction, and lateral spreading (*19*). In some instances, these failures cause more damage than the ground shaking itself. An M 8.6 earthquake in China's Gansu Province in 1920 triggered an extensive debris flow, which covered a region larger than 100 square kilometers and resulted in roughly 200,000 deaths. An immense rock and snow avalanche (60 million cubic meters) triggered by the 1970 Peru earthquake (M 8.0) buried the mountain towns of Yungay and Ranrahirca, killing 66,000 people (Figure 3.3). Many of those killed in the January 13, 2001, El Salvador earthquake were buried by a muddy landslide loosened from a slope in the capital's suburbs.

Liquefaction is the temporary conversion of water-saturated, unconsolidated soils into a medium that behaves like a fluid. It occurs when saturated sand or silty sand is shaken hard enough to mobilize individual grains. If the water cannot escape the granular soil matrix fast enough to

FIGURE 3.3 Destruction of the mountain towns of Yungay and Ranrahirca, Peru, which were buried by an avalanche triggered by the 1970 earthquake (M 8). SOURCE: Photo by Servicio Aerofotografico Nacional de Peru; available from the U.S. Geological Survey, <http://landslides.usgs.gov/>.

permit compaction, more of the overburden load becomes supported by the water, resulting in increased pore pressure. This process can progress relatively quickly to the point at which the pore-water pressure becomes equal to the overburden stress, creating quicksand-like conditions. The liquefaction potential of any particular saturated deposit depends primarily on the age and grain-size distribution of the deposit as well as the ampli-

FIGURE 3.4 Tilting of apartment buildings at Kawagishi-Cho, Niigata, Japan, produced by liquefaction of loose, water-saturated sediments caused a loss of load-bearing capacity during the 1964 Niigata earthquake (M 7.5). The losses from this earthquake exceeded $1 billion in 1964 dollars. SOURCE: National Oceanic and Atmospheric Administration, National Geophysical Data Center.

tude and duration of the ground shaking (*20*). The dangers of liquefaction are thus compounded in deep sedimentary basins, where the water table is often shallow and the shaking amplitude and duration tend to be increased by seismic-wave resonance within the basins. Liquefaction can severely damage foundations and other subsurface structures, causing large buildings to sink or tilt (Figure 3.4) and underground structures, such as pipelines and storage tanks, to float to the surface when they become buoyant in the liquefied soil. If the liquefied layer is close to the surface, it may break through dryer deposits overlying the water table, forming geysers that leave sandblows as postseismic evidence. In fact, the dating of such features has become an extremely useful tool for establishing prehistoric records of major earthquakes in the Charleston and New Madrid areas of the eastern and central United States (*21*).

Lateral spreading is a form of landsliding caused when liquefaction occurs on a sloping surface or adjacent to an embankment or excavation, typically resulting in the opening of fissures perpendicular to the surface gradient. Embedded structures are dragged by the flow, and the variable

FIGURE 3.5 Many apartment buildings on the fringe of the river delta at Golcuk, Turkey, slid into the sea during the Izmit earthquake of August 17, 1999. Land-use planning that accounts for the instability of young deltaic sediments could substantially reduce loss of life and property in future earthquakes. SOURCE: A. Barka, Istanbul Technical University.

displacements can literally rip structures apart. Lateral spreading tends to allow material to fill topographic depressions, such as streams and rivers, causing the channels to narrow and the flow to become dense—a major source of damage to bridges during earthquakes (22). A more recent case of lateral spreading occurred during the August 17, 1999, Izmit, Turkey earthquake (M 7.4) when unconsolidated, water-saturated deltaic sediments collapsed into the sea (Figure 3.5), resulting in numerous deaths. Lateral spreading or landsliding can also be caused by the shaking-induced loss of shear strength in certain types of "quick" or "sensitive" layers of salt-leached, clay-rich marine sediments. The spectacular damage to the Turnagain Heights district of Anchorage during the great 1964 earthquake (M 9.2) (Box 2.3) has been attributed to large (150- to 180-meter) displacements within a relatively thin zone of the Bootlegger Cove clay, 25 meters below the surface (23).

Empirical relations for predicting the extent and severity of liquefaction events have been developed through field studies, theoretical modeling, and laboratory experiments using geotechnical centrifuges. Less extensive research has been conducted on mass wasting in unsaturated conditions. However, postseismic mapping of landslides and rockfalls has shown that rock type, steepness of slope, and proximity to seismic source are the major contributors to failure. In 1990, California enacted the Seismic Hazards Mapping Act, which significantly broadened the responsibilities of the state geologist to include mapping liquefaction and landslide hazards. Mitigation money from the Federal Emergency Management Agency (FEMA) under the Stafford Act after the Northridge earthquake and fees from construction building permits provide funding for these mapping efforts (e.g., Figure 3.6).

FIGURE 3.6 Map of landslide and liquefaction hazards for the city of San Francisco. Areas of high landslide potential are shown in blue, and areas of high liquefaction potential are shown in green. SOURCE: California Division of Mines and Geology, <http://www.consrv.ca.gov/dmg/shezp/maps/m_sf.htm>.

Tsunamis

When a large earthquake occurs under the ocean, the vertical motion of the seafloor displaces the water column, causing gravitational instability; the potential energy change of the water column converts to kinetic energy, forming a tsunami, or seismic sea wave. (Although this tectonic mechanism is the most common cause of tsunamis, they can also be generated by submarine landslides or mass debris entering the sea from volcanic eruptions.) Tsunamis have wavelengths of tens to hundreds of kilometers, depending on the horizontal dimensions of the source, and travel over long distances with little attenuation. Their speed depends on ocean depth, increasing from 500 kilometers per hour in a 2-kilometer-deep ocean to about 900 kilometers per hour in a 6-kilometer-deep ocean (*24*). The waves are refracted away from regions of deep water and scattered from local bathymetric highs. Accurate bathymetric maps can be used to simulate tsunami propagation (Figure 3.7) and predict the arrival time and, if the details of the source are known, the amplitude. As the tsunami enters shallower water, the propagation speed and wavelength decrease, and the amplitude increases. Transgression of the shoreline by large tsunamis causes runup (measured as the water rise above the shoreline level, in meters), often as a very fast-rising tidal wave that floods well past the normal high-water level, sometimes as a turbulent bore. Considerable structural damage can occur to ports and other coastal installations from the exceptional currents generated during runup and withdrawal and the impact of debris entrained by these currents.

Tsunamis claimed more than 100,000 lives in the twentieth century. In some cases, the destruction occurred far from the earthquake epicenter, as on April 1, 1946, when an M 7.1 earthquake in the Aleutian Islands triggered a Pacific-wide tsunami. A runup of 8.1 meters occurred 4.9 hours later at Hilo, Hawaii, causing $26 million in damage and 159 deaths (Figure 3.8). This disaster led to the first Seismic Sea Wave Warning System, established in Hawaii on August 12, 1948. Additional systems have since been deployed, including those designed to provide rapid warnings of tsunami hazards from local earthquakes, when the runups occur soon after ground shaking (see below).

Although these warning systems can estimate tsunami arrival time accurately, their prediction of wave amplitude and coastal runup is much less precise. Major uncertainties are associated with the tsunami excitation process. Although tsunami amplitude depends on the total deformation of the seafloor (and should therefore be proportional to low-frequency seismic moment), it correlates poorly with standard earthquake magnitude determined at high frequencies (*25*). Moreover, seafloor deformation in the epicentral region depends on the depth and orientation of the fault-

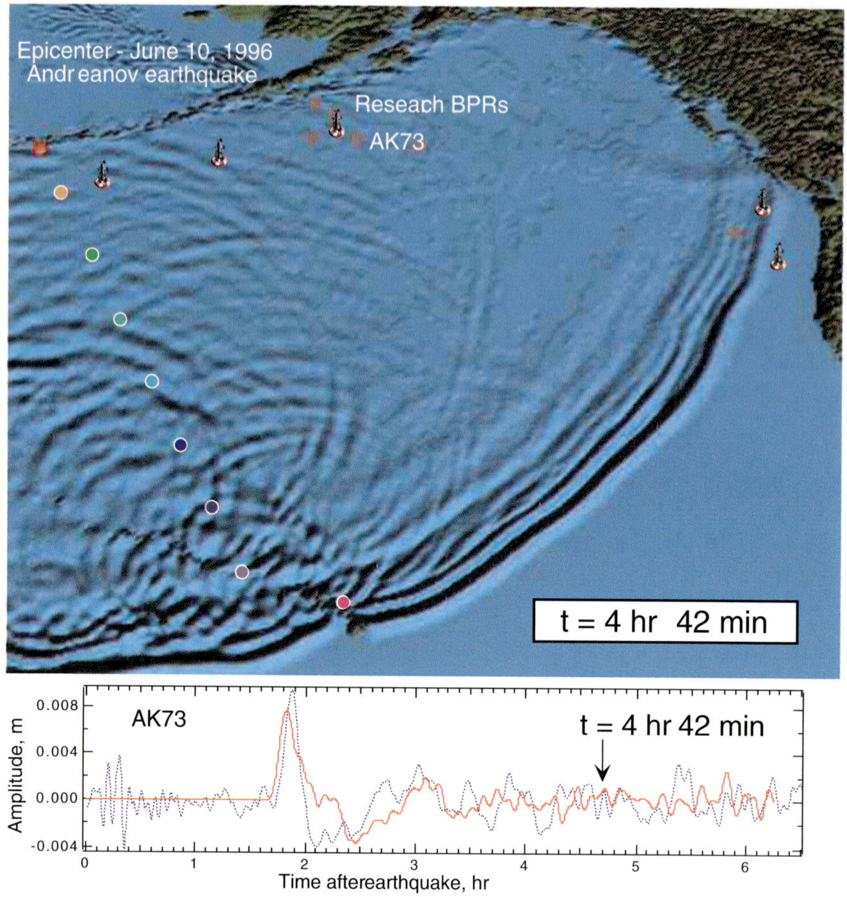

FIGURE 3.7 Numerical simulation of tsunami radiation into the Pacific Ocean by an M 7.7 earthquake of June 10, 1996, in the Andreanov Islands 50 miles southwest of Adak, Alaska. *Top:* Wavefield calculated at 4 hours and 42 minutes after the event. *Bottom:* Comparison of observed (blue line) and computed (red line) waveforms from bottom pressure recorder AK73. SOURCE: National Oceanic and Atmospheric Administration, Pacific Marine Environmental Laboratory.

ing. Subduction-zone earthquakes as large as the 1946 Aleutian event occur several times each year, for example, but they only rarely produce such a big tsunami. Fault ruptures that propagate at anomalously low velocity, and thus have a long source duration, appear to be responsible for some large tsunamis, whereas others are evidently due to landslides triggered by earthquakes or volcanic eruptions (*26*).

FIGURE 3.8 Tsunami generated by the April 1, 1946, Aleutian Islands earthquake, breaking over Pier No. 1 in Hilo Harbor, Hawaii. Man in the foreground was one of the 159 fatalities in the Hawaiian Islands. SOURCE: National Oceanic and Atmospheric Administration, National Geophysical Data Center.

Tsunami prediction is also limited by the complex hydrodynamics of tsunami propagation and runup. Considerable effort has been devoted to rigorously defining the effects of nonlinearity and dispersion for tsunamis that propagate over long distances (*27*). The 1992 Flores Island and the 1993 Okushiri Island tsunamis prompted investigation of the phenomenon of tsunami trapping near islands (*28*). Runup laws that relate the offshore waveform to maximum runup onshore have been derived theoretically, and these studies have been augmented by laboratory and numerical investigations of the runup associated with breaking waves and shallow beach slopes (*29*). Efforts are also under way to formulate site-specific inundation models near coastal population centers, which will require more accurate numerical methods that take into consideration topographic effects and bottom friction in the inundation region (*30*). Local tsunamis generated along continental margins pose a special problem for hazard mitigation (*31*). A particular issue is the excitation of edge waves, which can result in large-amplitude late arrivals, as observed in a tsunami generated by the 1992 Cape Mendocino, California, earthquake (*32*).

3.2 SEISMIC HAZARDS IN THE UNITED STATES

A major task for earthquake science is to characterize the geographical distribution of seismic hazards. At a particular location, the primary hazard of ground shaking can be quantified by probabilistic seismic hazard analysis (PSHA; see Section 2.7). Extending PSHA calculations to all points in a region generates a seismic hazard map, which conveys the spatial variations in the maximum expected intensity of shaking needed for building codes, loss estimation, and other risk mitigation applications. This section describes seismic hazards in the United States, using the national seismic hazard maps as a guide.

National Seismic Hazard Maps

Seismic hazard maps depict the ground-motion intensity that will be exceeded with a specified probability during a specified exposure time (See section 2.7). In 1996, the U.S. Geological Survey (USGS) released a new series of national seismic hazard maps (*33*) that contoured four measures of earthquake intensity—PGA and response-spectrum accelerations at frequencies of 1, 3, and 5 hertz—at three hazard levels, given by the 2 percent, 5 percent, and 10 percent probabilities of exceedance in 50 years. When used to determine the seismic-safety criterion in structural design, a lower exceedance probability specifies a higher shaking intensity and is assigned to structures whose failure would cause more severe consequences. The highest probability level depicted in the national maps (10 percent probability in 50 years) corresponds to a mean return period of 475 years (*34*) and is often used in the life-safety design of buildings, while the lowest (2 percent probability in 50 years) corresponds to a mean return period of 2475 years and is used in collapse-prevention design (*35*). Figure 3.9 compares the maps of PGA at these two hazard levels for firm-rock sites (B-C boundary in the NEHRP site classification) in the conterminous United States.

The 1996 maps incorporated a range of new information on earthquake hazards derived from NEHRP-sponsored research, such as the discovery of great prehistoric subduction-zone earthquakes in the Pacific Northwest, and recurrence rates and magnitudes of large earthquakes in the New Madrid and Charleston areas estimated from paleoliquefaction studies. The database for the western United States included catalogs of geologic slip rates, paleoseismic chronologies, and geodetic measurements comprising approximately 450 faults. In the calculation of expected intensities, earthquake recurrence times were estimated from geologic slip rates using two alternative hypotheses, a characteristic earthquake model and a truncated Gutenberg-Richter model. Results from competing models for fault recurrence, seismicity distribution, and attenuation relations were combined via a logic-tree formalism.

The process for developing the national seismic hazard maps involved an extensive dialogue between the geoscience and engineering communities, directed at improving the applicability of these maps as risk mitigation tools. This effort has paid off in greater utilization (see Section 2.7). Utilization has also been enhanced by the availability of the data and data products that went into the map calculations. An important data product for specialized applications is the disaggregation of the hazard at a map point into the individual earthquake sources (*36*). (Examples are shown for Knoxville, Tennessee, in Figure 3.10.) Such disaggregations identify which potential earthquakes dominate the hazard for a given site. These sources can then be used as scenarios for constructing ground-motion time histories needed to design critical facilities, conduct emergency management exercises, and estimate earthquake losses.

California

California has the highest levels of seismic hazard in the lower 48 states because more than 75 percent of the relative motion between the Pacific and North American plates occurs as active faulting within its borders. The cumulative fault slip at each latitude from the Mexican border to Cape Mendocino averages more than 3.5 meters each century. California seismic hazard is dominated by the San Andreas system. This "master fault" of the plate boundary has a documented history of earthquakes as large as M 8. Two of its four major segments have broken in large, historic events: a 420-kilometer segment (Cape Mendocino to San Juan Bautista) during the 1906 San Francisco earthquake and a 350-kilometer segment (Parkfield to Wrightwood) during the 1857 Fort Tejon earthquake. The frequency of large events on the 1906 segment remains poorly constrained, but paleoseismic investigations suggest a mean recurrence interval of about 250 to 300 years. The frequency of large events along the 1857 segment is better documented, varying by locality from about 50 to 300 years. These segments are separated by the "creeping section" of the San Andreas (San Juan Bautista to Parkfield), where most of the strain is taken up as aseismic creep and large earthquakes appear to be absent. According to paleoseismic data, the 200-kilometer-long southernmost segment (Wrightwood to Bombay Beach) has broken in at least four major earthquakes since A.D. 1000. The last rupture occurred circa 1680, so this segment appears to be overdue for another.

Auxiliary strike-slip faults of the San Andreas system are also hazardous. Over the past century alone, the Imperial, San Jacinto, Elsinore, and Newport-Inglewood faults of southern California have together produced more than a dozen earthquakes larger than M 6. In the San Francisco region, the San Andreas fault splays into several branches. One of the branches on the east side of San Francisco Bay, the Hayward fault, was

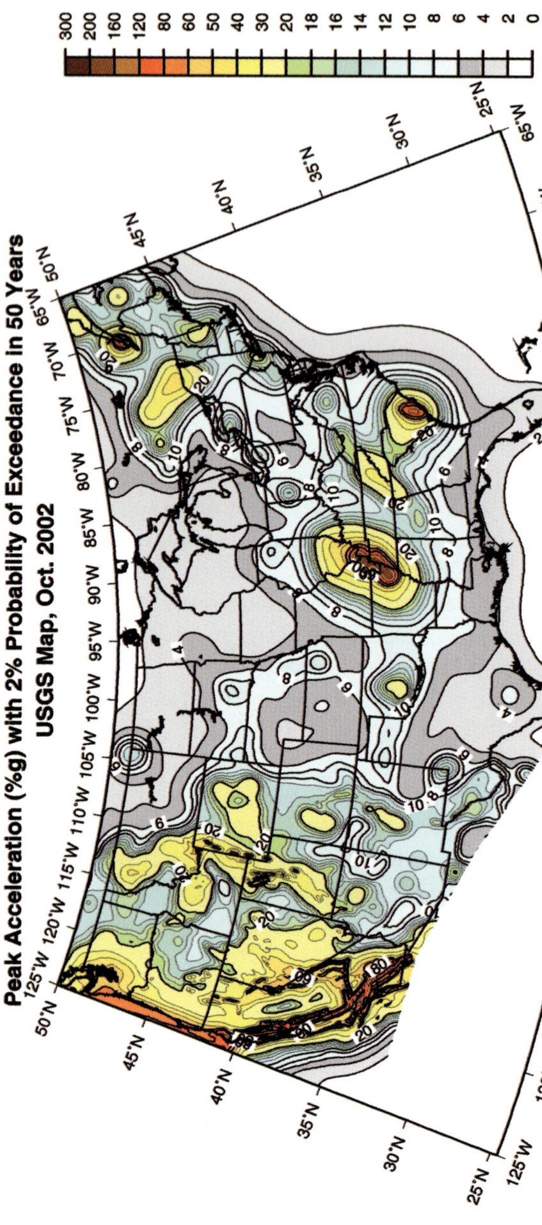

FIGURE 3.9 Probabilistic seismic hazard maps for the conterminous United States. *Left panel*: Peak ground acceleration with a 2 percent chance of exceedance in 50 years. *Right panel*: PGA with a 10 percent chance of exceedance in 50 years. The region of highest hazard lies along the San Andreas fault and the Transverse Ranges in California, with a branch extending into eastern California and west-

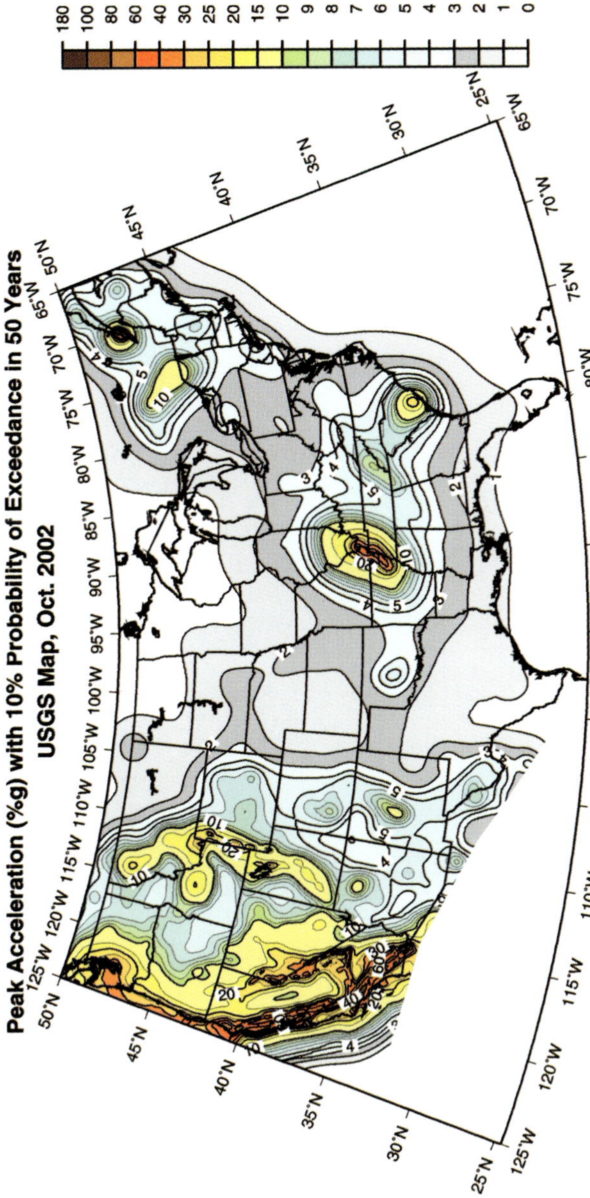

ern Nevada. High hazards are also found along the coast of the Pacific Northwest and in a zone following the intermountain seismic belt. In the central and eastern United States, the highest hazard areas are New Madrid, Missouri; Charleston, South Carolina; eastern Tennessee; and portions of the Northeast. SOURCE: U.S. Geological Survey, <http://geohazards.cr.usgs.gov/eq/>.

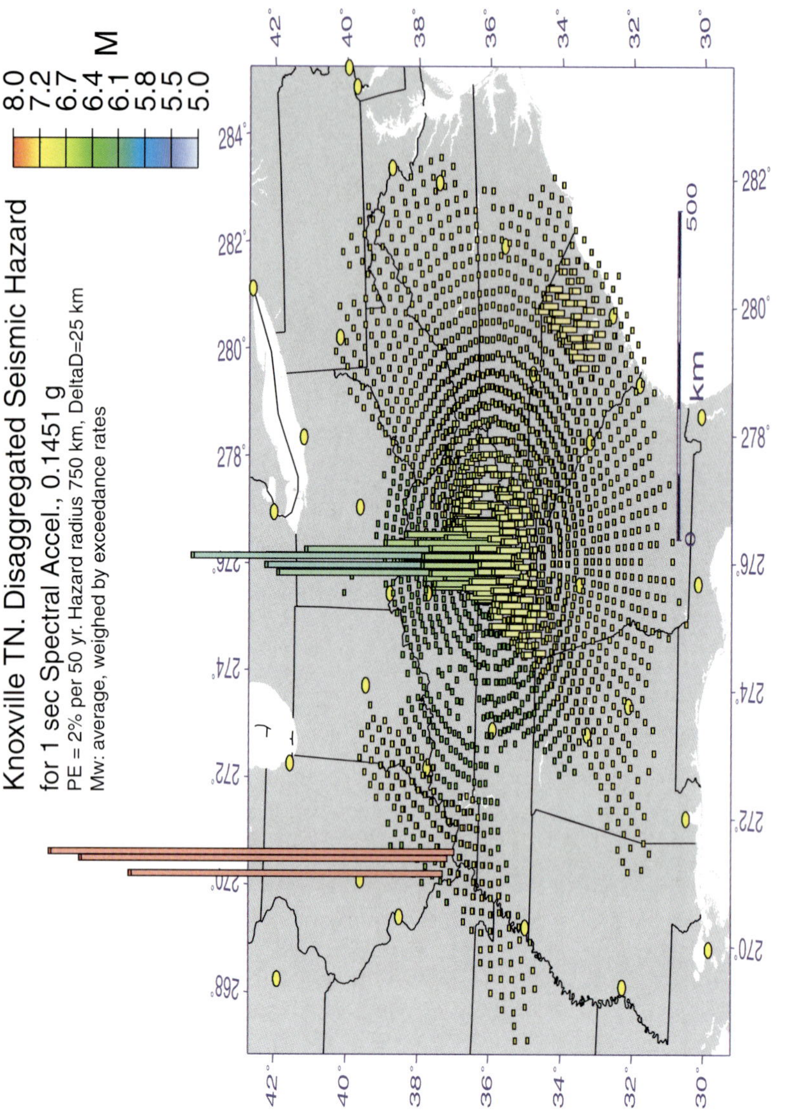

FIGURE 3.10 Disaggregated seismic hazards for Knoxville, Tennessee, for response-spectrum acceleration at 1 hertz (*left panel*) and 5 hertz (*right panel*). The height of the vertical bars is proportional to the level of hazard produced by individual earthquakes in that map cell. The color of each bar indicates the dominant magnitude contributing to the hazard. At 1 hertz, the greatest hazards come from very large earthquakes in the New Madrid seismic zone (red bars in southern Illinois) and from local intermediate-magnitude events (green bars near

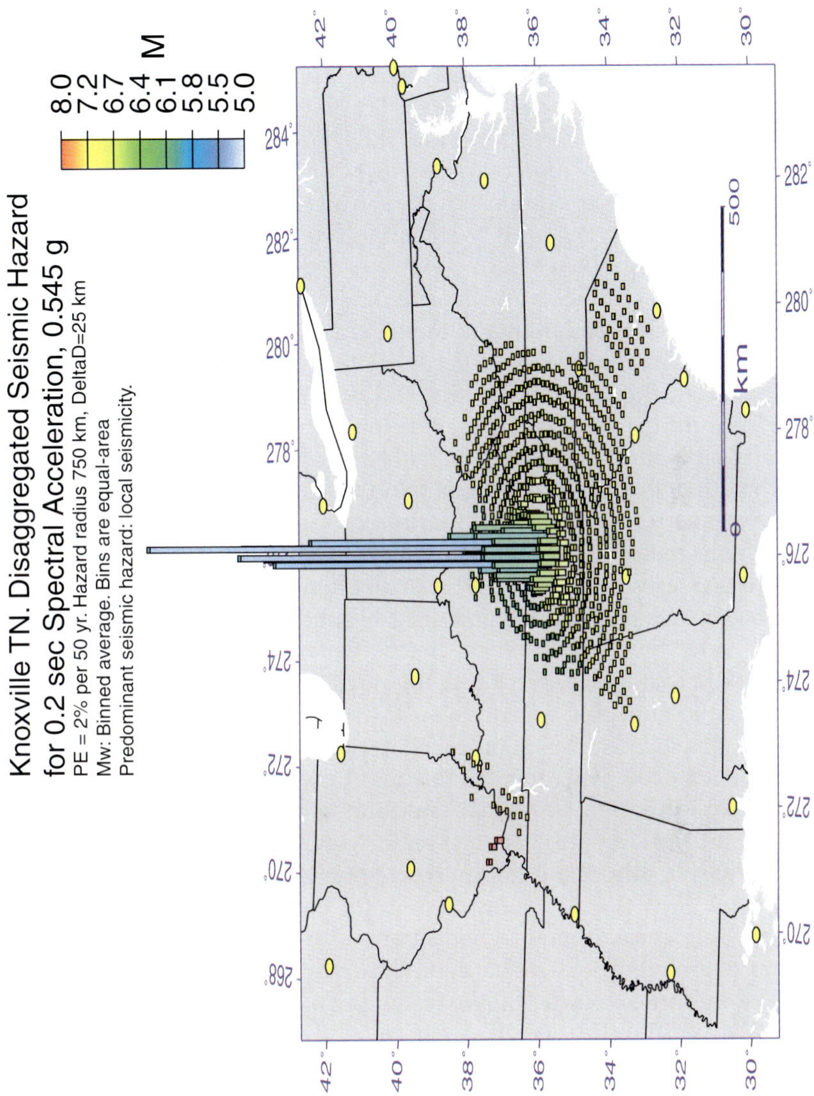

Knoxville); potentially large earthquakes near Charleston also make significant contributions (yellow bars in eastern South Carolina). At 5 hertz, the seismic waves from the more distant sources are more severely attenuated, and the hazard is dominated by the smaller, more local events (blue bars near Knoxville). Yellow circles are cities. SOURCE: U.S. Geological Survey, <http://geohazards.cr.usgs.gov/eq/>.

the source of a large (M about 6.8) earthquake in 1868. In 1990, the Working Group on California Earthquake Probabilities estimated that the chance of an M ≥ 6.7 earthquake on the Hayward-Rogers Creek fault system before 2030 is approximately 36 percent. The same group concluded, in a major reassessment, that the chance of an earthquake larger than M 6.7 in the Bay area over the next 30 years was about 70 percent (*37*). However, it is uncertain how much of the slip along the northern portion of the Hayward fault is accommodated by aseismic creep, rather than by large earthquakes (*38*).

The S-curve in the San Andreas fault north and east of Los Angeles is an example of a restraining bend that results in compressive deformation taken up by auxiliary reverse faults. These have produced damaging thrust-type earthquakes, including the 1952 Kern County (M 7.5), 1971 San Fernando (M 6.7), and 1994 Northridge (M 6.7) events. The latter occurred on a blind thrust (i.e., a shallow-dipping reverse fault that does not crop out at the Earth's surface). Blind thrusts are common in compressional regimes where thick sections of soft sediments cover active faulting in basement rocks. Rather than propagate to the surface, the faulting deforms the overlying sediments into distinctive fold structures (Figure 3.11). Because they lack surface scarps, blind thrusts are more difficult to identify, and they cannot be studied with paleoseismic trenching techniques; rather, their geometry and slip rates must be inferred using structural and geomorphic methods supplemented with seismologic and geodetic data. From this type of neotectonic analysis, geologists have estimated that the blind thrust fault that produced the 1994 Northridge earthquake is slipping at about 1.5 millimeters per year and that the recurrence interval for such events is on the order of 1700 years (*39*). In 1995, the Southern California Earthquake Center (SCEC) published a major assessment of earthquake hazards that, for the first time, merged results from geodetic measurements, neotectonic slip rates, and historic seismicity into a probabilistic seismic hazard analysis for southern California (*40*). That report concludes that earthquakes of M 7.2 to 7.6 have occurred and will recur in the Los Angeles region to relieve the contractional strain accumulating across the "Big Bend" of the San Andreas.

The eastern Mojave shear zone, which splays off the San Andreas system just east of the Big Bend, accommodates a portion of the Pacific-North American plate motion (0.7 to 1.2 meters per century) (*41*). It has produced a series of major earthquakes during the last decade, including the 1992 Landers (M 7.3) and 1999 Hector Mine (M 7.1) earthquakes. Along the eastern side of the Sierra Nevada, the Mojave shear zone has also been a source of high seismicity throughout history, including the 1872 Owens Valley earthquake (M 7.6). Other fault systems that contribute to seismic hazards in California include the Mendocino fracture zone

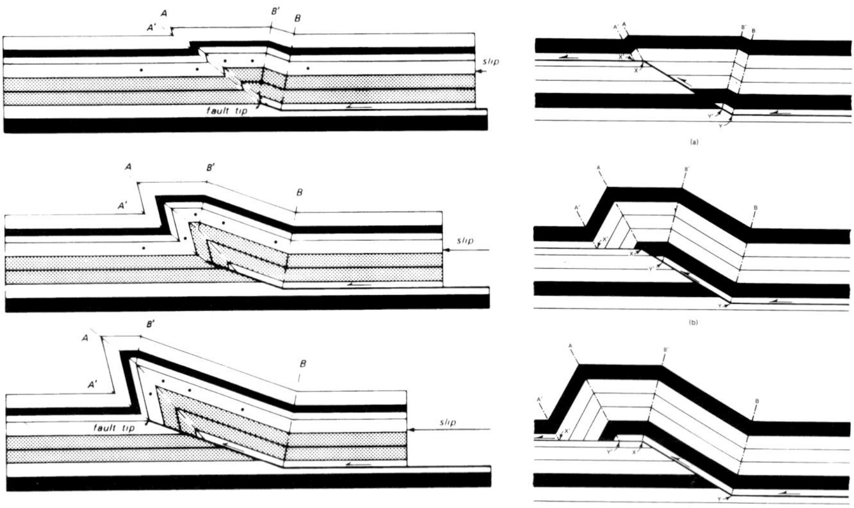

FIGURE 3.11 Two types of blind faults and the distinctive deformation of adjacent crustal blocks associated with them. In the case of a fault that steps from one flat surface to another across a shallow-dipping ramp, a symmetrical fault-bend fold forms and lengthens with increasing fault displacement. In the case of a fault that ramps up from a flat surface and is propagating upward from the top of a shallow-dipping ramp, a highly asymmetrical fault propagation fold forms. SOURCE: J. Suppe, Geometry and kinematics of fault-bend folding, *Am. J. Sci.*, **283**, 684-711, 1983; Reprinted by permission of the American Journal of Science; J. Suppe, *Principles of Structural Geology*, Prentice-Hall, Englewood Cliffs, N.J., p. 351, 1985.

in northern California and the Cascadia subduction zone off the coast north of the Mendocino triple junction.

Pacific Northwest

The band of high seismic hazard paralleling the Pacific coast, which includes the Portland and Seattle-Tacoma metropolitan areas, comes from three tectonically distinct sources: great earthquakes on the main thrust of the Cascadia subduction zone; shallow earthquakes in the upper crust above the subduction zone; and deeper earthquakes within the subducting lithosphere of the Juan de Fuca plate. The 1000-kilometer-long subduction interface that runs along the continental margin from Cape Mendocino to the northern tip of Vancouver Island appears to be locked and accumulating strain, rather than slipping aseismically as previously believed (42). Brian Atwater and his colleagues (43) have documented seven great subduction earthquakes during the last 3500 years, recorded as irregularly spaced episodes of sudden submergence in estuarine sediments. A combination of radiocarbon and tree-ring dating pinpointed the most recent submergence event to the winter of 1699-1700, and K. Satake correlated this event with a mysterious tsunami recorded along the Japanese coast on the night of January 26, 1700 (*44*). He estimated the size of this event at about M 9.

The Seattle fault is one of several shallow thrust faults accommodating north-south compression in the Puget lowlands (Figure 3.12). The

FIGURE 3.12 Photograph of uplifted beach from prehistoric earthquake on the Seattle fault, looking east across Puget Sound from Bainbridge Island toward Seattle, Washington. The broad, flat, grass-covered surface in the foreground is a prehistoric beach that was uplifted about 7 meters during a large earthquake (M > 7) about 1100 years ago. The earthquake was caused by slip on a portion of the Seattle fault, whose location beneath Puget Sound is indicated by the red line. SOURCE: U.S. Geological Survey.

frequency of large earthquakes on this fault is poorly determined, but paleoseismic data show that the most recent large earthquake occurred about 1100 years ago and was associated with as much as 7 meters of uplift, major landslides, and tsunamis in Puget Sound (*45*). Use of laser altimetry (e.g., light detection and ranging [LIDAR]) for precise mapping of topography has led to the identification of splays of the Seattle fault. Geologists who have trenched these splays find that earthquakes of M 6.5 or greater recur as often as about 1000 years.

Earthquakes damaging to Seattle, Tacoma, and Olympia in 1949 (M 7.1), 1965 (M 6.5), and 2001 (M 6.8) were caused by rupture at depths of 50 to 60 kilometers within the subducting slab of the Juan de Fuca plate. Owing to this greater depth, the shaking intensity is lower than that of comparable shallow events. For example, the PGA recorded for the February 28, 2001, Nisqually earthquake was only about 30 percent of gravity, compared with values more than 100 percent of gravity observed for the 1994 Northridge event (M 6.7). It nevertheless caused significant damage over a broad region.

Intermontane West

The seismic hazards in the intermontane regions of the western United States are dominated by high but relatively diffuse seismicity accommodating oblique crustal extension. The total extension rate between stable North America and the Sierra Nevada-Great Valley block is estimated by Global Positioning System (GPS) geodesy to be 1 ± 0.1 meter per century (*46*). Most is concentrated in the Basin and Range, a geologic province characterized by dozens of tilted, 10- to 30-kilometer-wide crustal blocks that form high mountain ranges alternating with deep basins. Earthquakes occur both on the normal faults bounding the mountain ranges and on the strike-slip faults that cut across the province.

The late Cenozoic normal faults are distributed relatively uniformly in the Basin and Range, but the historic and instrumental seismicity is concentrated in the central Nevada seismic belt, along the western margin of the province in eastern California and western Nevada, and the intermountain seismic zone, along the eastern edge of the province, from southern Nevada across central Utah to southwestern Montana and central Idaho. From 1915 to 1954, a sequence of five large earthquakes (M 6.8 to 7.7) ruptured adjacent segments of the central Nevada seismic belt. Another historically active area is the intermountain seismic zone centered on Yellowstone National Park in northwestern Wyoming, where a mantle hot spot is causing uplift and volcanism (*47*). The 1959 Hebgen Lake earthquake ruptured a normal fault adjacent to the Yellowstone area, and the 1983 Borah Peak earthquake (M 7.5) occurred on a range-front fault farther to the west in southeast Idaho. Paleoseismic studies show that even the most active faults in this region produce such big earthquakes only every few thousand years. Based on the Holocene average, the current level of activity appears to be abnormally high in the central Nevada seismic belt and abnormally low in the intermountain seismic belt. One area of future concern is the Wasatch front traversing the populated areas of Salt Lake City and Provo, where large, prehistoric earthquakes have been documented by paleoseismic techniques along six segments of front-bounding fault, but no large earthquakes have occurred in the historic period (Figure 2.4).

Central United States

Seismic activity decreases markedly in the stable continental interior, east of the Rockies. However, seismic waves propagate more efficiently through the colder, thicker lithosphere that underlies this region than through the hotter crust and upper mantle of the western United States (*48*). Earthquakes of comparable magnitude can consequently cause damage over larger areas. For instance, a large earthquake in southeastern Missouri on December 16, 1811, generated strong shaking (Modified Mercalli Intensity V or greater) over an area at least five times bigger than in the 1906 San Francisco earthquake, which had a larger magnitude. This was the first in a violent sequence of earthquakes that occurred for the next several months along an abandoned set of Paleozoic extensional faults called the Reelfoot Rift (Figure 3.13). The large magnitudes of these earthquakes, combined with the relatively short return periods for events of similar size found from paleoseismic studies, imply that the central Mississippi Valley has the highest seismic hazard east of the Rocky Mountains, although the level of this hazard remains controversial (Box 3.1).

Other areas of potential seismic hazard in the central United States include eastern Kansas and Nebraska, which has been the site of two M 5 earthquakes in the past 150 years. Liquefaction features indicate that M 6.5 to 7.5 earthquakes with recurrence times of a few thousand years have occurred in the Wabash Valley of southern Indiana and Illinois (*49*). The Meers fault in southern Oklahoma has generated two large earthquakes (M 7) in the past 3000 years (*50*), and the Cheraw fault in southeast Colorado has produced M 7 earthquakes in the past 10,000 years (*51*).

Eastern United States

Most people think of the eastern United States as seismically benign. At present, this region lies near the center of the North American plate, far from active plate boundaries to the east and west. The histori-

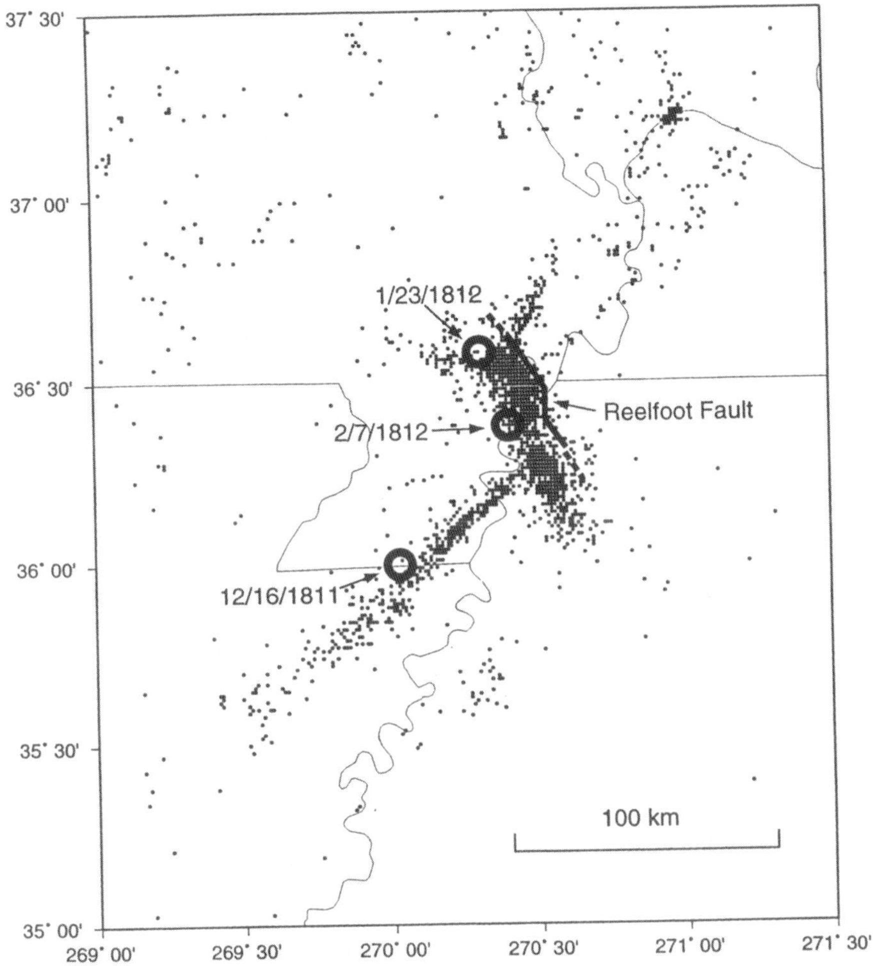

FIGURE 3.13 Structural setting of the New Madrid seismic zone showing the locations of historical earthquakes (M < 5) detected by regional seismic networks from 1974 to 1993. The circles show the estimated epicenters of the large New Madrid earthquakes of 1811-1812. SOURCE: S.E. Hough, J.G. Armbruster, L. Seeber, and J.F. Hough, On the Modified Mercalli intensities and magnitudes of the 1811-1812 New Madrid earthquakes, *J. Geophys. Res.*, **105**, 23,839-23,864, 2000. Copyright 2000 American Geophysical Union. Reproduced by permission of American Geophysical Union.

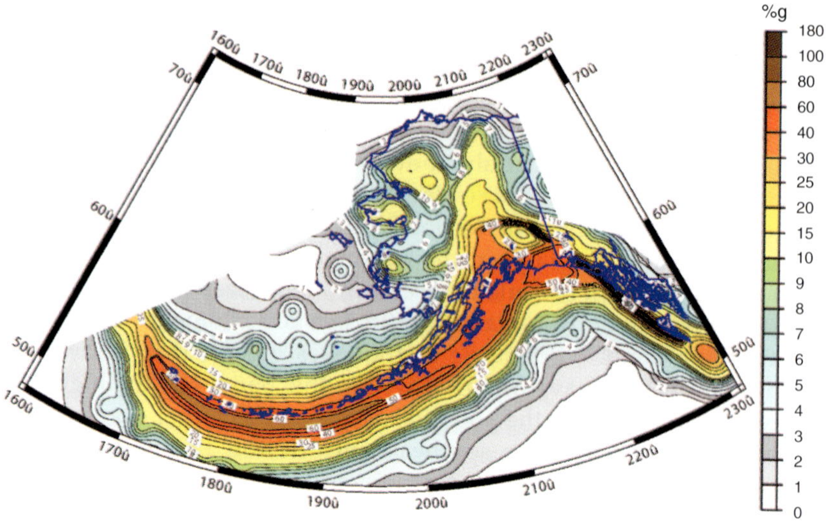

FIGURE 3.14 Alaska seismic hazard map showing peak ground acceleration with a 10 percent chance of exceedance in 50 years. SOURCE: U.S. Geological Survey, <http://geohazards. cr.usgs.gov/eq/>.

chatka to the Gulf of Alaska, and it accommodates most of the convergence between the Pacific and North American plates, which ranges from 6.3 meters per century near the Kenai Peninsula to 8.6 meters per century in the western Aleutian Islands. The largest historic earthquake in the United States (M 9.2) was generated in 1964; its ground shaking and liquefaction caused spectacular damage to Anchorage (Box 2.3). The fault plane responsible for this huge event was equivalent to the area of New York State moving an average of about 10 meters. The megathrust was also the source of great earthquakes in 1957 (M 9.1) and 1965 (M 8.7). Elsewhere in Alaska, major strike-slip faults, including the Denali and Fairweather faults, traverse the interior of Alaska and pose a substantial danger to Juneau. The Denali fault is about 1000 kilometers long and has a slip rate of 0.2 to 1.0 meter per century, but has not ruptured in the past few centuries. Earthquakes in the crust and subducted slab pose a significant hazard to Anchorage and Fairbanks.

Hawaii

The Hawaiian Islands were formed by the passage of the Pacific Plate over a mantle hot spot. This hot spot persists today, its eruptive centers on the southeast side of the "Big Island." Consequently, the largest earthquakes and highest level of seismic hazard in the Hawaiian Islands occur in the southeast portion of the Big Island. The seismicity generally decreases westward along the island chain, reflecting the increasing age of the islands (Figure 3.15). The M 7.2 Kalapana earthquake in 1975 ruptured a nearly horizontal surface at a depth of about 9 kilometers. This earthquake, and presumably the M 7.9 shock in 1868, moved the southern flank of the Big Island seaward in response to stress generated by intrusion of magma into the Kilauea and Mauna Loa volcanoes. Earthquake-induced landslides, the most recent of which occurred about 100,000 years ago, appear to have removed a sizable chunk of Mauna Loa's southwest flank, sending sea waves laden with coral blocks several hundred meters onto the coast of neighboring islands. Several M 6 to 7 earthquakes have occurred elsewhere in the Hawaiian Islands in the past 200 years, including events near Kilauea caldera and the Kona coast of the Big Island and in the Maui-Molokai region.

3.3 SEISMIC HAZARDS AROUND THE WORLD

Most of the world lacks the detailed information on fault activity comparable to that available in the United States. In many tectonic regions, especially in underdeveloped countries, the seismic stations for delineating active structures are sparse; neotectonic and paleoseismic data are not yet available even on the most dangerous faults; and the efforts to investigate and quantify seismic hazards are weak or nonexistent. Recognizing this need, the United Nations initiated a Global Seismic Hazard Assessment Program (GSHAP) as a demonstration project for the International Decade of Natural Disaster Reduction (IDNDR, 1990-1999). In December 1999, this program released the first global seismic hazard map based on a consistent probabilistic seismic hazard analysis (56), reproduced here as Figure 3.16. This map, in combination with the global seismicity map (Figure 2.10), furnishes the guide for a brief tour of seismic hazards around the world.

Convergent Environments

Most of the global seismic energy release is in subduction-zone earthquakes (Figure 3.17). Subduction zones are marked by intermediate- and deep-focus seismicity that defines the Wadati-Benioff zones of the subducted lithospheric slabs, which dip beneath volcanic arcs in places such as Japan, the Aleutian Islands, and the Cascadian province of the western United States (57). The interface between the subducting and overriding lithosphere is usually well expressed in seafloor bathymetry, commonly

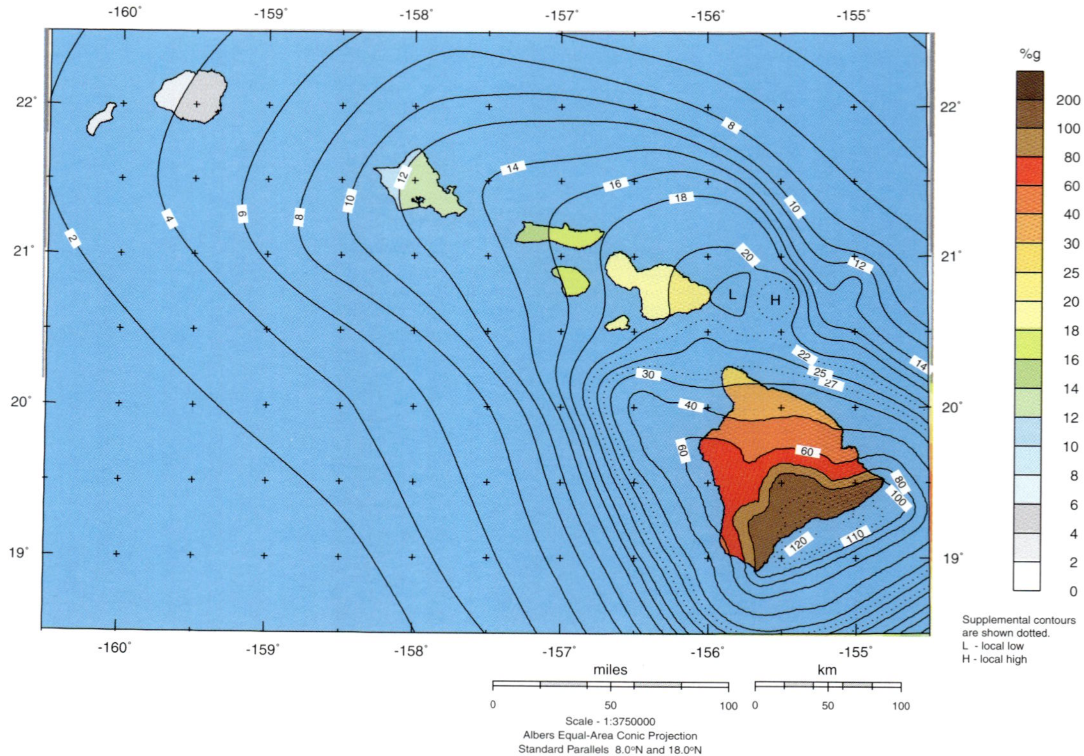

FIGURE 3.15 Hawaii seismic hazard map showing peak ground acceleration with a 10 percent chance of exceedance in 50 years. SOURCE: U.S. Geological Survey, <http://geohazards.cr.usgs.gov/eq/>.

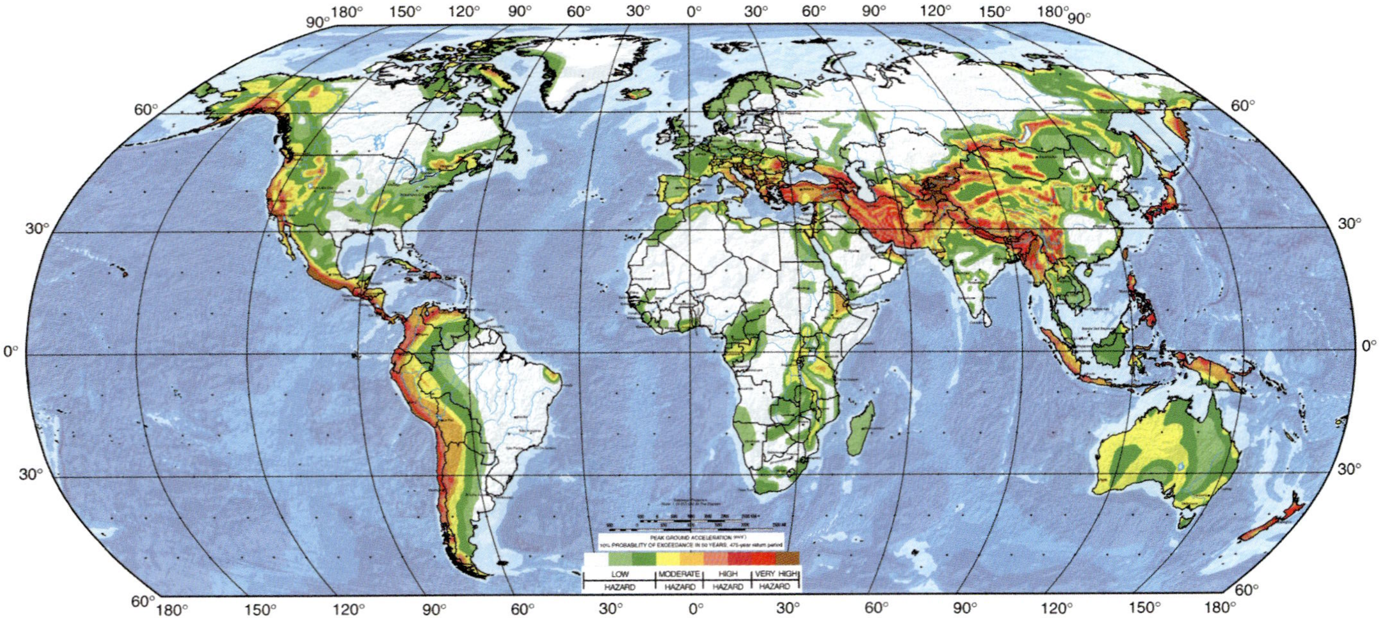

FIGURE 3.16 The GSHAP global seismic hazard map, depicting the peak horizontal ground acceleration for rock sites with a 10 percent probability of exceedance in 50 years (corresponding to a return period of 475 years). The map was compiled from existing national seismic hazard maps and new regional maps developed under the auspices of GSHAP. Except for the United States and parts of the former Soviet Union, the map is based on recurrence times derived entirely from historic seismicity, assuming a Gutenberg-Richter recurrence relation (Equation 2.5). SOURCE: D. Giardini, G. Grünthal, K. Shedlock, and P. Zhang, The GSHAP global seismic hazard map, *Ann. Geofisica*, **42**, 1225-1230, 1999.

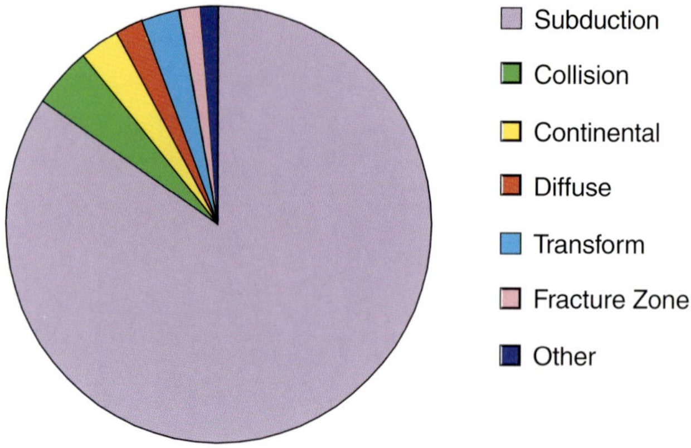

FIGURE 3.17 Comparison of seismic moment distribution for different tectonic environments, excluding southern Chile, from 1900 to 1989. Great earthquakes associated with circum-Pacific subduction dominate the recent energy release. In fact, about 30 percent of the global total for the last century was released in a single event—the great Chilean earthquake of 1960 (M 9.5). SOURCE: Modified from J.F. Pacheco and L.R. Sykes, Seismic moment catalog of large shallow earthquakes, 1900 to 1989, *Bull. Seis. Soc. Am.*, **82**, 1306-1349, 1992. Copyright Seismological Society of America.

by a deep ocean trench or a demarcation between undeformed oceanic-plate seafloor and highly deformed seafloor of the overriding plate. The dip of the subduction interface determined from seismicity and focal mechanisms ranges geographically from nearly flat to nearly vertical. Steeper subduction interfaces, such as that beneath the Mariana Island arc, usually produce smaller earthquakes (*58*); they commonly involve older, denser oceanic lithosphere and trench-normal extension of the overriding plate, sometimes in the form of back-arc spreading. Shallow-dipping subduction zones, like that beneath Chile, tend to produce larger earthquakes and commonly involve younger oceanic lithosphere. The percentage of the total long-term slip on a subduction interface expressed in earthquake ruptures, commonly referred to as the degree of seismic coupling, is believed to range widely (*59*). For example, the subduction zone that separates Indonesia from the Australian plate appears to be highly coupled offshore Sumatra, but poorly coupled offshore Java (*60*), and this contrast in reflected in the higher hazard for the Sumatran portion and the lower hazard for the Javan portion of the subduction zone.

In addition to reverse slip on the main plate interface and deeper seismicity within the Wadati-Benioff zones, large earthquakes take place in both the subducting slab (foot wall) and the overriding plate (hanging wall). Many of Japan's historic, destructive earthquakes, including the 1891 Nobi and 1995 Hyogo-ken Nanbu earthquakes, were caused by shallow faults within the overriding plate. In regions of oblique convergence, relative plate motion tends to separate into two parts, a subduction component taken up by reverse dip-slip faulting perpendicular to the trench and a lateral component taken up as strike-slip faulting parallel to the trench in the overriding plate (*61*); examples of such strike-slip structures include the Great Sumatran fault in Indonesia, the Philippine fault, and the Median Tectonic Line of Japan. Earthquakes that occur seaward of the trench are caused by flexing of the downgoing slab. These events may accommodate either stretching or contraction of the downgoing slab along either normal or reverse faults. Two of the largest historical events in this setting—the 1977 Sumba, Indonesia (M 8.3) and the 1933 Sanriku, Japan (M 8.4) earthquakes—were caused by normal faulting of the bending subducting plate.

Continental convergence zones have generated some of history's most destructive earthquakes. They are common throughout the broad Alpine-Himalayan belt that resulted from the closure of the ancient Tethys Ocean (Figure 3.18), but are also found behind ocean-continent subduction zones, such as the foreland fold-and-thrust belts in the eastern foothills of the Andes. The style of deformation ranges from block motions along large reverse faults that penetrate deep into the crystalline basement, as in the Pampean Ranges in northwestern Argentina and the Zagros of Iran and

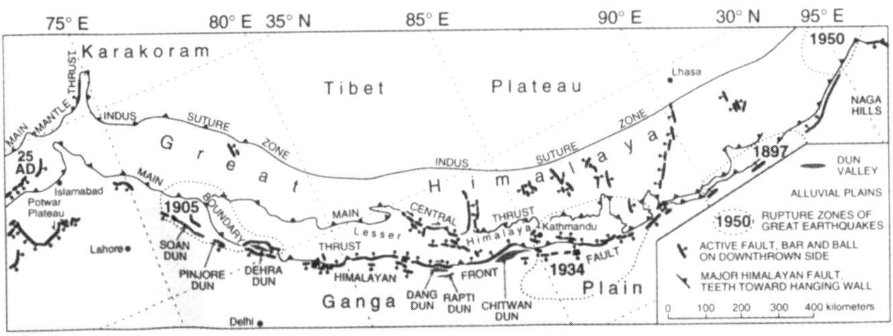

FIGURE 3.18 Four giant earthquakes—1897, 1905, 1934, and 1950—have resulted from slip on shallow thrust faults along the southern front of the Himalaya in India and Nepal. Although many active faults reach the surface in this 2000-kilometer-long region of continental convergence, the ruptures of these devastating earthquakes did not break to the surface. SOURCE: G. LeBlanc and F. Anglin, Induced seismicity at the Manic 3 Reservoir, Quebec, *Bull. Seis. Soc. Am.*, **68**, 1469-1485, 1978. Copyright Seismological Society of America.

in New Guinea, to the "thin-skinned" tectonics that characterizes crustal shortening in major sedimentary basins.

Divergent Environments

A predominance of normal faults, high heat flow, relatively low seismic-wave speed in the upper mantle, and volcanism usually characterize tectonic environments dominated by crustal divergence or extension. Oceanic spreading centers are the most common type, and their geometry and kinematics are usually fairly simple, conforming much more to the plate-tectonic ideal than their continental counterparts. Slow-spreading boundaries, such as the Mid-Atlantic Ridge, typically display a central rift valley or graben, while such features are usually absent on fast-spreading ridges like the East Pacific Rise. Most of the displacement on normal faults flanking the central rift valley occurs soon after crustal formation; as the crust moves away from the spreading center, the faults become inactive. Much of the deformation associated with seafloor spreading at the oceanic ridge crests appears to be aseismic, presumably because of the high temperatures generated by mantle upwellings, and large earthquakes in this environment are rare. Moreover, with the exception of a few places such as Iceland, most oceanic spreading centers are far removed from areas of human habitation, and their seismic activity poses little danger. Subaerially exposed spreading centers are rare and atypical, but examples in Iceland and Djibouti have offered particularly good opportunities to study this class of normal faults and their associated earthquakes (*62*).

Extension within continental crust can be localized in discrete rifts, such as those in East Africa (Figure 3.19) and the Baikal rift of eastern Russia, or distributed over broad regions, as in the extensional provinces of northeastern China and the Basin and Range Province of the western United States. While they share certain characteristics—high heat flow, volcanic activity, and normal faulting between horsts and graben—the issue of how these features relate to large-scale plate tectonics remains problematic. For example, the Basin and Range accommodates about 20 percent of the northwest-southeast motion between the Pacific and North American plates, but it does so along faults of north-to-northeast strike, which results in normal faulting with secondary strike-slip activity (*63*). This messy behavior is evidently associated with active mantle flow beneath the province.

Normal faulting is also found in association with hot spots, back-arc basins, high plateaus behind collision zones, and intercontinental rift zones. Although not very common, they pose a significant seismic hazard, and their underlying tectonic mechanisms are not well understood.

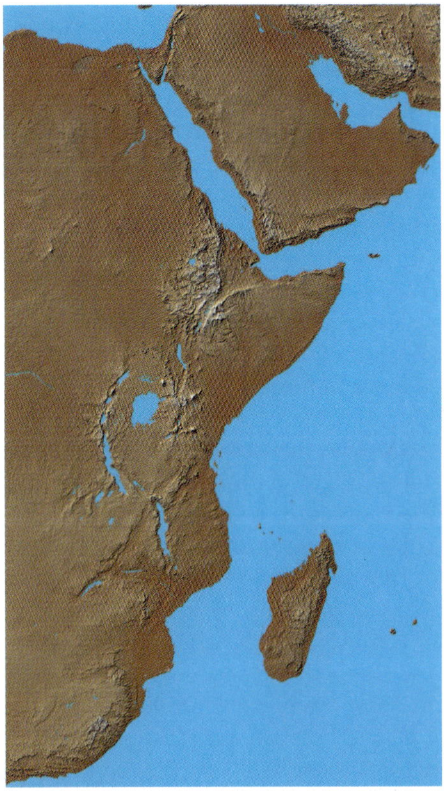

FIGURE 3.19 Spacecraft view of the African rift valleys from the Shuttle Radar Topography Mission. SOURCE: T. Farr, National Aeronautics and Space Administration.

Extension in the hanging-wall block above subduction zones, which is well documented, has produced significant earthquakes behind some circum-Pacific and Mediterranean arcs (*64*). In Tibet and the South American Altiplano, compression during plate collision has built large plateaus so high that they may be gravitationally collapsing toward their margins, resulting in active normal faulting within their interiors (*65*). At least a half-dozen north-south-trending fault systems are accommodating extension across an area of 1000 kilometers by 400 kilometers in southern Tibet (*66*). Other examples of active crustal extension behind collision zones include the lower Rhine graben, north of the Alpine orogenic belt. Recent paleoseismic excavations in the Roer Valley of eastern Belgium, southern Netherlands, and westernmost Germany have shown that the magnitudes of modern earthquakes (M about 5) may underestimate the maximum expectable magnitudes by at least one magnitude unit (*67*).

Strike-Slip Environments

Strike-slip faults release only a small percentage of total energy, yet they are found in great variety and have produced some of the most destructive and infamous earthquakes, including the 1906 San Francisco (Box 2.2) and the 1995 Hyogo-ken Nanbu (Box 2.5). The most common, but least seismic, of these lateral faults are the transform faults that connect oceanic spreading centers. Although this class includes some of the longest faults in the world—the Romanche transform in the equatorial Atlantic has been mapped as a nearly continuous fault for 950 kilometers—and the ones with the fastest slip rates (as much as 8.5 meters per century for the 480-kilometer-long Tharp transform in the southeastern Pacific), they generate earthquakes larger than M 7 rarely and M 8 almost never (*68*). Essentially all ridge-ridge transforms are under the oceans and therefore difficult to study, but some have been the subject of detailed oceanographic surveys (*69*). The low seismicity of ridge-ridge transforms is correlated with the high mantle temperature in these regions, although the peculiar mechanics of these faults may have something to do with the thinness of oceanic crust along ridge-ridge transforms and the abundance of weak, serpentinized upper mantle.

Transforms that run between trenches are far less common than ridge-ridge transforms, because they do not occur as closely spaced arrays separating trench segments. Most often, trench-trench transforms act as the lateral boundaries of large plates. They are typically many hundreds of kilometers long, involve continental lithosphere, and display complex and idiosyncratic geometry (*70*). Examples include the Alpine fault system of New Zealand (Figure 3.20) and the Macquarie Ridge, farther south. The latter structure generated an unusually large, enigmatic earthquake in 1989. Ridge-trench transforms connecting extensional and contractional zones also occur as lateral plate margins. The left-lateral Dead Sea fault, for example, connects the Red Sea spreading center between the African and Arabian plates to the oblique collisional Bitlis zone between the Arabian and Anatolian plates (*71*). The right-lateral Sagaing fault system runs through Myanmar (Burma) along the eastern edge of the Indian plate, between spreading in the Andaman Sea and convergence in the Himalayan system.

Many strike-slip faults are not transform faults; that is, they do not transform at their termini into other types of plate boundaries. A major class involves the trench-parallel strike-slip faults associated with the strain partitioning in oblique subduction zones, discussed above. Other strike-slip faults accommodate the considerable horizontal motions associated with the lateral advection of crust escaping continent-continent collision zones. The great faults in central Asia, north of India, are the most prominent cases of such indent-linked strike-slip faults, and they

FIGURE 3.20 Spacelab photograph of New Zealand's Alpine fault. The fault runs as a single structure for more than 500 kilometers and forms the sharp line separating the snow-covered southern Alps in the east from the low coastal plain bordering the Tasman Sea in the west. SOURCE: National Aeronautics and Space Administration.

have been the source of the Earth's greatest (greater than M 8) historic strike-slip earthquakes (*72*). Another spectacular example is the 2000-kilometer-long North Anatolian fault system, along which a big chunk of Turkey is extruding westward from the Arabian-Eurasian collision. This fault system is responsible for a remarkable sequence of earthquakes propagated from its eastern sections to the west, beginning with the Erzincan earthquake in 1939 and continuing to the Izmit-Düzce sequence in 1999 (Figure 3.21) (*73*).

Intraplate Earthquakes

Not all major earthquakes have occurred at plate boundaries or even within the broad plate boundary zones of distributed continental deformation (Figure 3.17). The Indian, Australian, and North American cratons, in particular, are well known for their infrequent but destructive intraplate earthquakes. The 1819 Rann of Cutch earthquake, highlighted in Lyell's *Principles of Geology*, occurred well within the Indian plate; meters of vertical deformation resulted from movement on a large fault in this flat, arid coastal region, causing spectacular flooding and uplift (*74*). The January 26, 2001, Bhuj earthquake (M 7.6) occurred in the same intraplate area, killing tens of thousands of people and causing extensive damage. A moderate (M 6.1) earthquake in 1993 in the middle of the Indian craton killed about 10,000 people (*75*). Between 1968 and 1988,

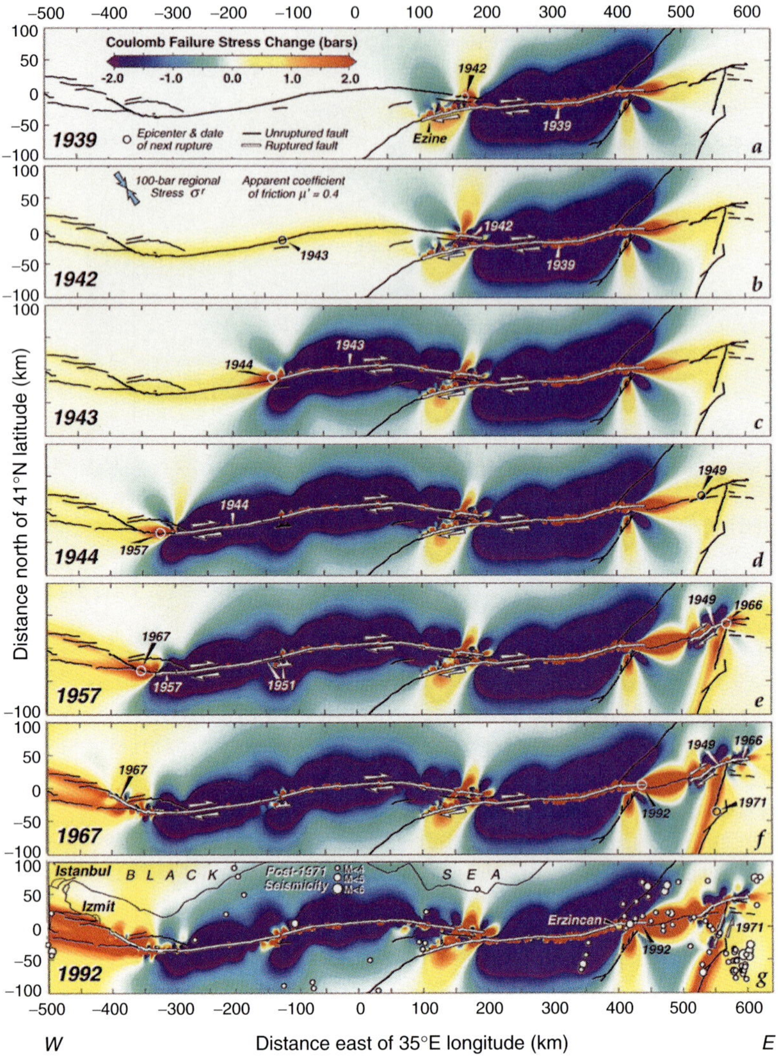

FIGURE 3.21 Map of North Anatolian fault system, showing the 53-year sequence of earthquakes and net Coulomb stress increments. Areas where the rupture-causing stresses were increased by the sequence of past events are shown in red; areas where they decreased in blue. This map, published in 1997, indicated that the city of Izmit, Turkey, was ripe for a future major earthquake. A large earthquake hit Izmit on August 17, 1999, killing over 17,000 people and causing $6 billion in direct economic losses. SOURCE: R.S. Stein, A.A. Barka, and J.H. Dieterich, Progressive failure on the North Anatolian fault since 1939 by earthquake stress triggering, *Geophys. J. Int.*, **128**, 594-604, 1997. Reproduced by permission of Blackwell Publishing.

fault rupture within the Australian plate caused several moderate earthquakes (*76*), and geologic reconnaissance has uncovered evidence of surface rupture associated with other intraplate earthquakes there within the past several thousand years. Paleoseismic investigations of the surficial fault ruptures associated with the earthquakes reveal that the fault had not moved for 50,000 to 100,000 years or more before the recent event (*77*).

Though they can occur far from plate boundaries, most intraplate earthquakes are still caused by plate-tectonic forces. The patterns of the tectonic stress that drive intraplate seismicity have been mapped using a variety of indicators—wellbore breakouts, volcanic alignments, and earthquake focal mechanisms—and their orientations are coherent over distances of 400 to 4000 kilometers (Figure 3.22). These observed trajectories generally match the predictions of intraplate stress from dynamic models of plate motions in which the primary driving forces are ridge push (compression due to gravitational sliding of newly formed lithosphere away from mid-ocean ridge highs) and slab pull (tension due to the gravitational sinking of the old subducting slabs). The spatial and temporal patterns of intraplate seismicity remain poorly understood, however. In some cases, the stress that causes these earthquakes may come from nontectonic sources, such as the withdrawal of large continental ice sheets. Reservoir loading and subsequent water infiltration are significant factors in generating some intraplate earthquakes.

3.4 ESTIMATING EARTHQUAKE RISK

Earthquake loss estimates are forecasts of damage and human and economic impacts that may result from future earthquakes. Seismic retrofitting and earthquake-resistant design involve substantial investments (Section 1.1), so it is necessary to measure the consequences of earthquakes in economic terms (i.e., dollars) that allow rational trade-offs between the known costs of preparation and the anticipated losses. The methods for constructing loss estimates have thereby become important tools for disaster preparation and decision making. Communities have begun to use such estimates in setting priorities for mitigation efforts (e.g., identifying specific structures for seismic retrofits) and developing contingencies for earthquake emergencies (e.g., alternative transport routes). Governments and insurance companies employ such estimates to anticipate the financial impact of earthquake damage. Rescue and response organizations are devising systems that make and revise damage projections in near real time based on seismic information received immediately after an earthquake, so they can focus their postseismic response where it will be most needed.

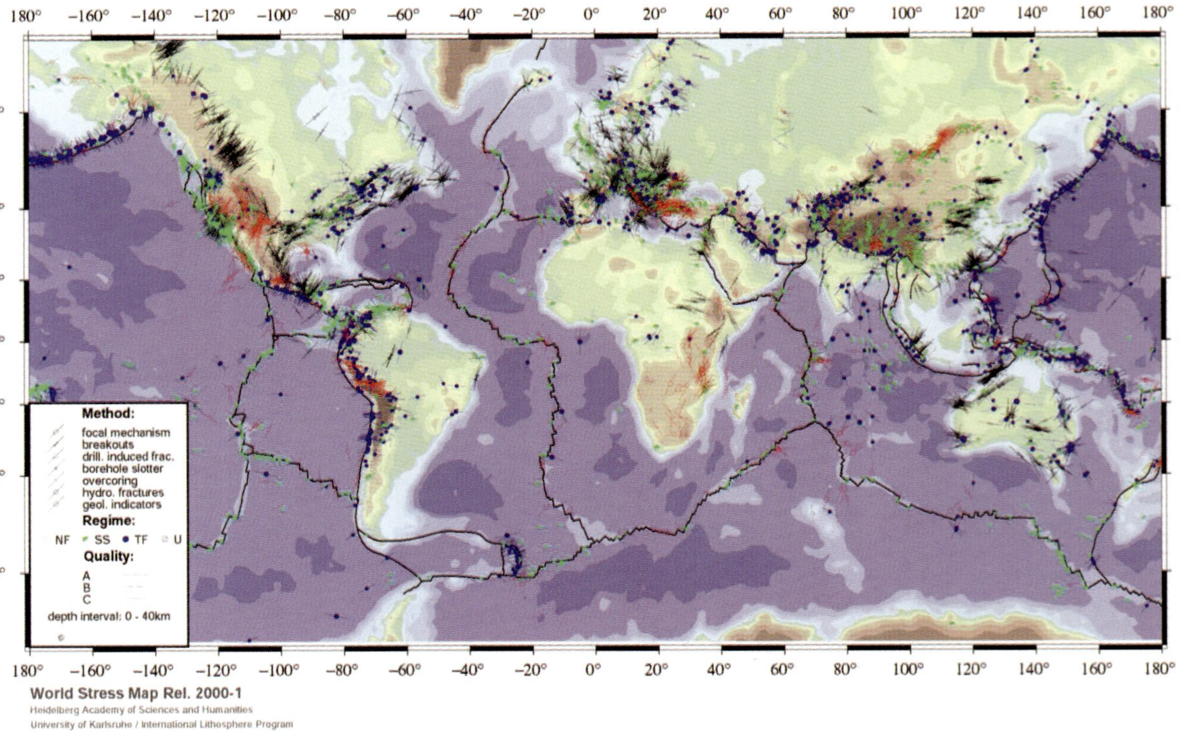

FIGURE 3.22 Global stress map. Shown are principal stress directions for normal faulting (red), strike-slip faulting (green), and thrust faulting (blue) regimes. SOURCE: B. Mueller, J. Reinecker, O. Heidbach, and K. Fuchs, The 2000 release of the world stress map, available on-line at <www.world-stress-map.org>.

Quantifying earthquake losses is difficult and uncertain, however. Earthquake forecasting is still in a primitive stage, and ground motions for a particular temblor can vary tremendously from site to site. The exposure of the built environment depends on the types and distribution of structures, and the vulnerability of each structure is a function of its construction type, age, and siting. Moreover, the true economic losses caused by earthquakes extend well beyond direct damage, owing to the consequent disruption of transportation and commerce.

In 1992, FEMA and the National Institutes for Building Standards (NIBS) initiated a program to improve the tools for calculating earthquake losses (*78*). The requirements included a nationally consistent inventory of buildings and infrastructure classified according to structural class; accurate estimates of the properties of potential earthquake hazards; a standardized methodology to describe and compute seismic damage to and replacement costs for buildings and lifelines, as well as secondary damage from floods, fire, and hazardous material release; and estimates of social and economic losses. This program has resulted in a public-domain software package and database termed Hazards U.S. (HAZUS) (*79*). HAZUS is built on a geographic information system platform and contains a nationwide inventory described by 36 model building types and 28 occupancy classes, as well as databases for populations, lifelines, and essential facilities, all except the last aggregated by census tract. In its current release, HAZUS output can include costs for repair and replacement of damaged buildings and lifelines; costs associated with loss of business revenue; casualties; people displaced from residences; quantity of debris; regional economic impacts; functionality losses (loss-of-function and restoration costs for buildings, critical facilities such as hospitals, and components of transportation and utility lifeline systems); and extent of secondary hazards (fire, flooding, and hazardous materials). As an example, Figure 3.23 shows the HAZUS estimates for direct economic losses from a repeat of the 1886 Charleston earthquake.

FEMA intends to develop HAZUS as its standard loss estimation tool for multihazard reduction efforts throughout the United States, including earthquakes, floods, and wind storms. Because the program is freely available and runs on a conventional personal computer, it offers the first opportunity to educate many computer users on the risks and vulnerabilities associated with earthquakes. Several HAZUS user groups have been formed at the state and local levels to educate community organizations about the use of this methodology to identify areas of high risk that can then be targeted for mitigation efforts (*80*). These educational efforts are important because users need to know the technical modules and options of this complex program sufficiently well to execute the analysis correctly (*81*). Moreover, there are still important limitations to the

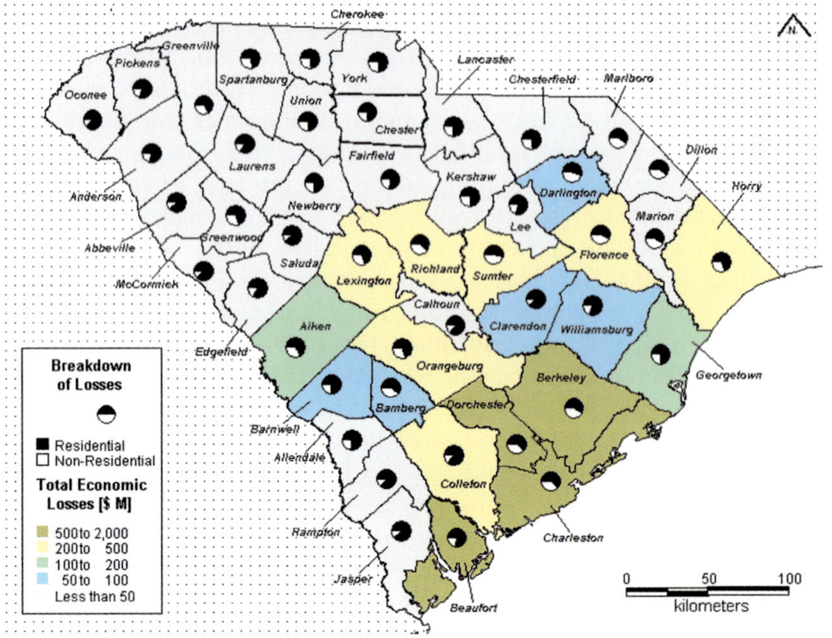

FIGURE 3.23 Map of South Carolina showing total expected economic losses (in millions of dollars) from a repeat of the 1886 Charleston earthquake, calculated using the HAZUS methodology. SOURCE: J. Bouabid, Durham Technologies, Inc., 2001.

HAZUS program that introduce uncertainties in its use. For example, HAZUS was originally designed to estimate urban losses primarily from building damage under particular earthquake scenarios. Modules for lifelines, such as transportation, water, and power systems, were added later but still need improvement, especially for estimating the effect losses from one system have on losses in other systems. HAZUS has also been expanded to estimate average annual loss at single locations, but it cannot be used to compute probabilistic aggregate loss for portfolios of properties spread over a large region and cannot take into account insurance variables, such as deductibles and coverage limits.

Requirements for improved loss estimation are driving a broad research agenda to understand the vulnerability of society to earthquakes. This will involve collecting data on the full inventory of vulnerable structures, characterizing the complete range of impacts from earthquake disasters, accurately characterizing the fragility of the built environment, and increasing the accuracy of hazard assessments for the probable earthquake sources. A particularly important issue is how to extend the methodology to evaluate losses to the total performance of extended infrastructure systems, including transportation, lifelines, and acute health care. Challenging problems emerge from the interactions among these systems, for example, how damage to the regional transportation system affects local acute health care. Assessing these system-level effects will require much improved capabilities for regional-scale simulations of earthquake-induced ground motions.

3.5 REDUCING EARTHQUAKE RISK

Earthquake risk, measured in expected losses over a given period of time, depends on the seismic hazard through the exposure and on the vulnerability of the built environment (Figure 1.4). A sound strategy for the reduction of earthquake risk has four basic components: better characterization of seismic hazards; land-use policies to limit exposure to seismic hazards; preparation of the built environment to withstand future earthquakes; and rapid response to earthquake disasters.

Seismic Hazard Characterization

Characterizing and mapping seismic hazards has progressed substantially in recent years, but much work remains to be done in collecting information on active faults and incorporating new results from earthquake research into national and global seismic hazard maps. In many regions of the world, there are few data on active faults, and hazard estimates (Figure 3.16) have too little resolution and accuracy to be useful. In many areas covered with thick vegetation, even the location of these faults is unknown. Earthquake forecasting information, such as fault slip rates and dates of past earthquakes, is badly needed. Regional deformation measurements made with GPS can provide useful constraints on the expected rate of large earthquakes across zones of deformation. The time dependence of seismic hazard can be evaluated based on the dating of past earthquakes and calculations of fault interaction. The synthesis of these research efforts will lead to a detailed global map of ground-motion forecasts.

Another key goal is the development of urban seismic hazard maps for populous centers in active seismic zones. These maps would integrate the latest models of time-dependent earthquake hazard with estimates of site response and the effects of sedimentary basins, which typically underlie major cities. Such urban seismic hazard maps would have much more spatial detail than national or regional seismic hazard maps. The urban maps would show variations of seismic hazard maps over a few city blocks. Maps depicting global ground-motion forecasts and detailed seismic hazard in high-risk urban areas are critically needed tools for mitigating the loss of life and property from future large earthquakes.

Land Use

Although little can be done to alter the seismic hazard—which is geologically fixed by proximity to potential earthquake sources, local rock or soil type, and exposure to secondary ground failures—it is possible to limit the exposure to earthquake destruction by land-use policies. This approach is feasible when the hazard is localized, as in fault rupture. Unless specially designed, any structures built across an active fault will be forced to follow the ground deformation caused by fault rupture. The slip across faults in major earthquakes can be as large as 10 meters, and it is usually not practical or possible to design structures to withstand this type of displacement without severe damage. Therefore, restricting land use in active fault zones is the primary strategy for mitigating fault rupture as a seismic hazard (Box 3.2; Figure 3.24). These policies are less effective where the hazard is distributed across a broad area and all sites face significant seismic hazards; here the primary mechanisms for reducing risk are good engineering design and construction standards and effective emergency management.

Life Safety

U.S. building codes have been successful largely in achieving a high degree of life safety during earthquakes. From 1983 to 2001, in the western United States, 129 people died in eight severe earthquakes (*82*), while more than 160,000 people worldwide were killed in earthquakes (*83*). Some of this success is attributable to revisions in building codes prompted by vigorous postearthquake investigations of structural failure and improved understanding of ground motion and dynamic structural behavior (Section 2.7). One notable example is the change in codes after the near collapse of Olive View Hospital during the 1971 San Fernando earthquake. To allow for more open space on the first floor, many buildings of that era, including the hospital, were designed with a major discontinuity in the structural system between the first and successive stories, with the upper stories much stiffer and stronger. This condition was usually referred to as the "soft first story," and the codes were modified to require that buildings have no significant discontinuities in stiffness and strength from the top floor to the foundation.

The San Fernando earthquake also exposed a deficiency in bridge design that resulted in improvements to the design standards for these structures. This earthquake also spawned a major seismic retrofit pro-

BOX 3.2 Fault Rupture and the Alquist-Priolo Act

Surface breaks did not receive serious attention as a seismic hazard until the 1971 San Fernando earthquake, when a fault ruptured a densely populated area of Los Angeles, destroying almost 100 structures. California responded in 1972 with the Alquist-Priolo Special Studies Act, which prevented construction of new buildings for human occupancy across an active fault. For existing residences on a fault, real estate agents were required to disclose the information to potential buyers. Notably, the act did not cover publicly owned facilities, critical facilities and lifelines, or industrial facilities, many of which contain hazardous materials.

The act required the state geologist to initiate a broad, and continuously updated, program of seismic hazard mapping to define the fault zones. "The earthquake fault zones shall ordinarily be one-quarter mile or less in width, except in circumstances which may require the State Geologist to designate a wider zone . . . The State Geologist shall continually review new geologic and seismic data and shall revise the earthquake fault zones or delineate additional earthquake fault zones when warranted by new information." The California Division of Mines and Geology issues fault-zone maps at a scale of 1:24,000.

In 1990, California enacted the Seismic Hazards Mapping Act, which significantly broadened the responsibilities of the state geologist beyond the Alquist-Priolo zones by requiring that landslide and liquefaction hazards be mapped throughout the state. Funds for this program are provided by FEMA from mitigation resources authorized by the Stafford Act and fees from construction building permits.

gram of highway bridges in California that was greatly accelerated by the dramatic bridge failures (e.g., the Interstate 880 Cypress Viaduct in Oakland, the upper deck [east-to-west] crossing of the Oakland-San Francisco Bay Bridge, and the Struve Slough Bridge near Monterey) caused by the 1989 Loma Prieta earthquake.

Bridge seismic retrofits in California from 1971 to 1989 were limited to a cable-restrainer installation at hinges between adjoining bridge segments to prevent the ends of the spans from falling off the support seats. After the Loma Prieta earthquake, the retrofit program was broadened to include bridge columns and footings. Based on focused research, novel yet practical schemes, such as wrapping columns with steel collars or fiber-reinforced composite materials, were implemented on many vulnerable bridges to improve their seismic performance.

A notable aspect of the U.S. experience in achieving life safety is the consensus on the standards and practice for earthquake-resistant design established through the efforts of private, nonprofit organizations. Among the most important are the Structural Engineers Association of California, the Building Seismic Safety Council (BSSC) (*84*), the Applied Technology Council (ATC) (*85*), the Earthquake Engineering Research Institute (*86*), and the American Society of Civil Engineers (*87*). Backed by state and

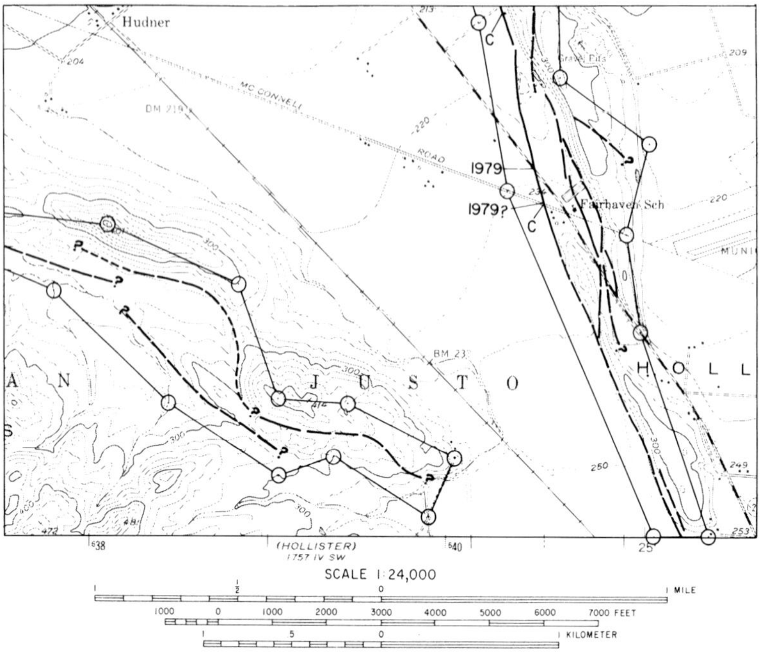

MAP EXPLANATION

Active Faults

1906 C — Faults considered to have been active during Holocene time and to have a relatively high potential for surface rupture; solid line where accurately located, long dash where approximately located, short dash where inferred, dotted where concealed; query (?) indicates additional uncertainity. Evidence of historic offset indicated by year of earthquake-associated event or C for displacement caused by creep or possible creep.

Earthquake Fault Zone Boundaries

These are delineated as straight-line segments that connect encircled turning points so as to define earthquake fault zone segments.

Seaward projection of zone boundary.

FIGURE 3.24 Alquist-Priolo zones for active faults in a region south of Hollister, California, mapped by the California Division of Mines and Geology. SOURCE: E.W. Hart and W.A. Bryant, *Fault Rupture-Hazard Zones in California*, California Division of Mines and Geology, Special Publication 42, Sacramento, 38 pp., revised 1997.

federal government as well as private funding, these groups have sponsored hundreds of workshops and published scores of reports, which have become the basis for revisions to the seismic provisions in building codes.

Achieving the proper balance between risk to human life and construction costs has been an ongoing objective of engineers for decades. As the codes are evolving to more risk-based design, research will be directed to improve the probabilistic definitions of ground motion, ground movement, and structural vulnerability. At the same time, the search for improved construction materials and devices to limit the seismic forces will continue, with the objective of enhancing seismic performance at lower cost. This research will also include the investigation of new and more cost-effective techniques to stabilize poor soils prone to catastrophic failure. As in the past, this research program will be guided or redirected based on observations during future earthquakes.

Advanced Engineering Technologies

Engineering technologies that enhance protection against seismic hazards have advanced considerably over the last two decades, especially for high-value facilities. Sophisticated systems have been developed for retrofits and new construction to isolate structures from seismic shaking (base isolators) (*88*), to dissipate seismic energy without structural damage (hysteretic and viscoelastic devices) (*89*), and to dynamically modify the response of the structure in real time to counteract earthquake forces (active control systems) (Table 3.1) (*90*). Base isolation was introduced in the United States in the mid-1980s as a technique for seismic retrofits of historic monuments. The first application was the City Hall building in Salt Lake City, soon followed by City Halls in Oakland, San Francisco, and Los Angeles; the U.S. Appeals Court in San Francisco; and the School of Mines Building at the University of Nevada. In recent years, it has increasingly been used in new construction, where the additional costs are significantly less than for retrofitting.

The primary challenge of base isolation and other such advanced technologies is that they require sufficient understanding of the seismic response of the structure, and in some cases the soil foundation system, to identify the appropriate placement and capacity of the engineering systems. Realistic estimates of the earthquake ground motion are also required to optimally design these systems. For example, for base isolated structures, the emphasis is on the prediction of long-period ground motions; thus, fault-rupture directivity and basin response are potentially important effects that should be considered in the estimation of these motions.

Performance-Based Engineering

The seismic provisions in most current U.S. building codes are designed "to safeguard against major failures and loss of life, not to limit damage, maintain functions, or provide for easy repair" (*91*). The increas-

TABLE 3.1 Advanced Earthquake Engineering Technologies for Improving Seismic Performance of Structures

Technology	Description
Base isolation	Flexible mounting systems (e.g., elastomeric bearings, friction pendulums, coil springs) are inserted between the building foundation and the ground. "Moats" are constructed around the structure to allow displacements of the entire building without damaging shear strain
Jackets	High-strength jackets (e.g., steel, fiber-reinforced materials) are placed over structural columns to increase the lateral strength
Active control systems	A mass is displaced in a structure to counteract and cancel earthquake-induced acceleration. These systems can be used for both wind and earthquake hazards
Energy dissipaters	Hysteretic or viscoelastic devices are inserted in the structure above the foundation to absorb energy from earthquake vibrations

ing exposure of urban regions to significant economic loss, even in moderate earthquakes, was underscored in 1994, when the Northridge earthquake caused direct damages amounting to $20 billion and surprised engineers with extensive brittle failure of welds in nominally ductile steel moment-frame buildings. Northridge accelerated the U.S. effort to improve building practices that limit damage in future earthquakes. Implementation of what is now called "performance-based earthquake engineering" got a boost with the publication of *Guidelines for the Seismic Rehabilitation of Buildings* in 1997 (FEMA 273), the U.S. government's first published consensus report on the subject (*92*) and, two years later, with the release of a major study on design guidelines for new steel moment-frame buildings (*93*).

In FEMA 273, performance-based engineering is implemented by matching the desired performance objectives, such as the four NEHRP categories described in Table 3.2, with different levels of ground motion along a seismic hazard curve, ranging from weak and frequent to intense and rare. For example, standard buildings might be required both to withstand a frequent earthquake (e.g., 50 percent probability of exceedance in 50 years) with limited economic damage to contents and negligible structural damage, ready for immediate reoccupancy, and to survive a maximum considered earthquake (2 percent probability of exceedance in 50 years) without collapsing. Although the target levels of performance for various types of structures can be specified relative to earthquake intensity (Figure 3.25), the actual procedures are still evolving. This iterative process, for which increasingly quantitative approaches are being devel-

TABLE 3.2 Building Performance Levels

Level	Performance
	High ↑
Operational Very little damage, backup utility services maintain functions	
Immediate Occupancy Building receives a "green tag" (safe to occupy) inspection rating; any repairs are minor	
Life Safety Structure remains stable and has significant reserve capacity; hazardous nonstructural damage is controlled	
Collapse Prevention Building remains standing, but only barely; any other damage or loss is acceptable	
	↓ Low

SOURCE: Building Seismic Safety Council and Applied Technology Council, *Guidelines for the Seismic Rehabilitation of Buildings*, Federal Emergency Management Agency Report FEMA-273, Washington, D.C., 400 pp., October, 1997.

oped, attempts to balance the costs of construction against the willingness to accept risk and uncertainty about future losses.

Conceptually, performance-based seismic engineering begins with the identification of appropriate measures of building deformation, a common choice being the maximum interstory drift (largest of the peak displacements between successive floors of the building). Each performance objective, from operational to collapse prevention, is assigned a tolerable annual probability of occurrence (with increasingly undesirable performance decreasing in annual probability of occurrence) and mapped into maximum allowed values of interstory drift. The latter step is based on experience in actual earthquakes, engineering judgment, or computer analyses of buildings. In more complete applications, structural deformation is also being used to predict the performance both of nonstructural building components, such as elevators and heating-ventilating-air conditioning systems, and of building contents.

This performance-based approach increases demands on accuracy and completeness of earthquake forecasting and strong-motion seismology. As input, it utilizes the probabilities of exceedance over a range of ground-motion intensities—the entire hazard curve, not just a single intensity value corresponding to, say, the maximum considered earthquake. Moreover, ground-motion intensity must be specified in a way that is compatible with the method chosen for analyzing building response.

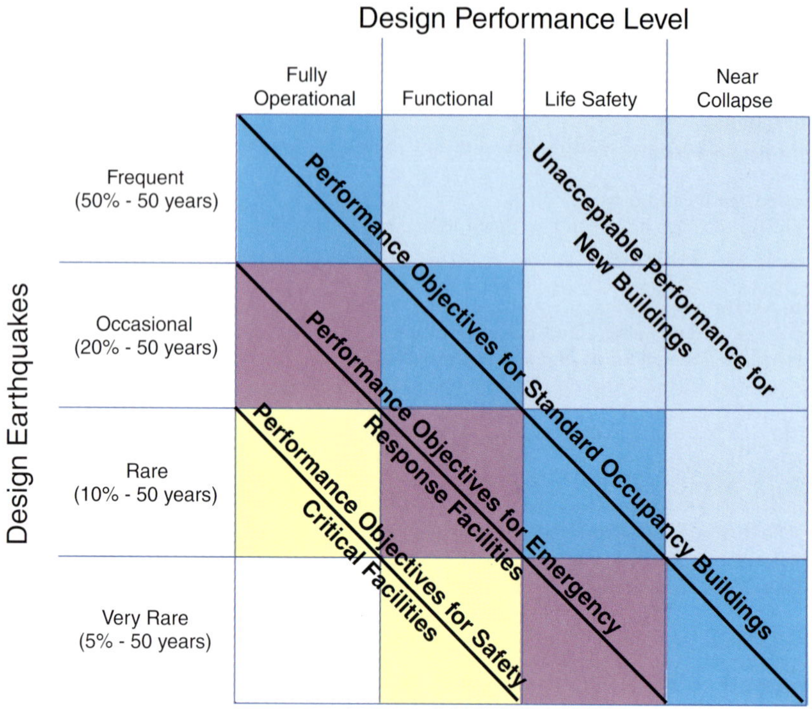

FIGURE 3.25 Performance objectives as a function of hazard level for different building types proposed in the Vision 2000 report, which uses performance objectives that differ slightly from those in FEMA 273 (Table 3.2). For each building type, increasingly rare (large) ground-motion levels are associated with increasingly unacceptable building performance. SOURCE: Vision 2000 Committee, *Vision 2000—A Framework for Performance Based Design*, Structural Engineers Association of California, Sacramento, Calif., Volumes I, II, III, 1995.

The performance-based approach increases the demands on accuracy on the engineering side of the equation as well. To achieve performance objectives, the calculation must go well beyond the standard pseudostatic lateral-force considerations prescribed by traditional building codes. Ground-motion intensity measures may be needed for several probabilities of exceedance. For example, the collapse prevention performance level may be specified by the intensity of the maximum considered earthquake at the frequency of greatest structural fragility (e.g., related to the low-order modes of building response), whereas the operational performance level may be determined by the much lower intensity of a maximum frequent earthquake at the high frequency responsible for nonstructural damage. The ground-motion intensity may have to be described by vector quantities that are combinations of parameters, such as spectral acceleration at a particular period and duration, or the peak velocity and period of the near-fault rupture directivity pulse, to predict building response more optimally.

The most complete treatments abandon response-spectrum analysis, which is based on linear modal superposition, and use suites of ground-motion time histories to drive simulations of the fully nonlinear response, the most complete treatment for capturing the essence of building damage and failure. If ground motions within the specified probabilities are found by structural analysis to cause building drifts that exceed those required to meet the associated performance objective, the design of the building has to be modified (e.g., by increasing its stiffness, strength, or ductility capacity) until the performance objective is met. (Figure 3.26 shows the results of nonlinear dynamic analyses of a model of a five-story steel moment-resisting building frame.) Because the suites of observed strong-motion seismograms are limited, there is an increasing need for realistic simulations of wavefields at frequencies important for building performance (0.3 to 10 hertz). Current physics-based simulations of ground motions for relatively well-characterized regions, such as the Los Angeles metropolitan area, are now feasible only at the lower end of this frequency range (*94*).

Major efforts will be required to improve the accuracy of predictions of nonlinear structural behavior, to link the displacements to cost and injury statistics, and to simplify the procedures for routine practice. Progress toward the last objective will require new types of intensity measures more general than response spectral values (e.g., nonlinear functionals of the seismogram), judiciously derived as predictable aspects of the time histories and explicitly formulated for a simplified nonlinear analysis. This generalization of seismic hazard analysis to seismic performance assessment analysis holds great promise (*95*) and is well matched with the new capabilities of earthquake science in earthquake forecasting, source characterization, and ground-motion simulation. However, this research will demand a greater degree of collaboration between the scientific and engineering communities.

Warning and Rapid Response

John Milne anticipated the current enthusiasm for seismic warning systems when he wrote in 1899 (*96*):

> An earthquake which traveled at the rate of four seconds to a mile might, if it were allowed to close a circuit which fired a gun at a station fifteen miles distant, give the inhabitants at that place a minute's warning to leave their houses. The inhabitants of Australia and the western shores

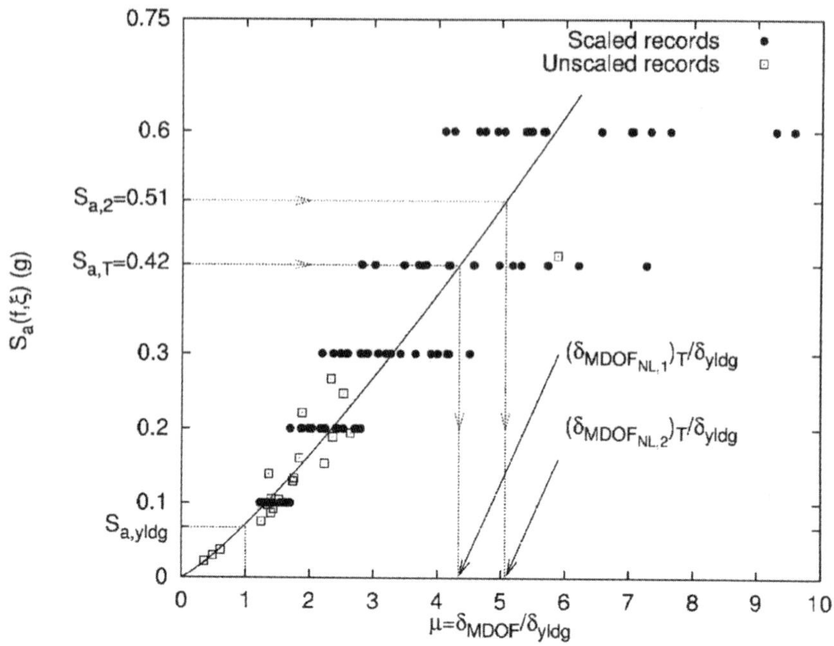

FIGURE 3.26 Predicted maximum nonlinear interstory drift ductility displacements in a five-story steel building frame subjected to recorded accelerograms of various intensities S_a. Recorded accelerograms of various intensity levels, as measured here by the spectral acceleration at the 1-hertz fundamental mode of the structure (in the elastic range), have been used as input to the analyses. The response is measured by a maximum story ductility, the maximum interstory drift (denoted δ_{MDOF}) normalized by δ_{yldg}, the drift level required to induce yield or the onset of nonlinear behavior. As a point of reference, in central Los Angeles the annual probability of exceeding 0.42*g* 1-hertz spectral acceleration, which is the intensity measure being used here, is 0.1 percent. If this level of shaking were to occur, the maximum story ductility expected is about 4, implying displacements four times the elastic limit, major damage to nonstructural elements, and permanent distortions or possibly fracture of local beam-column connections in the building. SOURCE: P. Bazzurro, C.A. Cornell, N. Shome, and J.E. Carballo, Three proposals for characterizing MDOF nonlinear seismic response, *J. Struct. Engr.*, **124**, 1281-1289, 1998. Reproduced by permission of the American Society of Civil Engineers.

> of the Pacific might, by telegraphic communication, receive eighteen to twenty-five hours' warning of the coming of destructive sea waves resulting from earthquakes in South America.

Based on Milne's observation that seismic waves propagate much more slowly than electromagnetic waves, single-seismometer warning systems were first deployed in Japan in 1950 to stop trains after earthquakes. New communication and computing technologies have advanced these warning systems over the past 50 years (*97*). Real-time seismic systems have also been deployed in Taiwan, Mexico, and the United States, involving dedicated digital seismic networks that are automatically analyzed by computers and broadcast to interested users (e.g., emergency service organizations, utilities, train operators). In Mexico, a seismic alert system, operating since 1991, warns of subduction-zone earthquakes several hundred kilometers west of Mexico City. During the September 14, 1995, Guerrero earthquake (M 7.3), its warning arrived 72 seconds before the strong ground motion (*98*). An early warning system set up by the USGS after the 1989 Loma Prieta earthquake gave workers on the collapsed freeways in Oakland (about 100 kilometers from the rupture zone) as much as 20 seconds' notification before the shaking from strong aftershocks, allowing them to evacuate hazardous areas (*99*).

The requirements for a real-time monitoring system are: reliable components that will perform during an earthquake, broad bandwidth communications facilities, and seismic instrumentation with high dynamic range and broad band-recording capabilities. With this approach, real-time systems in the United States are being developed in southern (Caltech USGS Broadcast of Earthquakes) and northern (Rapid Earthquake Data Integration Project) California (*100*). With partnerships between university researchers, private corporations, and government (federal, state, and local), these systems have largely evolved by upgrading the communications and data-processing infrastructure of existing seismic networks. Significant improvements in the southern California seismic network are currently under way (*101*).

The recent deployment of high-performance systems in California has motivated an active dialogue regarding operational goals and practices. This has focused attention on the differing data delivery needs before, during, and after the start of ground shaking, the type of information required for different purposes (e.g., simple warning, magnitude, hypocenter, extent and location of strong ground shaking), the tolerance for false alarms, and the policies for broadcasting data. Consideration of these questions has revealed a much broader range of applications for real-time data than originally anticipated. At first, the systems were expected to deliver warnings before strong ground shaking began so that evasive actions could be taken, such as evacuation, duck-and-cover procedures, and automated shutdown of sensitive facilities. However, collaboration between developers and users of real-time information, together with the experience from the Northridge and Hyogo-ken Nanbu earthquakes, has revealed that real-time broadcasts will still be valuable even if they arrive after the earthquake. For example, hypocenter determinations can be combined with spatial maps of ground shaking and real-time loss modeling to predict the areas of greatest damage before response and rescue teams are deployed.

A key recent development is ShakeMap, which provides maps of observed ground shaking and intensity within a few minutes of an earthquake's occurrence. This product was developed by the USGS and was originally implemented as part of the Tri-Net seismic network in southern California (*102*). ShakeMap is now operational in other areas with high seismic risk and relatively dense seismograph coverage, including the San Francisco Bay area, the Puget Sound region, and the Wasatch Front. ShakeMap interpolates between observed ground motions recorded at seismograph sites using empirical ground-motion relations and maps of site conditions. As more dense instrumentation is deployed in urban areas under the Advanced National Seismic System, ShakeMap will be implemented in other areas. ShakeMap is an effective tool for rapidly communicating shaking and intensity information to emergency managers, government officials, the private sector, and the general public. The ShakeMap software can also be linked to HAZUS to provide rapid estimates of losses. From this information, impacts on critical facilities can be estimated quickly, followed by rapid decisions as to response and mitigation strategies for the facility, the occupants, and the community (*103*).

In detail, the terminology for these efforts shifts from real time or early warning, to rapid ground-motion estimation. However, as the time scales for these data deliveries shrink to within tens of seconds of the events, the distinction between the systems begins to blur. Early estimates of ground shaking (provided as warnings) can be updated with time as more and more stations report the entire history of shaking at their sites, permitting interpolation of the entire network's direct shaking observations, rather than more primitive, forward ground-motion predictions based only on location and magnitude estimates. Prior and posterior information about the likely effects of local site conditions can be factored in at still later stages.

Utilization of and demand for this rapid information will be user specific, indicating that the costs and benefits of these systems will vary among regions. For example, in postearthquake emergency services, it is more important to know the distribution of strong motion than the rupture initiation point. This motivates a need for data on the center of energy radiation (ground-motion centroid), the conventional hypocenter, and the spatial distribution of ground motion. With these requirements, detection and location algorithms for real-time warning will differ from standard procedures at traditional seismographic stations.

3.6 CLOSING THE IMPLEMENTATION GAP

For the past 25 years, the NEHRP has sponsored a combination of basic and applied research on the causes of earthquakes, their distribution, and their damage to the built environment (Appendix A). The review in this chapter demonstrates that this information is paying off in the ability to anticipate and reduce earthquake destruction. However, the degree to which the knowledge gained through research has been utilized in coordinated programs of risk mitigation has been the focus for criticism of the NEHRP strategy. In a 1995 report to the U.S. Congress, the Office of Technology Assessment stated (*104*):

> NEHRP has made significant contributions toward improving our understanding of earthquakes and strategies to reduce their impact. However, much of the United States remains at risk for significant earthquake losses. Risk-reduction efforts lag far behind the knowledge base created by research; this lag or implementation gap, reflects the limitations of NEHRP's information-based strategy for encouraging nonfederal action.

Considerable controversy surrounds which types of loss reduction measures should be implemented by government and the private sector through regulatory policies, economic incentives, long-term investments, and public education. Part of the debate concerns the role of scientific research in closing the implementation gap. The key question is: how can researchers participate more effectively in translating the technical understanding of earthquake phenomena into civic actions?

This chapter has explored the great utility of combining seismic hazard analysis with engineering-based performance estimates to produce measures of earthquake risk, in terms of either the damage expected during the lifetime of an individual structure or the annualized economic loss expected for a specified region. Three further steps are necessary to reduce potential losses:

1. Mitigation options must be identified and evaluated. These options include improved building codes, design enhancements, retrofitting, land-use planning, and insurance. Evaluation must include assessments of the cost and effectiveness of each option to reduce risk.

2. The public, elected officials, property owners, and other decision makers must understand the nature of the risks, their mitigation options, and the costs of action and inaction.

3. Mitigation decisions must be made and implemented. Setting priorities for action is imperative, since the need for improvement will always vastly exceed the available resources.

Experience gained during NEHRP demonstrates that a consensus to invest in risk reduction is best achieved through an active collaboration among scientists, engineers, government officials, and business leaders, working together with an informed populace. A corollary is that earthquake research will contribute to risk reduction more effectively when it is carried out in a context that recognizes the problem's engineering, economic, and political dimensions (*105*).

An example of a public-private partnership designed to bridge the implementation gap is the lifelines program conducted by the Pacific Gas and Electric Co., Caltrans, the California Energy Commission, and the Pacific Earthquake Engineering Research Center (*106*). This program identifies and funds research on how to make lifelines less vulnerable to earthquake shaking and follows up this research with steps to implement research findings within the sponsoring organizations.

FEMA's Project Impact is another notable example. Project Impact fostered public-private partnerships in selected communities to prepare for the occurrence of earthquakes and other natural hazards. In Seattle, for example, Project Impact involved a partnership between local, state, and federal agencies, universities, and the private sector and included programs for seismic retrofit of schools and homes, as well as a disaster mitigation plan for businesses (*107*). Although Project Impact was terminated as a federal program in 2001, it helped minimize damage and injuries during the February 28, 2001, Nisqually earthquake. As a result, the city of Seattle is continuing the program.

Scientists and engineers have lacked effective organizational and programmatic mechanisms to exploit the synergy between the two fields and to educate each other about implementation issues. The federal government has established an elaborate array of research programs focused on the engineering aspects of earthquake safety, which include three National Science Foundation (NSF) earthquake engineering centers (*108*), the NSF initiative for a Network for Earthquake Engineering Simulation (NEES) (*109*), and the NEHRP component of the National Institute of Standards and Technology (Appendix A). The participants have been mainly engineers, with earthquake scientists engaged only on the periphery of these efforts. The USGS and NSF Earth Science Division sponsor large, diverse programs in basic and applied earthquake research, including major centers to promote interdisciplinary collaborations and deliver new products for seismic hazard analysis (*110*). The primary participants in these programs have been geoscientists, with engineers relegated to a relatively minor role. The Federal Emergency Management Agency leads NEHRP and holds the federal responsibility for seismic risk mitigation. Although FEMA is deeply involved in both risk assessment and emergency management, it is primarily a user, not a coordinator, of earthquake research. At present, no agency or organization is responsible for ensuring an integrated approach to earthquake science and engineering. This vacuum could be filled through structured collaborations between the science and engineering research centers, explicitly funded and frequently reviewed by NEHRP agencies. These collaborations should involve economists and social scientists with expertise in mitigation issues.

Scientists also need better organizational and technological support for communicating with all levels of society about earthquake hazards, mitigation measures, and the appropriate use of earthquake information. The challenges are to select the right kinds of educational activities, target the appropriate audiences, and present them at the right places and times. Appropriately, NEHRP agencies are now placing more emphasis on efforts to interpret scientific research and reduce the results to understandable, usable products. Even if well-packaged, however, such products cannot be "simply thrown over the wall" for public consumption. Effective communication between researchers and end users requires a two-way, continuing dialogue with repeated opportunities for the exchange of ideas and plans. Likewise, effective public education requires interactive mechanisms that can engage an audience at an appropriate level. The new technologies of the Internet—interactive web pages backed by powerful, simple-to-use query languages and digital libraries with up-to-the-minute earthquake information—offer considerable promise. However, their utilization will depend on support structures with more financial and human resources than a typical research group.

NOTES

1. As described in Section 1.2, the warning times for destructive tsunamis that cross wide ocean basins can be several hours or more.
2. Secondary hazards also include fires, dispersal of nuclear materials, and other threats indigenous to the built environment.
3. The displacement across the fault scales with the cube root of seismic moment for earthquakes of magnitude less than about 6.5 and with the square root of moment for larger events.
4. A complete discussion of these complexities is given in R.S. Yeats, K. Sieh, and C.R. Allen, *The Geology of Earthquakes*, Oxford, New York, 568 pp., 1997.
5. Throughout the engineering literature, the redundant term "time history" is used to describe ground motion as a function of time and is thus synonymous with seismogram,

accelerogram, and seismic waveform. The spectral response method is described in Section 2.7.

6. N.A. Abrahamson and P.G. Somerville, Effects of the hanging wall and foot wall on ground motions recorded during the Northridge earthquake, *Bull. Seis. Soc. Am.*, **86**, S93-S99, 1996.

7. See summary of ground motion models described by N.A. Abrahamson and K.M. Shedlock (Overview of ground motion attenuation models, *Seis. Res. Lett.*, **68**, 9-23, 1997).

8. E.H. Field and the SCEC Phase III Working Group, Accounting for site effects in probabilistic seismic hazard analyses of Southern California: Overview of the SCEC Phase III report, *Bull. Seis. Soc. Am.*, **90**, S1-S31, 2000. This major study of regional attenuation relations and local site effects concluded that "any model that attempts to predict ground motion with only a few parameters will have substantial intrinsic variability. Our best hope for reducing such uncertainties is via waveform modeling based on the first principles of physics."

9. P.G. Somerville and J. Yoshimura, The influence of critical Moho reflections on strong ground motions recorded in San Francisco and Oakland during the 1989 Loma Prieta earthquake, *Geophys. Res. Lett.* **17**, 1203-1206, 1990. The ground motions recorded in San Francisco and Oakland were actually stronger than those for some closer sites with similar geology.

10. S.K. Singh, E. Mena, and R. Castro, Some aspects of source characteristics of the 19 September 1985 Michoacan earthquake and ground motion amplification in and near Mexico City from strong motion data, *Bull. Seis. Soc. Am.*, **78**, 451-477, 1988.

11. See S. Gao, H. Liu, P.M. Davis, and L. Knopoff (Localized amplification of seismic waves and correlation with damage due to the Northridge earthquake: Evidence for focusing in Santa Monica, *Bull. Seis. Soc. Am.*, **86**, S209-S230, 1996) for an example of focusing during the 1994 Northridge earthquake.

12. H.B. Seed and I.M. Idriss, Analyses of ground motions at Union bay, Seattle during earthquakes and distant nuclear blasts, *Bull. Seis. Soc. Am.*, **60**, 125-136, 1970; M. Zeghal and A.-W. Elgamal, Analysis of site liquefaction using earthquake records, *J. Geotech. Engr.*, **120**, 996-1017, 1994; E.H. Field, P.A. Johnson, I.A. Beresnev, and Y.H. Zeng, Nonlinear ground-motion amplification by sediments during the 1994 Northridge earthquake, *Nature,* **390**, 599-602, 1997; J. Aguirre and K. Irikura, Nonlinearity, liquefaction and velocity variation, of soft soil layers in Port Island, Kobe, during the Hyogo-ken Nanbu earthquake, *Bull. Seis. Soc. Am.*, **87**, 1244-1258, 1997.

13. NEHRP site classifications are rated on a five-level scale ranging from A (hard rock with measured shear-wave velocity [v_S] more than 5000 feet per second) to E (soft soil with v_S less than 600 feet per second); see Building Seismic Safety Council, 1997 Edition NEHRP Recommended Provisions for Seismic Regulations for New Buildings and Other Structures, FEMA 302/303, Part 1 (Provisions) and Part 2 (Commentary), developed for the Federal Emergency Management Agency, Washington, D.C., 337 pp., 1998.

14. Some have argued (W.D.L. Finn, Geotechnical engineering aspects of microzonation, in *Proceedings of the Fourth International Conference on Seismic Zonation*, August 25-29, 1991, Stanford, California, Vol. I, pp. 199-259, 1991; K. Aki, Local site effects on weak and strong ground motion, *Tectonophysics*, **218**, 93-111, 1993) that a very low but approximately constant shear modulus and site resonance at Mexico City can explain the ground motions without appeal to nonlinear effects. Others (S.K.E. Singh, E. Mena, and R. Castro, Some aspects of source characteristics of the 19 September 1985 Michoacan earthquake and ground motion amplification in and near Mexico City from strong motion data, *Bull. Seis. Soc. Am.*, **78**, 451-477, 1988; C. Lomnitz, Mexico 1985; The case for gravity waves, *Geophys. J. Int.*, **102**, 569-572, 1990) argue for a strong nonlinear shear modulus reduction during the strong shaking. Recent efforts to measure dynamic strains at depth in the Valley of Mexico from

other earthquakes (S.K. Singh, M.A. Santoyo, P. Bodin, and J. Gomberg, Dynamic deformations of shallow sediments in the valley of Mexico; Part II, Single-station estimates, *Bull. Seis. Soc. Am.*, **87**, 540-550, 1997; P. Bodin, S.K. Singh, M. Santoyo, and J. Gomberg, Dynamic deformations and shallow sediments in the Valley of Mexico, I: Three-dimensional strains and rotations recorded on a seismic array, *Bull. Seis. Soc. Am.*, **87**, 540-550, 1997) suggest that the Mexico City clays exhibit only a mildly nonlinear response, even up to strains of 1 percent. This strain is one to two orders of magnitude beyond the strains at which more common, more granular sediments begin to respond inelastically. Even though this may not increase strong-motion amplitudes, it often leads to other ground failure problems, the most common and most serious of which is liquefaction.

15. See, for example, A.-W. Elgamal, M. Zehal, and E. Parra, Liquefaction of reclaimed island in Kobe, Japan, *J. Geotech. Engr.*, **122**, 39-49, 1996; M. Zeghal and A.-W. Elgamal, Analysis of site liquefaction using earthquake records, *J. Geotech. Engr.*, **120**, 996-1017, 1994.

16. For a concise summary of nonlinear site response, see E.H. Field and SCEC Phase III Working Group, Accounting for site effects in probabilistic seismic hazard analysis of southern California; An overview of the SCEC Phase III report, *Bull. Seis. Soc. Am.*, **90**, S1-S31, 2000, and references therein.

17. G. Plafker, *Tectonics of the March 27, 1964, Alaska Earthquake*, U.S. Geological Survey Professional Paper 543-I, U.S. Government Printing Office, Washington, D.C., 74 pp., 1969.

18. J.C. Savage and L.M. Hastie (Surface deformation associated with dip-slip faulting, *J. Geophys. Res.*, **71**, 4897-4904, 1966) showed how a dislocation model could be used to fit Plafker's observations of uplift and subsidence following the 1964 Alaska earthquake; see also S.R. Holdahl and J. Sauber, Coseismic slip in the 1964 Prince William Sound earthquake: A new geodetic inversion, *Pure Appl. Geophys.*, **142**, 55-82, 1994.

19. National Research Council, *Liquefaction of Soils During Earthquakes*, National Academy Press, Washington, D.C., 240 pp., 1985.

20. H.B. Seed and I.M. Idriss, *Ground Motions and Soil Liquefaction During Earthquakes*. Earthquake Engineering Research Institute, Engineering Monograph on Earthquake Criteria, Structural Design, and Strong Motion Records 5, El Cerrito, Calif., 134 pp., 1982.

21. P. Talawani and W.T. Shaeffer, Recurrence rates of large earthquakes in South Carolina coastal plain based on paleoliquefaction data, *J. Geophys. Res.*, **106**, 6621-6642, 2001; M.P. Tuttle and E.S. Schweig, Recognizing and dating prehistoric liquefaction features; Lessons learned in the New Madrid seismic zone, central United States, *J. Geophys. Res.*, **101**, 6171-6178, 1996.

22. During the 1964 Alaska earthquake, compression resulting from such dense flows buckled or skewed spans and damaged abutments on more than 250 bridges. See National Research Council, *The Great Alaska Earthquake of 1964*, National Academy Press, Washington, D.C., 15 volumes, 1972-1973.

23. W.R. Hansen, Effects of the Earthquake of March 27, 1964, at Anchorage, Alaska, U.S. Geological Society Professional Paper 542-A, Washington, D.C., 68 pp. + 2 plates, 1966.

24. Tsunami propagation can be treated by the theory of shallow-water waves, which states that the propagation speed varies as the square root of water depth. An elementary discussion of the tsunami physics is given by T. Lay and T.C. Wallace, *Modern Global Seismology*, Academic Press, San Diego, pp. 147-153, 1995.

25. Early in their history, tsunami warning systems generated many false alarms because they relied on earthquake size determined from high-frequency magnitude scales, such as m_b. See H. Kanamori, Mechanism of tsunami earthquakes, *Phys. Earth Planet. Int.*, **6**, 346-359, 1972.

26. Kanamori and Kikuchi (The 1992 Nicaragua earthquake; A slow tsunami earthquake associated with subducted sediments, *Nature*, **361**, 714-716, 1993) suggested that there are two types of tsunami earthquakes: those that arise from slow rupture, such as the 1992

and A.L. Moore, A tsunami about 1,000 years ago in Puget Sound, Washington, *Science*, **258**, 1614-1617, 1992; R.E. Karlin and S.E.B. Abella, Paleoearthquakes in the Puget Sound region recorded in sediments from Lake Washington, *Science*, **258**, 1617-1620, 1992; R.L. Schuster, R.L. Logan, and P.T. Pringle, Prehistoric rock avalanches in the Olympic Mountains, Washington, *Science*, **258**, 1620-1621, 1992; G.C. Jacoby, P.L. Williams, and B.M. Buckley, Tree ring correlation between prehistoric landslides and abrupt tectonic events in Seattle, Washington, *Science*, **258**, 1621-1623, 1992.

46. R.A. Bennett, B.P. Wernicke, and J.L. Davis, Continuous GPS measurements of contemporary deformation across the northern Basin and Range Province, *Geophys. Res. Lett.*, **25**, 563-566, 1998; W. Thatcher, G.R. Foulger, B.R. Julian, J. Svarc, E. Quilty, and G.W. Bawden, Present-day deformation across the Basin and Range Province, western United States, *Science*, **283**, 1714-1718, 1999.

47. R.B. Smith and L. Siegel, *Windows into the Earth: The Geologic Story of Yellowstone and Grand Teton National Parks*, Oxford University Press, New York, 242 pp., 2000.

48. T.C. Hanks and A.C. Johnston, Common features of the excitation and propagation of strong ground motion for North American earthquakes, *Bull. Seis. Soc. Am.*, **82**, 1-23, 1992.

49. S.F. Obermeier, P.J. Munson, C.A. Munson, J.R. Martin, A.D. Frankel, T.L. Youd, and E.C. Pond, Liquefaction evidence for strong Holocene earthquake(s) in the Wabash Valley of Indiana-Illinois, *Seis. Res. Lett.*, **63**, 321-336, 1992. Although small, nondestructive earthquakes are relatively common in the Wabash River Valley of southeastern Illinois and southern Indiana, no large earthquakes have struck the region in 200 or so years of historical record. Nevertheless, paleoliquefaction features indicative of very large earthquakes have been identified, including clastic dikes ranging up to 2.5 meters in width, are widespread throughout a region of about 200 kilometers by 250 kilometers. The most widespread set of these formed during a large event about 6100 years ago. The fact that the largest dikes cluster within a region about 50 kilometers in diameter suggests that the source of the earthquake was there, near the Illinois-Indiana border with a size of M_W about 7.5. More restricted sets of dikes formed in this same region during an event about 12,000 years ago and in a smaller region within Indiana about 3000 years ago.

50. A.J. Crone and K.V. Luza, Style and timing of Holocene surface faulting on the Meers fault, southwestern Oklahoma, *Geol. Soc. Am. Bull.*, **102**, 1-17, 1990.

51. A.J. Crone, M. Machette, L. Bradley, and S. Mahan, *Late Quaternary Surface Faulting on the Cheraw Fault, Southeastern Colorado*, U.S. Geological Survey Map I-2591, Reston, Va., 1997.

52. D. Amick and R. Gelinas, The search for evidence of large prehistoric earthquakes along the Atlantic seaboard, *Science*, **251**, 655-658. 1991. Near Charleston, prehistoric sandblow craters, similar to those that formed in 1886, formed four times in the 5000 years before 1886. Small twigs and bark that fell into these ancient craters soon after they were formed yield radiocarbon ages of about 600, 1250, 3200, and 5150 years. Judging from the size and geographic extent of the craters formed 600 and 1250 years ago, the magnitude of the causative earthquakes was at least M_W 7.5.

53. R.T. Marple and P. Tawani, The Woodstock lineament; A possible surface expression of the seismogenic fault of the 1886 Charleston, South Carolina, earthquake, *Seis. Res. Lett.*, **63**, 153-160, 1992.

54. C. Powell, G. Bollinger, M. Chapman, M. Sibol, A. Johnston, and R. Wheeler, A seismotectonic model for the 300-kilometer-long eastern Tennessee seismic zone, *Science*, **264**, 686-688, 1994.

55. A. Frankel, C. Mueller, T. Barnhard, D. Perkins, E. Leyendecker, N. Dickman, S. Hanson, and M. Hopper, *National Seismic-Hazard Maps: Documentation June 1996*, U.S. Geological Survey Open-File Report 96-532, Denver, Colo., 111 pp, 1996.

56. D. Giardini, G. Grünthal, K.M. Shedlock, and P. Zhang, The GSHAP Global Seismic Hazard Map, *Ann. Geofisica*, **42**, 1225-1230, 1999.

57. The volcanism in subduction zones is caused primarily by water that is carried down with the subducted slab to depths on the order of 100 kilometers. This water eventually fluxes into the mantle wedge above the slab, lowering the melting temperature of the rocks and producing a small fraction of melt that rises into shallow magma chambers, which erupt to form the andesitic volcanoes characteristic of the island arcs.

58. S. Uyeda and H. Kanamori, Back-arc opening and the mode of subduction, *J. Geophys. Res.*, **84**, 1049-1061, 1979.

59. L. Ruff and H. Kanamori, Seismicity and the subduction process, *Phys. Earth Planet. Int.*, **23**, 240-252, 1980.

60. W.R. McCann, S.P. Nishenko, L.R. Sykes, and J. Krause, Seismic gaps and plate tectonics: Seismic potential for major boundaries, *Pure Appl. Geophys.*, **117**, 1082-1147, 1979. Offshore of Java, the subduction zone has produced only major earthquakes (M ≤ 7.5) in the 450-year-long historic record, whereas offshore of Sumatra, giant earthquakes (M 8.5 to 8.8) have occurred. Future geodetic measurements across subduction zones will enable better quantification of the degree of coupling and, hence, better estimates of seismic potential.

61. T.J. Fitch, Plate convergence, transcurrent faults and deformation in Asia and Pacific, *J. Geophys. Res.*, **77**, 4432-4460, 1972.

62. A.Y. Le Dain, B. Robineau, and P. Tapponnier, The tectonic effects of the seismic and volcanic event of November 1978 in the Asia-Ghubbet Rift, *Soc. Gèol. France Bull.*, **22**, 817-822, 1979; T. Forslund and A. Gudmundsson, Crustal spreading due to dikes and faults in southwest Iceland, *J. Struct. Geol.*, **13**, 443-457, 1991; R.S. Stein, P. Briole, J.-C. Ruegg, P. Tapponnier, and F. Gasse, Contemporary, Holocene, and Quaternary deformation of the Asal Rift, Djibouti: Implications for the mechanics of slow spreading ridges, *J. Geophys. Res.*, **96**, 21,789-21,806, 1991.

63. Several million years of normal faulting in the million-square-kilometer Basin and Range Province have led to a northwest-southeast extension of more than 100 kilometers and the creation of dozens of tilted, 10- to 30-kilometer-wide crustal blocks that form the alternating basins and ranges. Although late Cenozoic normal faults are distributed relatively uniformly across this region, historical and instrumental seismicity is concentrated in two zones: the central Nevada seismic zone, which extends along the western margin of the province in eastern California and western Nevada, and the intermountain seismic zone, along the eastern edge of the province from southern Nevada across central Utah to southwestern Montana. The large (>M 7) earthquakes of 1872, 1915, and 1954 occurred within the former zone and the large events of 1959 and 1983 within the latter. Paleoseismic studies of normal faults in the Basin and Range Province suggest that many of the faults produce such big earthquakes only every few thousand years and that the current level of activity is abnormally high in the central Nevada seismic zone and abnormally low in the intermountain seismic belt. This possibility is of particular importance to Carson City and Salt Lake City, the capitals of Nevada and Utah, respectively, which sit on the edges of the province.

64. Examples include the 1987 Edgecomb earthquake (M 6.6) caused by failure of several normal faults within the volcanic arc of North Island, New Zealand, as well as dozens of historically important earthquakes in Greece and western Turkey that have occurred in the broad extensional back-arc setting of the Aegean Sea, for example, in the Bay of Corinth in 1861 and 1981, on the Peloponnese near Kalamata in 1981 and 1998, and probably an earthquake that destroyed Sparta in 464 B.C. (R. Armijo, H. Lyon-Caen, and D. Papanastassiou, A possible normal-fault rupture for the 464 BC Sparta earthquake, *Nature*, **351**, 137-139, 1991).

65. An alternative explanation for this type of normal faulting involves behind-the-arc divergence associated with changes in the curvatures of the plate boundary thrust faults.

66. R. Armijo, P. Tapponnier, J.L. Mercier, and T.-L. Han, Quaternary extension in southern Tibet: Field observations and tectonic implications, *J. Geophys. Res.*, **91**, 13,803-13,872, 1986. A combination of right-lateral and normal faulting in the eastern part of this extensional region resulted in the great Beng Co earthquake of 1951 and its large aftershock.

67. T. Camelbeeck and M. Meghraoui, Large earthquakes in northern Europe more likely than once thought, *Eos, Trans. Am. Geophys. Union*, **77**, 405-409, 1996.

68. R.S. Yeats, K. Sieh, and C.R. Allen, *The Geology of Earthquakes*, Oxford University Press, Oxford, U.K., **568** pp., 1997.

69. These oceanographic surveys that have included high-resolution bathymetric mapping, seismicity studies using portable ocean-bottom seismographs, and visual investigations involving bottom photography and manned submersibles (e.g., P. Lonsdale, Structural geomorphology of the Eltanin fault system and adjacent transform faults of the Pacific-Antarctic plate boundary, *Marine Geophys. Researches*, **17**, 105-143, 1994).

70. N.H. Woodcock, The role of strike-slip fault systems at plate boundaries, *Phil. Trans. Roy. Soc. Lond.*, **A317**, 13-29, 1986.

71. The 56,000 year paleoseismic record of N. Porat, A.G. Wintle, R. Amit, and Y. Enzel (Late Quaternary earthquake chronology from luminescence dating of colluvial and alluvial deposits of the Arava Valley, Israel, *Quaternary Res.*, **46**, 107-117, 1996) is also discussed by G. Leonard, D.M. Steinberg, and N. Rabinowitz (An indication of time-dependent seismic behavior—An assessment of paleoseismic evidence from the Arava Fault, Israel, *Bull. Seis. Soc. Am.*, **88**, 767-776, 1998). Fragments of the seismic history of the Dead Sea transform are known from four millennia of recorded history and from archeological evidence (A. Ben-Menahem, Four thousand years of seismicity along the Dead Sea Rift, *J. Geophys. Res.*, **96**, 20,195-20,216, 1991). A major earthquake destroyed the ancient city of Jericho, on the northern edge of the Dead Sea, in the sixteenth century, B.C. This earthquake may have influenced the Old Testament writer who described the collapse of the walls of Jericho with the sounding of Joshua's trumpet (Joshua 6:20). Careful analysis of historical accounts (N.N. Ambraseys and M. Barazangi, The 1759 earthquake in the Bekaa Valley; Implications for earthquake hazard assessment in the eastern Mediterranean region, *J. Geophys. Res.*, **94**, 4007-4013, 1989; N.N. Ambraseys and C.P. Melville, An analysis of the eastern Mediterranean earthquake of 20 May 1202, in *Historical Seismograms and Earthquakes of the World*, W.H.K. Lee, H. Meyers, and K. Shimazaki, eds., Academic Press, San Diego, Calif., pp. 181-200, 1988) suggest that the northern 350 kilometers of this system, in Lebanon and Syria, ruptured in a series of eight major destructive earthquakes during the past millennium. These occurred in three temporal clusters in the twelfth, fifteenth, and seventeenth centuries.

72. P. Tapponnier and P. Molnar, Active faulting and Cenozoic tectonics of the Tien Shan, Mongolia, and Baykal regions, *J. Geophys. Res.*, **84**, 3425-3456, 1979.

73. A.A. Barka, Slip distribution along the North Anatolian fault associated with the large earthquakes of 1939-1967, *Bull. Seis. Soc. Am.*, **86**, 1238-1254, 1996; R.S. Stein, A.A. Barka, and J.H. Dieterich, Progressive failure on the North Anatolian fault since 1939 by earthquake stress triggering, *Geophys. J. Int.*, **128**, 594-604, 1997.

74. K. Rajendran and C. Rajendran, Paleoseismological investigations in Runn of Kachch, India, the site of the large 1819 earthquake, in *Summer School in Active Faulting and Paleoseismology*, M. Meghraoui, ed., European Centre for Geodynamics and Seismology, Luxembourg, pp. 123-124, 1998.

75. L. Seeber, G. Ekström, S.K. Jain, C.V.R. Murty, N. Chandak, and J.G. Armbruster, The 1993 Killari earthquake in central India: A new fault in Mesozoic basalt flows? *J. Geophys. Res.*, **101**, 8543-8560, 1996.

76. A. Crone, M. Machette, and R. Bowman, *Geologic Investigations of the 1988 Tennant Creek, Australia, Earthquakes: Implications for Paleoseismicity in Stable Continental Regions*, U.S. Geological Survey Bulletin 2032A, Reston, Va., p. 51, 1992.

77. M.N. Machette and J.R. Bowman, Episodic nature of earthquake activity in stable continental regions revealed by paleoseismicity studies of Australian and North American Quaternary faults, *Austr. J. Earth Sci.*, **44**, 203-214, 1997.

78. Proprietary software packages for loss estimation were first developed in the private sector. An example was the Early Post-Earthquake Damage Assessment Tool (EPEDAT) developed by EQE International, Inc., which applies intensity-damage relationships over specified zip codes. EPEDAT was developed for the California Office of Emergency Services and has been customized for five southern California counties (Los Angeles, Orange, Riverside, San Bernardino, and Ventura). HAZUS, developed by FEMA, uses PGA, PGV, Sa, and Sv inputs and building capacity-pushover analysis for census tracts.

79. HAZUS was developed by a consortium of university and private sector researchers and is maintained through NIBS. Its trial release in 1994 was followed by extensive efforts to collect input data for the software, to refine the algorithms for calculating damage, to educate state and local planning officials about the need for accurate information to support loss modeling, and to validate the methodology against well-characterized earthquakes. The final report on the methodology was delivered in fall 1997. The currently released version is HAZUS ®99, SR2, available at <http://www.fema.gov/hazus>, and further improvements are continuing.

80. See <www.hazus.org>.

81. The user should also have a good technical understanding of seismic hazards and the earthquake vulnerability of the modeled facilities to create realistic representations of the urban inventory and its fragilities, as well as the geologic hazards. The default data for some of these inputs, which a less skilled user might be tempted to employ, are crude and can lead to unrealistic loss estimates.

82. Western U.S. earthquakes resulting in loss of life were the 1987 Whittier Narrows (M 5.9, 8 deaths), 1989 Loma Prieta (M 7.0, 63 deaths), 1992 Landers (M 7.3, 1 death), and 1994 Northridge (M 6.7, 57 deaths). Other severe earthquakes with no loss of life were 1983 Coalinga (M 6.5), 1992 Petrolia (M 6.9), 1999 Hector Mine (M 7.1), and 2001 Nisqually (M 6.8).

83. A worldwide total of 160,000 is a lower bound from conservative USGS estimates for the 21 earthquakes resulting in more than 1000 deaths each that occurred from October 1983 to January 2001. See <http://neic.usgs.gov/neis/eqlists/eqsmajr.html>.

84. The BSSC (<http://www.bssconline.org>) is an independent, voluntary membership body representing a wide variety of building community interests. It was established in 1979 under the auspices of the NIBS to enhance public safety by providing a national forum to foster improved seismic safety provisions for use by the building community in the planning, design, construction, regulation, and utilization of buildings. The BSSC promotes the development and adoption of seismic safety provisions in building codes suitable for use throughout the United States.

85. The ATC (<http://www.atcouncil.org>) is a nonprofit, tax-exempt corporation established in 1971 through the efforts of the Structural Engineers Association of California. Its mission is to develop state-of-the-art, user-friendly engineering resources and applications for use in mitigating the effects of natural and other hazards on the built environment. ATC also identifies and encourages needed research and develops consensus opinions on structural engineering issues in a nonproprietary format. ATC is guided by a Board of Directors consisting of representatives appointed by the American Society of Civil Engineers, the Structural Engineers Association of California, the Western States Council of Structural Engineers Associations, and four at-large representatives concerned with the practice of structural engineering. Funding for ATC projects is obtained from government agencies and from the private sector in the form of tax-deductible contributions.

86. The Earthquake Engineering Research Institute (<http://www.eeri.org>) was founded in 1949 as an outgrowth of the Advisory Committee on Engineering Seismology of the U.S. Coast and Geodetic Survey. A founding purpose of the institute was to encourage research in the field of earthquake engineering. It is a national, nonprofit, technical society of engineers, geoscientists, architects, planners, public officials, and social scientists with the mission of advancing the science and practice of earthquake engineering and the solution of national earthquake engineering problems to protect people and property from the effects of earthquakes.

87. The American Society of Civil Engineers (<http://www.asce.org/>) is a not-for-profit organization actively involved in developing seismic standards and codes.

88. J.E. Bevers, ed., Theme issue: Seismic isolation, *Earthquake Spectra*, **6**, 161-430, 1999.

89. R.D. Hanson, ed., Theme issue: Passive energy dissipation, *Earthquake Spectra*, **9**, 319-641, 1993; R.D. Hanson and T.T. Soong, *Seismic Design with Supplemental Energy Dissipation Devices*, Earthquake Engineering Research Institute, Monograph MNO-8, Oakland, Calif., 135 pp., 2001.

90. T.T. Soong and M.C. Constantinou, eds., *Passive and Active Structural Vibration Control in Civil Engineering*, Springer-Verlag, Vienna, 380 pp., 1994.

91. The quote is from Seismology Committee, *Recommended Lateral Force Requirements—Commentary*, Structural Engineers Association of California, Sacramento, 203 pp., 1990.

92. *Guidelines for the Seismic Rehabilitation of Buildings* (Building Seismic Safety Council and Applied Technology Council, Federal Emergency Management Agency Report FEMA-273, Washington, D.C., 400 pp., October, 1997). Important precursors were two reports published soon after the Northridge earthquake, *Performance Based Seismic Engineering of Buildings* (Vision 2000 Committee, J. Soulanges, ed., Structural Engineers Association of California, Sacramento, 2 vols., 115 pp., 1995) and *Guidelines for the Seismic Rehabilitation of Buildings* (Building Seismic Safety Council and Applied Technology Council, Federal Emergency Management Agency Report FEMA-273, Washington, D.C., 400 pp., October, 1997).

93. SAC Joint Venture, *Recommended Seismic Design Guidelines for New Steel Moment-Frame Buildings*, FEMA 350, U.S. Government Printing Office: Washington, D.C., 207 pp., 2000.

94. K. Olsen, Site amplification in the Los Angeles Basin from three-dimensional modeling of ground motion *Bull. Seis. Soc. Am.*, **90**, S77-S94, 2000.

95. A. Cornell and H. Krawinkler, Progress and challenges in seismic performance assessment, Pacific Earthquake Engineering Research Center, *PEER Center News*, **3**, 1-3, 2000.

96. J. Milne, *Earthquakes and Other Earth Movements,* D. Appelton and Company, New York, p. 304, 1886.

97. National Research Council, *Real-Time Earthquake Monitoring*, National Academy Press, Washington, D.C., 52 pp., 1991. See also H. Kanamori, E. Hauksson, and T.H. Heaton, Real-time seismology and hazard mitigation, *Nature*, **390**, 461-464, 1997.

98. J.M. Espinosa Aranada, A. Jiménez, G. Ibarrola, F. Alcantar, A. Aguilar, M. Inostroza, and S. Maldanado, Mexico City seismic alert system, *Seis. Res. Lett.*, **66**, 42-53, 1995.

99. W.H. Bakun, F.G. Fischer, E.G. Jensen, and J. VanSchaack, Early warning system for foreshocks, *Bull. Seis. Soc. Am.*, **84**, 359-365, 1994.

100. In California, warning systems were first used to alert rescue workers about aftershocks following the Loma Prieta earthquake.

101. The centerpiece of this upgrade is the new installation of the TriNet network of 170 broadband sensors and 700 strong-motion instruments in southern California. The operational goal is that the system will detect all events to $M > 1.8$ and that all of the data will be available in real time for hazard mitigation purposes.

102. D.J. Wald, V. Quitoriano, T.H. Heaton, H. Kanamori, C.W. Scrivener, and B. Worden, Trinet "shakemaps": Rapid generation of peak ground motion and intensity maps for earthquakes in Southern California, *Earthquake Spectra*, **15**, 537-555, 1998.

103. Such capabilities could be of great value following an earthquake when communication and transportation are difficult. Examples include improved coordination of the response of firefighting and medical efforts, as well as routing and prioritizing the overload of telephone calls in the critical hours after an earthquake.

104. Office of Technology Assessment, *Reducing Earthquake Losses*, OTA-ETI-623, U.S. Government Printing Office, Washington, D.C., 162 pp., 1995.

105. One example of this holistic approach to hazard mitigation is the Natural Hazards Center of the University of Colorado, which is a national clearinghouse for information on natural hazards mitigation, with emphasis on social and political aspects (<www.colorado.edu/hazards>).

106. See <http://peer.berkeley.edu/lifelines/>.

107. See <http://www.cityofseattle.net/projectimpact/>.

108. The NSF Directorate of Engineering funds the Pacific Earthquake Engineering Research Center (<http://peer.berkeley.edu/>), the Multidisciplinary Center for Earthquake Engineering Research (<http://mceer.buffalo.edu/>), and the Mid-America Earthquake Center (<http://mae.ce.uiuc.edu/>).

109. NEES will provide real-time remote access to a complete set of testing and experimental facilities, making them widely available to earthquake engineers. The on-line network, or "collaboratory," will furnish researchers across the country with shared-use access to advanced equipment, databases, and computer modeling and simulation tools (<http://www.eng.nsf.gov/nees/>).

110. The NSF and USGS sponsor the Southern California Earthquake Center, which involves 40 universities, government laboratories, and other public and private research organizations. The USGS also maintains major centers for earthquake research in Menlo Park, California, and Golden, Colorado.

4

Observing the Active Earth: Current Technologies and the Role of the Disciplines

Not long ago, seismologists worked in rooms filled with drum recorders and big tables for hand-measuring seismograms. They now use digital monitoring systems that integrate high-performance seismometers, real-time communications, and automatic processing to produce high-quality information on seismic activity in near real time. Geodesists have replaced the theodolite and spirit level with space-based positioning and deformation imaging that can map crustal movements precisely and continuously, and they can hunt for slow, silent earthquakes with arrays of sensitive, stable strainmeters. Geologists have learned to decipher the subtle features of the rock record that mark pre-historic earthquakes, and they can date these events precisely enough to reconstruct the space-time behavior of entire fault systems. Laboratory and field scientists who study the microscopic processes of rock deformation are now formulating and calibrating the scaling laws that relate their reductionistic approach to the nonlinear dynamics of macroscopic faulting in the real Earth.

In each of these four domains—seismology, geodesy, geology, and rock mechanics—key technological innovations and conceptual breakthroughs were made within the last decade. The Global Seismic Network (GSN), initiated with the founding of the National Science Foundation (NSF)-sponsored Incorporated Research Institutions for Seismology (IRIS) in 1984, is reaching its design goal of 128 broadband, high-dynamic-range stations (as of December 2001, 126 stations had been installed and 122 were operational). The first continuously recording network of Global Positioning System (GPS) stations for measuring tectonic deformation was installed in Japan in 1988 by the National Research Institute for Earth Science and Disaster Prevention (*1*), and the first image of earthquake faulting using interferometric synthetic aperture radar (InSAR) was constructed in 1992. Paleoseismologists produced a preliminary 1000-year history of major ruptures on the San Andreas fault in 1995 and discovered a prehistoric moment magnitude (M) 9 earthquake in the Cascadia subduction zone in 1996. The first three-dimensional simulations of dynamic fault ruptures using laboratory-derived, rate- and state-dependent friction equations were run in 1996.

The unprecedented flow of new information opened by these advances is stimulating research on many fronts, from fault-system dynamics and earthquake forecasting to wavefield modeling and the prediction of strong ground motions. This chapter summarizes the state of the art in the main observational disciplines; it focuses on new technologies for observing the active Earth, and it highlights through a few examples the richness of the data sets now becoming available for basic and applied research.

4.1 SEISMOLOGY

Seismology lies at the core of earthquake science because its main concern is the measurement and physical description of ground shaking. The central problem of seismology is the prediction of ground motions from knowledge of seismic-wave generation by faulting (the earthquake source) and the elastic medium through which the waves propagate (Earth structure). In order to do this calculation (forward problem), information must be extracted from seismograms to solve two coupled inverse problems: imaging the earthquake source, as represented by its space-time history of faulting, and imaging Earth structure, as represented by three-dimensional models of seismic-wave speeds and attenuation parameters. Because seismic signals can be recorded over such a broad range of frequencies—up to seven decades (2)—seismic signals can be used to observe earthquake processes on time scales from milliseconds to almost an hour, and they provide information about elastic structure at dimensions ranging from centimeters to the size of the Earth itself.

Seismometry

Seismic waves span a wide range of amplitude, as well as frequency. The ground motions in the vicinity of a large earthquake can have velocities greater than 1 meter per second and accelerations exceeding the pull of gravity ($1g = 9.8$ m/s^2). The lower limit of seismic detection is typically eight orders of magnitude smaller, set by the level of the ambient ground noise (3). No single sensor has yet been developed that can faithfully record the violent displacements close to an earthquake and still be capable of detecting small events at the background noise level. For this reason, instruments historically have been divided into weak-motion and strong-motion seismometers. The former have been the principal sensors for studies of Earth structure and remote earthquakes by seismologists (4), while the latter have provided the principal seismological data to earthquake engineers. Technology is closing this gap. Modern force-feedback systems (5) can faithfully record ground motions from the lowest ambient noise at quiet sites to the largest earthquakes at teleseismic distances and achieve a bandwidth that extends from free oscillations with periods of tens of minutes to body waves with periods of tenths of seconds (Figure 4.1).

Seismic Monitoring Systems

Seismic monitoring systems comprise three basic elements: a network of seismometers that convert ground vibrations to electrical signals, communication devices that record and transmit the signals from the stations to a central facility, and analysis procedures that combine the signals from many stations to identify an event and estimate its location, size, and other characteristics. Monitoring systems are multiple-use facilities; they furnish information about earthquakes and nuclear explosions to operational agencies in near real time, and they also function as the basic data-gathering mechanisms for long-term research and education. With current technology, seismic networks of different types and spatial scales must be deployed to register the Earth's seismicity over its complete geographic and magnitude range (Table 4.1, Figure 4.2). Since this coverage is typically overlapping, monitoring systems can be effectively organized into nested structures.

Global Seismic Networks State-of-the-art seismic stations for global seismic networks comprise three-component sensors with high dynamic

TABLE 4.1 Scales of Seismic Monitoring

Type	Typical Network Size	Typical Station Spacing	Detection Threshold[a]
Global	Global	1000 km	4.5
Regional	500 km	25-50 km	2.0
Local	10 km	<1000 m	–1.0

[a]Magnitude of smallest event with a high probability of detection; see examples in Figure 4.2.

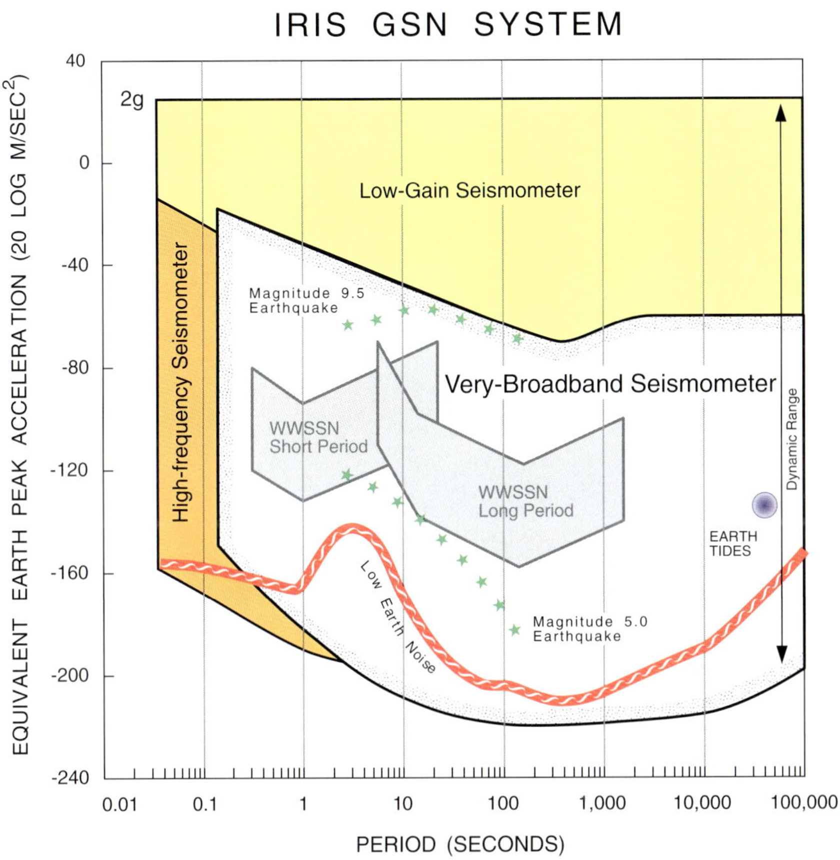

FIGURE 4.1 Plot of acceleration amplitude versus frequency, showing effective bandwidth and dynamic range for current broadband, high-frequency, and low-gain (strong-motion) digital seismometers, as well as for the older short-period and long-period analog seismometers of the World Wide Standardized Seismographic Network. Broadband instruments are capable of faithfully recording ground motions ranging from ambient noise at quiet sites (line labeled *low Earth noise*) to the peak accelerations generated by an M 9.5 earthquake at an epicentral distance of 3000 kilometers (upper line of stars). Low-gain seismometers are needed to record ground accelerations in the damage zones of large earthquakes, which can exceed 1*g*, and high-frequency seismometers are needed for periods less than about 0.1 second. SOURCE: R. Butler, IRIS.

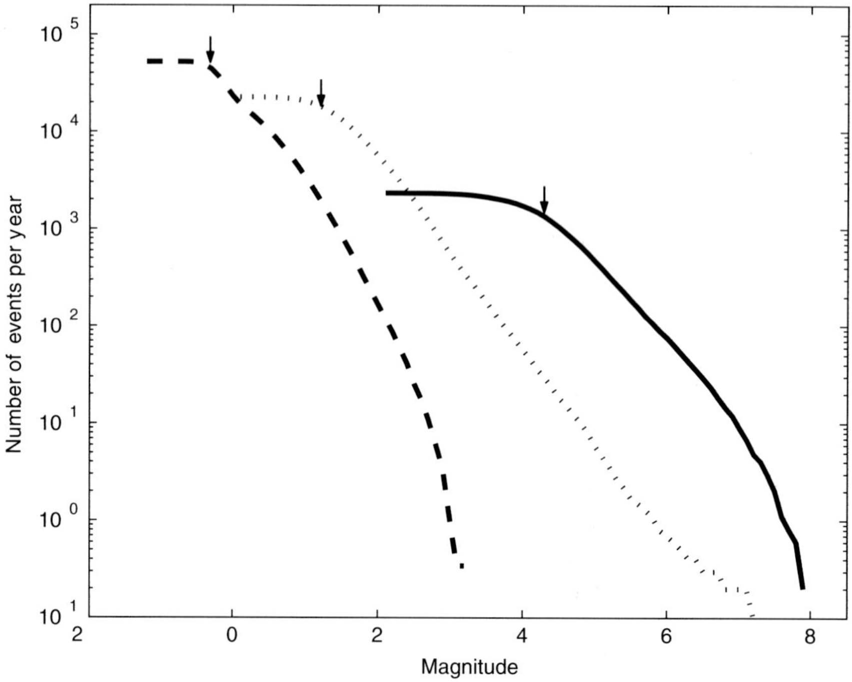

FIGURE 4.2 Number of earthquakes per year greater than a specified magnitude recorded by three networks of the types described in Table 4.1. Solid line shows the seismicity of the entire Earth from the global network of seismic stations cataloged by the International Seismological Centre during the decade 1990-1999. Dotted line is for events in southern California during the same interval recorded by the Southern California Seismic Network. Dashed line is for mining-induced seismicity recorded during 1997-1999 by a local network in the Elandrands gold mine, South Africa. Arrows show the approximate detection thresholds for the three networks, below which the sampling of seismicity is incomplete. All magnitudes are moment magnitudes. SOURCE: M. Boettcher, E. Richardson, and T.H. Jordan, University of Southern California.

range (up to 140 decibels) and broadband response (0.0001–10 hertz). Since 1984, more than 300 such stations have been installed at permanent locations worldwide, as elements of global and regional networks (Figure 4.3). Close to half of these stations are part of the GSN, which has been constructed and operated under a cooperative agreement between the U.S. Geological Survey (USGS) and IRIS (*6*). The GSN is coordinated with other international networks through the Federation of Digital Seismographic Networks (FDSN), and the data are archived and made available

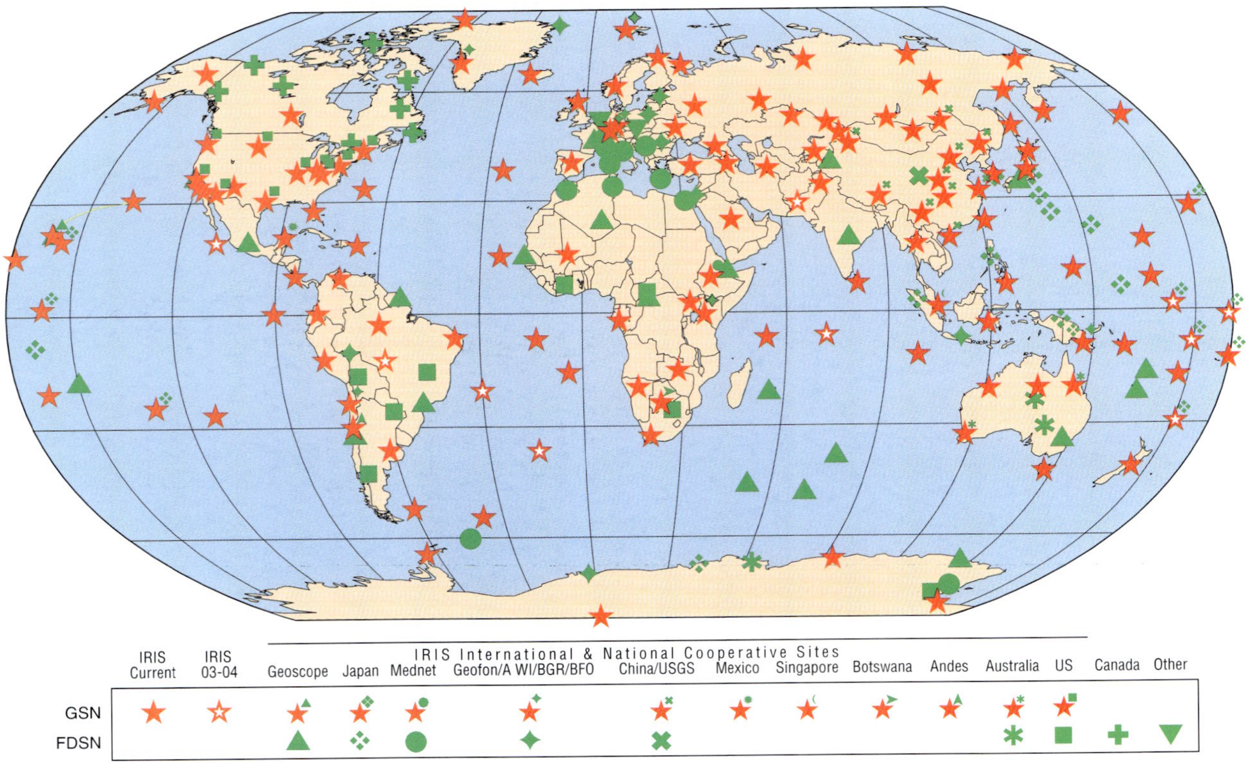

FIGURE 4.3 Current global distribution of broadband high-performance seismic stations. The Global Seismic Network is part of the Federation of Digital Seismographic Networks (FDSN). SOURCE: IRIS-FDSN.

on-line by the IRIS Data Management Center (DMC) in Seattle, Washington (7). Some stations still record on local magnetic or optical media that are shipped periodically to the DMC, but direct telemetry is being deployed as communication with remote sites becomes cheaper. At many locations with telephone access, the data can be retrieved via telephone dial-up or Internet connection (108 stations in 2001). The five-year goal is to have all stations on-line all the time. Achieving this goal, especially at remote sites, will depend in part on the cost of satellite communications.

The GSN data acquired over the last 15 years have facilitated many advances in the study of global Earth structure and earthquake sources. Seismic tomography has provided dramatic images of subducting slabs, plume-like upwellings, and other features of the mantle convective flow responsible for plate-tectonic motions (Figure 4.4). The GSN data have also improved the plate-tectonic framework for understanding earthquake hazards through better earthquake locations and centroid moment tensor (CMT) solutions (Figure 4.5). Seismologists have used the broadband waveforms to elucidate the details of rupture processes during large earthquakes from a variety of tectonic settings, shedding new light on the geologic and dynamic factors that govern the configuration of seismogenic zones and how earthquakes start and stop.

These successes have in no way diminished the need for continued monitoring. Discoveries based on data now being collected by the GSN will undoubtedly continue into the indefinite future. On the rapidly slipping plate boundaries, large earthquakes recur at intervals ranging from decades to centuries, while the recurrence times for significant intraplate events can extend to many millennia. With each passing year, GSN data will thus add new information to the evolving pattern of global seismicity by the direct observation of large, rare events and the delineation of low-level seismicity that may mark the eventual occurrence of such events. The densification of seismic sources through time will also improve tomographic mapping of features in the crust and mantle that control seismicity and may be indicative of the forces causing lithospheric faulting.

Global seismological monitoring could be further enhanced by increasing the spatial resolution on land with permanent and temporary deployments of seismometers, expanding the coverage of global networks to the ocean floor, and upgrading the present networks as new technologies become available. However, sustained funding of the global networks will present a continuing challenge. In terms of annualized expenditures, the operation and maintenance of the GSN is projected to be comparable to its initial capitalization. Under current arrangements, the USGS shares a portion of the costs of GSN operations with the NSF. Stable support of the GSN from a federal agency that embraces the mission of global seismic monitoring is essential to the long-term health of earthquake science.

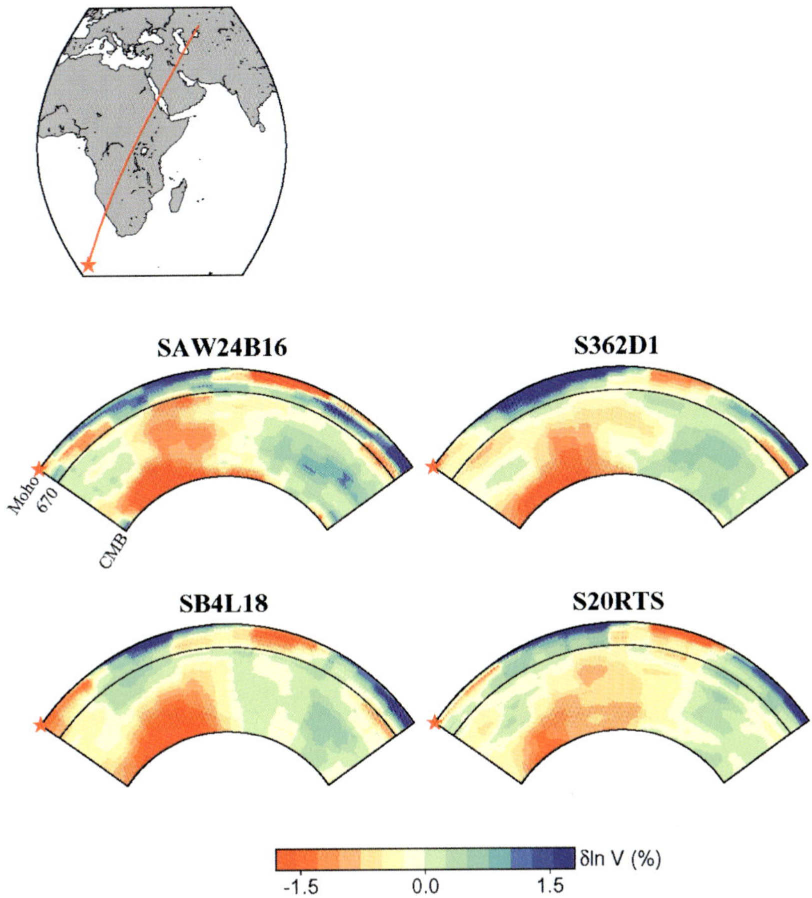

FIGURE 4.4 Depth cross section showing the African plume as seen in four recent global tomographic models of *S* velocity: SAW23B16 (Megnin and Romanowicz, 2000), SB4L18 (Masters et al., 1999), S362D1 (Gu et al., 2001), and S20RTS (Ritsema et al., 1999). Although the models differ in detail, they are in agreement on the broad characteristics of this major upwelling. SOURCE: Y.J. Gu, A.M. Dziewonski, W.-J. Su, and G. Ekström, Models of the mantle shear velocity and discontinuities in the pattern of lateral heterogeneities, *J. Geophys. Res.*, **106**, 11,169-11,199, 2001; G. Masters, H. Bolton, and G. Laske, Joint seismic tomography for P and S velocities: How pervasive are chemical anomalies in the mantle? *Eos, Trans. Am. Geophys. Union*, **80**, S14, 1999; C. Megnin and B. Romanowicz, The 3D shear velocity structure of the mantle from the inversion of body, surface and higher mode waveforms, *Geophys. J. Int.*, **143**, 709-728, 2000; J. Ritsema, H. van Heijst, and J. Woodhouse, Complex shear wave velocity structure imaged beneath Africa and Iceland, *Science*, **286**, 1925-1928, 1999. SOURCE: B. Romanowicz and Y. Gung, University of California, Berkeley.

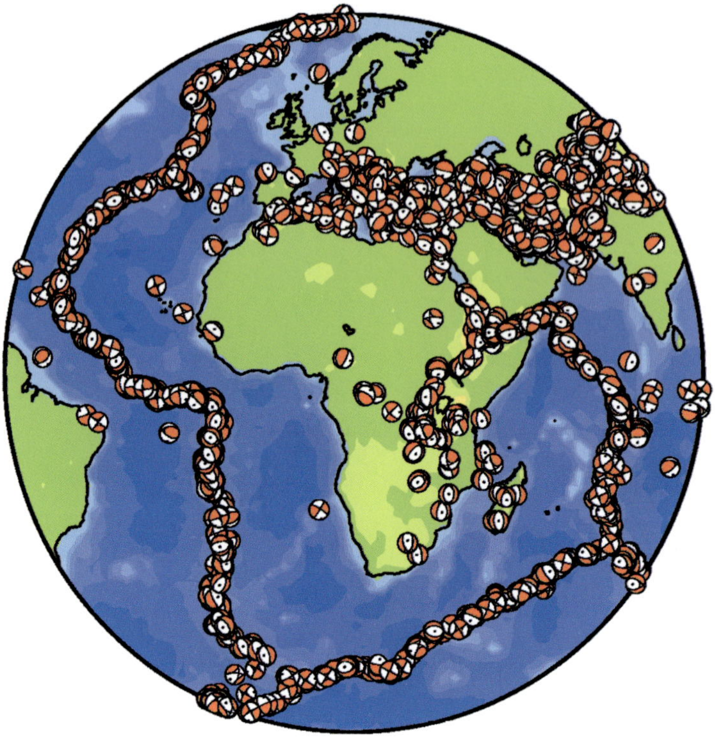

FIGURE 4.5 25 years of CMT solutions (1976-2000) for the region surrounding Africa. The availability of broadband data has made it possible to describe global seismicity in terms not only of location and magnitude, but also of fault mechanism, thereby greatly enhancing our view of active tectonics. SOURCE: Harvard CMT group.

Regional Seismic Networks Owing to their sparse station coverage, global networks do a poor job of detecting and locating events with magnitudes less than about 4.5 (Figure 4.2), and their sampling is too crude for investigating how waves are produced by fault ruptures, especially the near-fault radiation that generates the complex patterns of strong ground motions observed in large earthquakes. To deal with these problems, seismologists have densified station arrays in areas of high (or otherwise interesting) seismicity. Regional networks are collections of seismographic stations distributed over tens to hundreds of kilometers, usually as permanent facilities. The information supplied by regional networks services three overlapping but distinct communities: (1) scientists and engineers engaged in basic and applied research; (2) engineers, public officials, and other decision makers charged with the management of earthquake risk and emergency response; and (3) public safety officials, news media, and the general public. As information technology has transformed the regional networks into integrated monitoring systems, they have become centers for educating the general public about earthquake hazards, as well as key facilities for training graduate students in seismology (*8*).

The short-period, high-gain instruments historically used in regional networks (*9*) brought seismicity patterns into much clearer focus (Figure 4.6), but the dynamic range of these instruments was too low to furnish useful recordings of large regional events. In the last decade, deployments of broadband, high-dynamic-range seismometers have begun to transform the regional networks into much more powerful tools for investigating the basic physics of the earthquake source, the detailed structure of the Earth's crust and deep interior, and the patterns of potentially destructive ground motions. With these data, seismologists can now map the patterns of slip during earthquakes using seismic tomography, just as they map Earth structure. Images of fault ruptures during the more recent earthquakes in the Los Angeles, San Francisco, and Seattle regions have all been captured by high-performance networks (Figure 4.7).

Long-term funding has been a persistent problem for regional network operators, and new investments in equipment are badly needed (*10*). In particular, the implementation of new broadband technologies in regional monitoring has been lagging in the United States, especially when compared to the investments made by other high-risk countries such as Japan (Box 4.1) and Taiwan. Two exceptions are the Berkeley Digital Seismic Network in northern California and Caltech's TERRAscope Network in southern California. Both are equipped with a combination of three-component broadband seismometers and three-component strong-motion accelerometers; they have digital station processors and feed continuous data streams via real-time telemetry to central processing sites. Although these networks have developed independently, a major effort is under way, with some support from the State of California, to modernize the earthquake monitoring infrastructure throughout the region by integrating the regional networks into a California Integrated Seismic Network monitoring in the United States.

Local Networks Networks have been deployed with seismometers distributed over a few tens of kilometers or less for specialized purposes such as seismic monitoring of critical facilities (e.g., dams and nuclear power plants) or localized source zones (e.g., volcanoes or geothermal reservoirs). Local networks are important instruments for the study of natural earthquake laboratories such as deep mines. Digital arrays of very

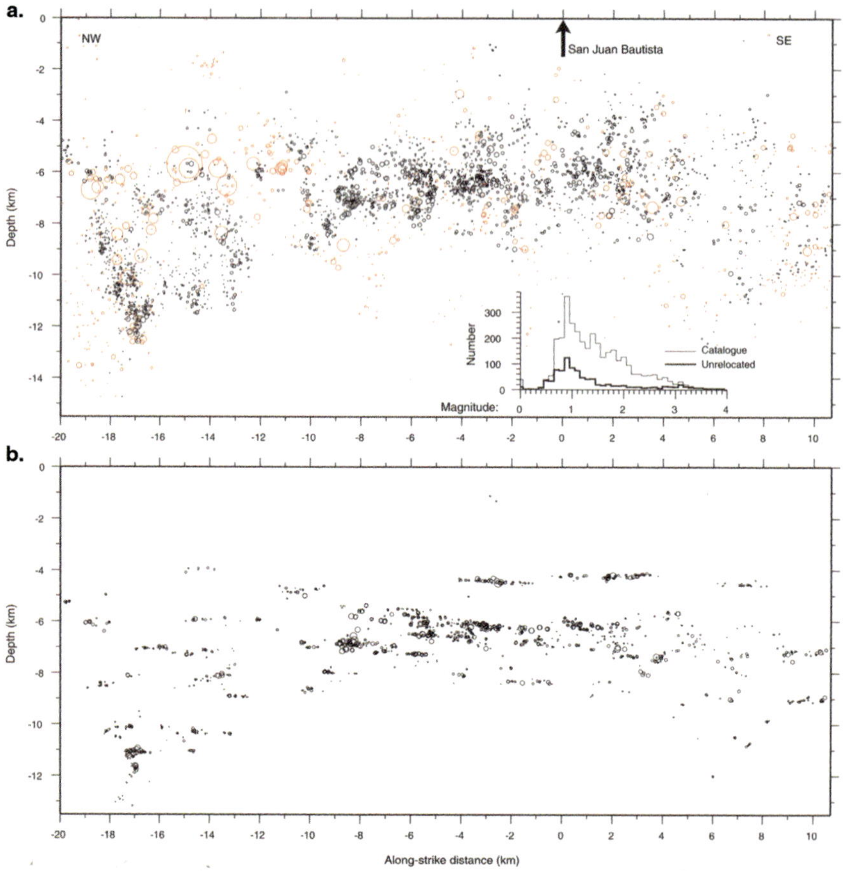

FIGURE 4.6 Vertical along-strike cross sections of earthquakes located along a 30-kilometer section of the San Andreas fault. Circles are an estimate of event size. (a) Locations of earthquakes between May 1984 and December 1997 reported in the Northern California Seismic Network catalog. (b) Relocated earthquakes show a linear, nearly horizontal pattern of seismicity. SOURCE: A.M. Rubin, D. Gillard, and J.-L. Got, Streaks of microearthquakes along creeping faults, *Nature*, **400**, 635-641, 1999. Reprinted by permission from Nature copyright 1999 Macmillan Publishers Ltd.

high frequency sensors have been deployed in deep mines in Canada, Poland, and South Africa to monitor mine tremors and rock bursts induced by mining activities (*11*), and they have furnished unique, close-in observations of earthquakes as large as M 5 and at depths as great as 4 kilometers. Recent research has shown that in the deep gold mines of South Africa, mine tremors caused by friction-controlled slip on faults

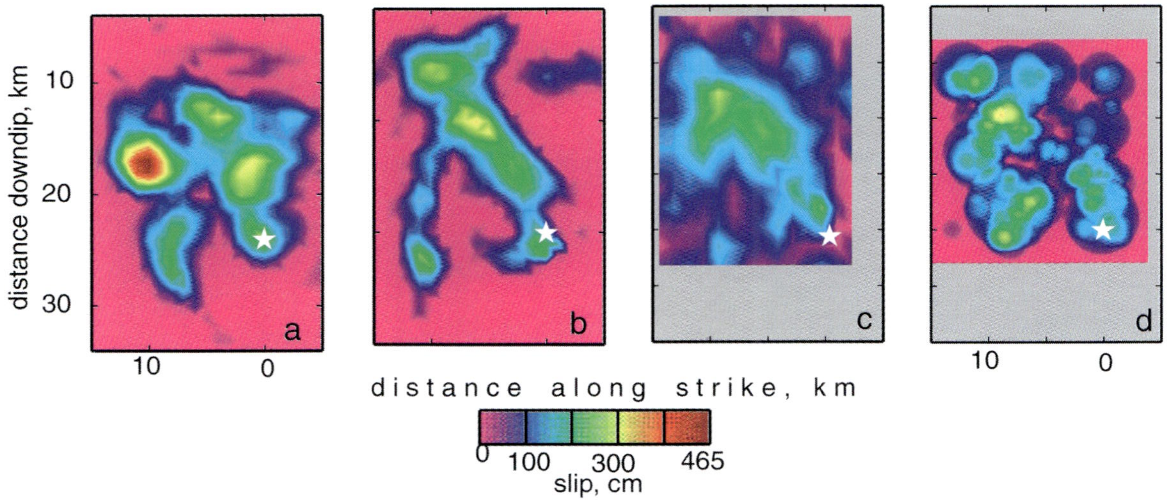

FIGURE 4.7 Comparison of fault slip maps for the 1994 Northridge earthquake derived from different methods. (a) Inversion of regional distance broadband data (Dreger, 1997). (b) Inversion of seismic moment rate functions derived from empirical Green's function deconvolution (Dreger, 1994). (c) Inversion of local strong motion, teleseismic waveform data, and geodetic and leveling data (Wald et al., 1996). (d) Inversion of strong motion data alone (Zeng and Anderson, 1996). Each map has the same scaling, and the white star identifies the hypocenter. SOURCE: D. Dreger, Empirical Green's Function study of the January 17, 1994 Northridge, California earthquake, *Geophys. Res. Lett.*, **21**, 2633-2636, 1994; D. Dreger, The large aftershocks of the Northridge earthquake and their relationship to mainshock slip and fault zone complexity, *Bull. Seis. Soc. Am.*, **87**, 1259-1266, 1997; D.J. Wald, T.H. Heaton, and K.W. Hudnut, The slip history of the 1994 Northridge, California, earthquake determined from strong-motion, teleseismic, GPS, and leveling data, *Bull. Seis. Soc. Am.*, **86**, S49-S70, 1996; Y. Zeng and J.G. Anderson, A composite source modeling of the 1994 Northridge earthquake using Genetic Algorithm, *Bull. Seis. Soc. Am.*, **86**, 71-83, 1996. SOURCE: D. Dreger, University of California, Berkeley.

BOX 4.1 Seismic Infrastructure in Japan

Japan has been at the forefront of seismic monitoring since instrumental observations of earthquakes began in the 1870s. The Japan Meteorological Agency (JMA) operates a national network that provides essential data for the study of earthquake sources and seismotectonics throughout the Japanese islands.[1] A number of local networks are operated by universities and other institutions, such as the National Research Institute for Earth Science and Disaster Prevention of the Science and Technology Agency, primarily for research on microseismicity and earthquake prediction. The Earthquake Prediction Data Center of the Earthquake Research Institute, University of Tokyo, receives hypocenter and arrival-time data from member universities and compiles them into two databases, one for real-time analysis and a revised one for archival purposes. More than 700 stations are currently operational, making the detection and location of all earthquakes of $M > 2$ possible almost everywhere in the country.

A distinctive feature of earthquake monitoring in Japan has been the systematic collection of observations on the intensity of seismic shaking, a tradition that dates back to 1884. For many years, intensities were estimated by the observers on duty at meteorological stations, but this procedure had several problems: the observations were too subjective and inconsistently reported, they often disagreed with ground motions reported by the public, and they were not suitable for rapid dissemination. The inadequacies of this system were made clear during the 1995 Hyogo-ken Nanbu earthquake (see Box 2.4). After that disaster, the intensity scale used in Japan was revised and redefined on the basis of instrumental measurements,[2] and suitable strong-motion instruments were deployed at 600 sites with approximately 20-kilometer spacing. The high density of this new national system provides adequate sampling of the rapid geographic variations in the ground motions typically observed for large earthquakes. Immediately after an event the instruments automatically send out parametric data to a central computer, which combines them and rapidly produces intensity maps of the seismic shaking. A number of counties, cities, and private organizations are also deploying arrays of digital strong-motion instruments; at last count, there were more than 1000 such instruments linked to central sites by real-time telemetry.

Over the last several years, a Japanese initiative has focused on the deployment of a dense network of state-of-the-art broadband, high-dynamic-range instruments for the purpose of research on earthquake source processes and global Earth structure. Begun as an unofficial collaboration among several university groups, the Ocean Hemisphere Project (OHP) initiative was officially inaugurated in 1997. The OHP includes provisions for seismic, gravity, and geomagnetism observations. Its goal is to deploy ocean-bottom stations as well as land-based instruments not only in Japan but, in cooperation with neighboring countries, throughout the western Pacific region.

---

[1]The information on Japanese seismic networks has been summarized by M. Ohtake and Y. Ishikawa, Seismic observation networks in Japan, *J. Phys. Earth*, **43**, 563-584, 1995. Additional material for this box came from *Report on the Investigation of System of Seismic Observation and Dissemination of Seismic Information in the Pacific Countries*, Japan Meteorological Agency, Tokyo, 80 pp., 1997 (courtesy of Dr. Kohichi Uhira).

[2]Japan Meteorological Agency, *Note on the JMA Seismic Intensity*, JMA Report, *Gyosei*, 1996.

have a lower cutoff near M 0, consistent with the minimum nucleation size of earthquakes implied by laboratory data (*12*).

The USGS and other institutions maintain special arrays of surface and borehole instrumentation on the San Andreas fault at Parkfield, California, as part of a long-term, multidisciplinary program for the study of earthquake processes at the transition between the creeping and locked sections of the fault (*13*) (Figure 2.15). Special arrays of surface and borehole instrumentation have furnished insight into seismogenic processes at scales much smaller than typical seismological investigations (Figure 4.6). For example, results from microearthquake and controlled-source Vibroseis studies using data from the High-Resolution-Seismic-Network provide a picture of a fault zone that is highly heterogeneous in seismic velocity structure (*14*), in the distribution and spatial clustering of microearthquakes (*15*) (Figure 4.6), and in the generation of fault-zone trapped seismic waves. These studies reveal structural detail at depth that is highly correlated with the transition from creeping to locked behavior inferred from surface observation, and they indicate temporal changes in propagation, seismicity, and slip rate at depth that correlate with deformation and water-level changes observed at and near the surface (*16*). On a finer scale, precise relative relocations of the microseismicity using waveform correlation techniques are revealing constellations of earthquakes and the detailed distribution of fault slip at depth (*17*). They have also yielded a surprising and strikingly detailed picture of the strength, strength distribution, and evolution of the deep San Andreas fault; the scaling of the earthquake source (*18*); and the strain accumulation on the Parkfield locked zone at depth. The discovery of numerous characteristically repeating microearthquake sequences at Parkfield has contributed significantly to the development of earthquake recurrence models currently being used to estimate earthquake hazard in California (*19*). Owing to the enhanced understanding of earthquake processes achieved through these observations, the Parkfield natural laboratory has been chosen as the site for the San Andreas Fault Observatory at Depth (SAFOD), a component of the EarthScope initiative that will use deep drilling to conduct in situ investigations of the San Andreas fault zone at seismogenic depths of 3 to 4 kilometers.

U.S. National Seismic Network A notable advance in earthquake monitoring has been the construction of a new U.S. National Seismic Network (USNSN), managed by the USGS National Earthquake Information Center in Golden, Colorado. A central objective is to transform the regional networks into highly automated seismic information systems, capable of broadcasting refined information about seismic ruptures and shaking in near real time to a wide audience concerned with emergency response to earthquake disasters. The idea for the USNSN dates back nearly 30 years (*20*); the concept was to complement the relatively dense coverage provided in selected areas by the regional seismic networks with a well-distributed but sparse permanent network of three-component, broadband stations. The USNSN currently maintains 32 complete broadband stations and some equipment at 96 cooperative broadband stations in North America (7 in Canada) from which it acquires real-time data. It also acquires real-time data from 82 short-period stations, 30 foreign broadband stations, and another 62 stations worldwide. Through participation of the Advanced National Seismic System (ANSS) and the planned EarthScope program, the USNSN will be expanded to 100 permanent broadband stations in North America and will serve as the "backbone" for both programs. Ten of the new stations will be built to GSN standards and, thus, be capable of high-quality recording at the low frequency of the Earth's free oscillations.

Strong-Motion Seismology

Accurate recordings of strong motions near earthquake sources are crucial to both earthquake engineering and science, because they provide the forcing functions for structural design and testing, as well as valuable information on earthquake source processes. The motions are registered by triggered, three-component, low-gain accelerographs located at free-field sites and housed in important structures, such as dams, bridges, and high-rise buildings. Accelerographs are capable of recording 2*g* accelerations in the frequency band from 0.1 to 10 hertz. The attenuation relations derived from the free-field data are key components of seismic hazard analysis and mapping, while the housed recordings furnish ground truth for structural performance during earthquakes.

The USGS oversees a national network of about 900 strong-motion accelerographs through the National Strong-Motion Program (NSMP). The NSMP coordinates data collection by a variety of federal, state, and local agencies, companies, and academic institutions (*21*). The California Geological Survey (CGS) operates the California Strong-Motion Instrument Program with basic funding provided by state tax on permits for new construction; it comprises 910 analog and digital accelerographs in California, 255 of which are in extensively instrumented structures. Strong-motion databases are maintained by both the USGS and the CGS, as well as by the Southern California Earthquake Center (SCEC) and the Pacific Earthquake Engineering Research (PEER) Center (*22*). Coordination of the various organizations that collect, process, and distribute strong-motion data has been a long-standing issue (*23*), but the situation has benefited substantially from on-line access now offered by all data centers and the virtual strong-motion database (in fact, a meta-database) recently set up by the Consortium of Organizations for Strong-Motion Observation Systems. However, as in broadband regional seismology, the U.S. effort falls short of the Japanese, who have created a database system called Kyoshin Net (K-Net), managed by the National Research Institute for Earth Science and Disaster Prevention, to archive and distribute data from the dense array (25-kilometer spacing) of 1000 digital strong-motion stations deployed throughout Japan (*24*). In Taiwan, a strong-motion network of 614 stations provided unprecedented strong ground-motion data during the 1999 Chi-Chi earthquake.

The 1999 Izmit, Turkey, and Chi-Chi, Taiwan, earthquakes (M 7.4 and 7.6, respectively) have substantially increased the number of strong-motion records for large earthquakes, allowing detailed mapping of the ruptures in time and space (Figure 4.8). Yet, despite more than 70 years of strong-motion seismology, the data coverage remains poor. There are few strong-motion recordings for subduction-zone earthquakes with magnitude greater than 8 and none for magnitude greater than 9. Intraslab earthquakes of M 7 and larger are also poorly sampled, yet they pose a substantial hazard to major cities around the world, as evidenced in the 2001 El Salvador earthquake (M 7.6) and the 1949 (M 7.1), 1965 (M 6.5), and 2001 (M 6.8) events beneath the Seattle-Tacoma metropolitan area. Likewise, there are no close-in recordings (closer than 50 kilometers) of intraplate earthquakes in the central and eastern United States for magnitudes greater than about 5.2 and few worldwide for interplate earthquakes with magnitudes greater than about 7.3. The improved national monitoring structure planned in the framework of the ANSS is clearly needed to remedy this situation (see Chapter 6).

Portable-Array Studies

Portable arrays of seismometers augment the data from permanent monitoring networks by increasing the recording of seismicity in reconnaissance studies and during periods of anomalous activity, including aftershock sequences and swarms. They are also used to image the architecture of fault systems and other aspects of crustal structure, such as sedimentary basins, that affect the amplitude and duration of strong motions. Until recently, this mode of operation was limited to short-period seismometers with low dynamic range, but large pools of broadband instruments are now efficiently organized within the IRIS Program of Array Seismic Studies of the Continental Lithosphere (PASSCAL) (*25*) and the USGS (*26*). Subsets are available for deployment after a major earthquake in a coordinated effort called the Rapid Array Mobilization Program (RAMP). These deployments have been used to determine the

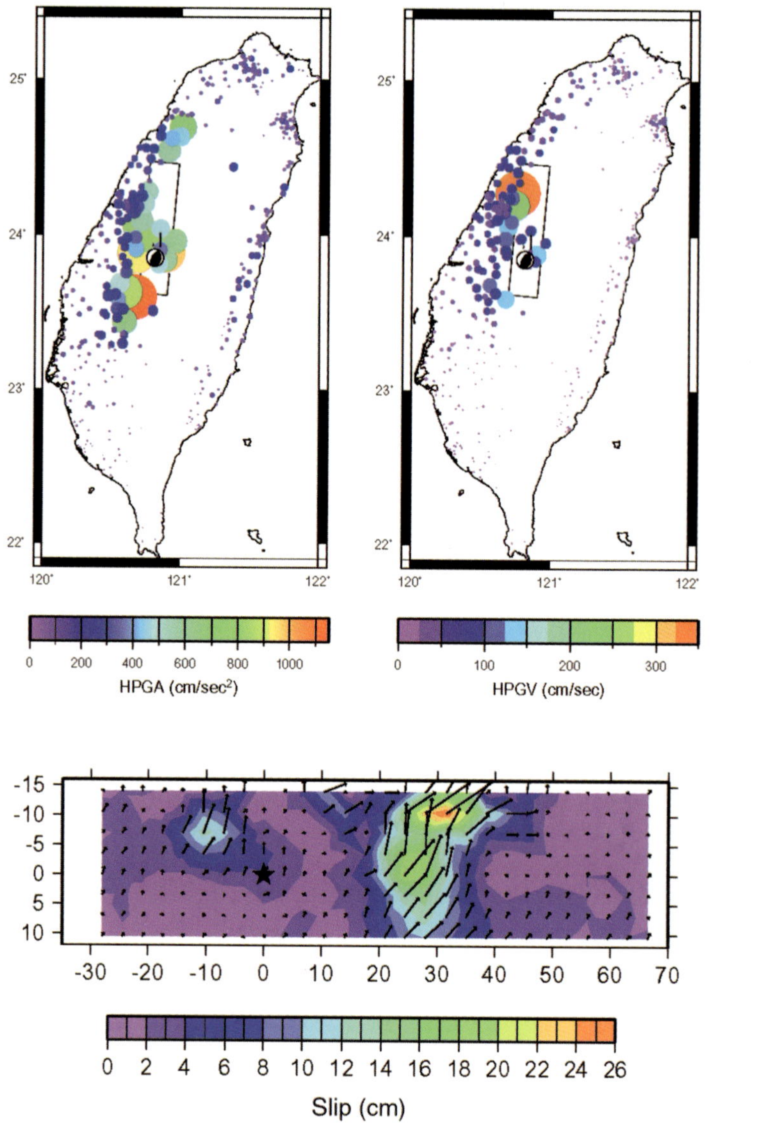

FIGURE 4.8 Peak ground acceleration (*top left*) and peak ground velocity (*top right*) of the 1999 Chi-Chi earthquake. The station coverage of the Central Weather Bureau Seismic Network is still poor in relatively inaccessible central mountainous areas. *Bottom*: Depth cross section showing the slip distribution derived from this data set using a finite-fault inversion methodology. SOURCE: W.-C. Chi, D. Dreger, and A. Kaverina, Finite source modeling of the 1999 Taiwan (Chi-Chi) earthquake derived from a dense strong motion network, *Bull. Seis. Soc. Am.*, **91**, 1144-1157, 2001. Copyright Seismological Society of America.

source parameters of aftershocks and their relationships to the main shocks—important data for studies of rupture propagation, postseismic relaxation, and stress transfer. Recordings of aftershocks have also begun to elucidate the causes of anomalous ground shaking and damage concentration, including basin resonance, basin-edge effects, and Moho reflection (see Section 3.1). Various forms of telemetry are making it possible to monitor state of health and to retrieve ground-motion data in near real time, allowing portable arrays to be integrated with permanent seismic monitoring systems for a wide range of seismic applications.

Imaging the Earth

Investigations of Earth structure have always figured prominently in the study of earthquakes because they frame the interpretation of seismograms in terms of source processes. Indeed, the problems of discovering the space-time structure of faulting and the three-dimensional variations in the Earth's elastic properties are strongly coupled and must be worked out together, either through joint inversion of the seismograms or iteratively through successive approximations. The primary seismological parameters needed to specify Earth structure are the local speeds of the two basic types of seismic waves, compressional (v_p) and shear (v_s), their associated attenuation factors, and the mass density (27). The variations in Earth structure that can be resolved are limited by the size and spacing of the seismic array and the distribution of seismic sources used to illuminate the array. Global networks can therefore determine worldwide structure at relatively low spatial resolution (Figure 4.4), whereas regional and local networks give finer details but only within more limited volumes of the Earth (Figure 4.9).

Portable arrays are useful in enhancing the structural resolution at spatial scales below the station spacing of permanent arrays. They can be deployed in two basic modes of observation: (1) to record artificial sources—explosions, mobile ground-shaking devices such as Vibroseis, or marine air guns—by high-frequency sensors (active-source experiments) and (2) to record signals from natural events, either regional or teleseismic earthquakes (passive experiments). PASSCAL experiments use both modes. Shallow structure (in the upper 2000 meters) can be imaged with highly portable, multichannel systems that record waves reflected from subsurface discontinuities, using hammer blows or small charges as sources. For example, in the Los Angeles Region Seismic Experiment (LARSE), researchers used air guns and explosions to construct images of the subsurface structure that may lead to a better understanding of earthquake hazards in southern California (Figure 4.10). These systems have proved very effective in delineating fault planes within sedimentary ba-

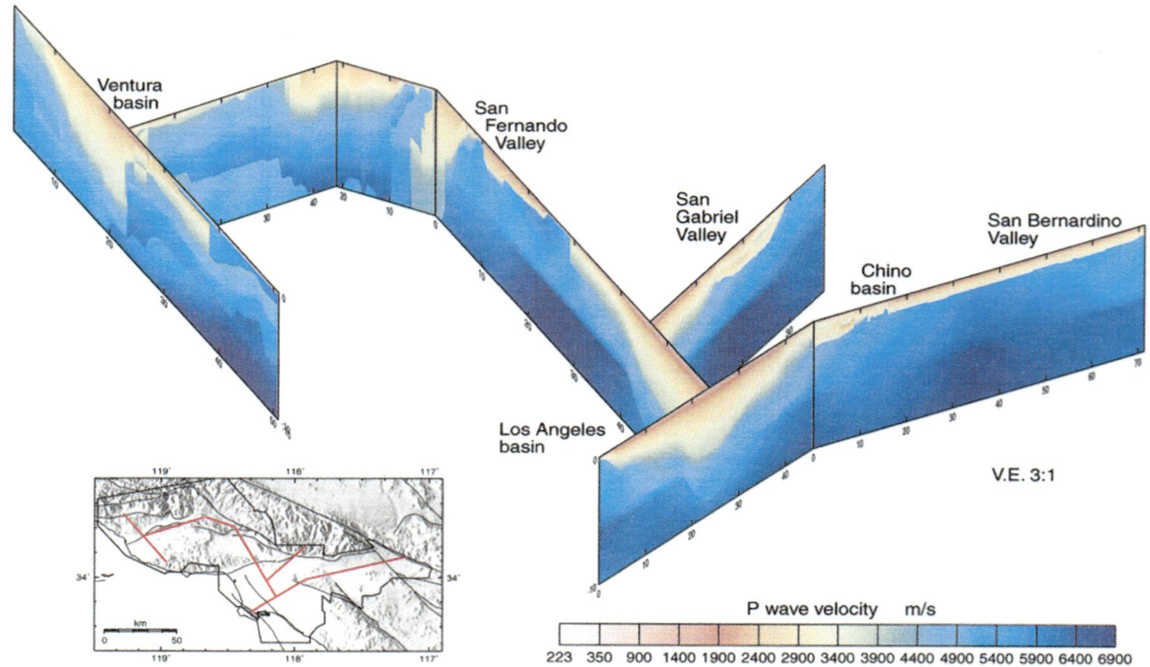

FIGURE 4.9 Fence diagram showing three-dimensional variation of shear-wave speeds in the Los Angeles region determined from seismic tomography. Cross sections are located as red lines in the lower left panel. Low speeds indicated by warm colors show deep sedimentary basins that trap seismic waves, thus amplifying and extending the shaking in regional earthquakes. SOURCE: H. Magistrale, S. Day, R.W. Clayton, and R. Graves, The SCEC southern California reference three-dimensional seismic velocity model version 2, *Bull. Seis. Soc. Am.*, **90**, S65-S76, 2000. Copyright Seismological Society of America.

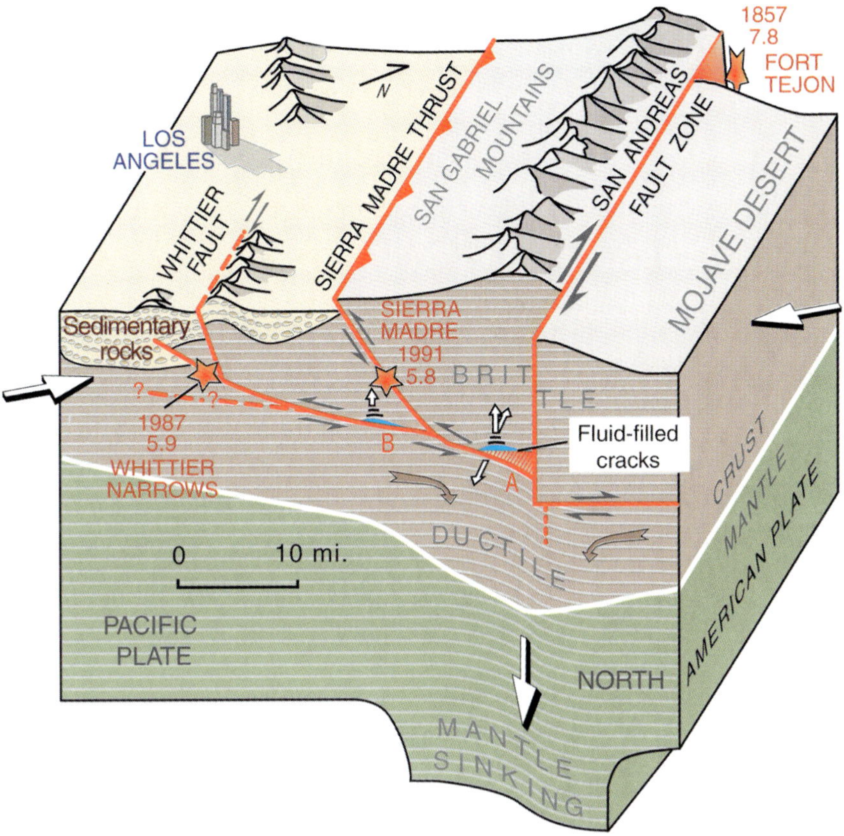

FIGURE 4.10 An interpretation of the geologic structure along Line 1 of the Los Angeles Region Seismic Experiment line, showing the interaction of strike-slip and thrust faults. "Blind" thrust faults, such as the one responsible for the 1987 Whittier Narrows earthquake, do not break the surface, but can be imaged by seismic methods. Relative movement on faults is shown by pairs of small arrows. Large white arrows show the olique direction in which the Pacific Plate (left) and the North American Plate (right) are converging. Red stars with dates and numbers (magnitudes) are earthquakes. SOURCE: U.S. Geological Survey, Fact Sheet 110-99, <http://geopubs.wr.usgs.gov/fact-sheet/fs110-99/>.

sins. Deep structure (down to the base of the crust at 30- to 40-kilometer depth) can be imaged using larger multichannel systems in conjunction with explosion or large Vibroseis sources.

Seismicity Catalogs

The basic product of seismic monitoring is the seismicity catalog, a sequential listing of all earthquakes, explosions, and other localized seismic disturbances, natural or man-made. In modern monitoring systems, the detection, association, and inversion of seismic arrivals are done automatically from continuous digital data streams, although seismic analysts are still employed to review, evaluate, and often modify the results. The output may include the event's origin time, hypocentral location (latitude, longitude, and depth), magnitude, and other source parameters, such as seismic moment and focal mechanism (usually in the form of a moment tensor) and a measure of rupture duration. Improving the completeness and accuracy of these seismicity catalogs is a major objective of seismic hazard analysis, which often depends on small earthquakes to identify the potential for damaging fault ruptures, and of earthquake physics, which relies on catalogs as the basic space-time record of fault-system behavior.

The USGS National Earthquake Information Service (NEIS) operates an Earthquake Early Alerting Service to determine as rapidly as possible the location and magnitude of significant earthquakes in the United States (M ≥ 4.5) and around the world (M ≥ 6.5, or known to be damaging) (*28*). The International Seismological Centre (ISC), based at Thatcham in Berkshire, United Kingdom, is a nongovernmental organization charged with producing a standard global catalog (*29*); it provides the most comprehensive compilations of short-period arrival times and amplitudes from the largest, most globally distributed set of seismic stations (approximately 3000), including earthquake reports from a number of regional seismic monitoring agencies. It currently processes about 5000 events per month worldwide. The International Monitoring System (IMS) currently operates 36 primary stations and arrays and collects data from 38 auxiliary stations. It produces an earthquake bulletin, the Reviewed Event Bulletin, within seven days, aiming at completeness down to M 3.5 (see Box 4.2). Specialized catalog services are rendered by university observatories and laboratories. Harvard University produces a global catalog of centroid locations, centroid times, and moment tensors for most large earthquakes (M ≥ 5.5), primarily from the broadband data provided by FSDN stations (*30*). Though operated on a very modest budget through a private university, this centroid-moment tensor service has proven to be immensely useful in earthquake research, be-

BOX 4.2 Nuclear Monitoring

Since the first underground nuclear tests in the late 1950s, underground test monitoring and test ban treaty verification have motivated the development of better seismic networks (see Section 2.3). With the breakup of the former Soviet Union and the increased number of emerging nuclear nations, the emphasis has shifted from a bilateral superpower test ban treaty to a global comprehensive test ban treaty (CTBT) and the Nuclear Nonproliferation Treaty. In the current plans, seismic networks represent one of four main technologies for monitoring the CTBT (along with infrasonic, hydroacoustic, and radionuclide techniques). The seismic component of the IMS will utilize 170 stations and reduce the global detection threshold to around m_b 4.0. The primary stations (alpha stations) are mostly dense arrays of high-quality, short-period sensors, located at carefully selected sites around the globe, with equipment for continuous telemetry to the International Data Center (IDC) for the primary purpose of detecting seismic events on a global scale. Auxiliary stations (beta stations) are meant to support rapid, on-demand, automatic retrieval of data for use in improving the location of events detected by the primary network. Most of the beta stations will be drawn from established three-component, broadband stations of the FDSN, ensuring a strong partnership between the CTBT monitoring community and earthquake scientists. Approximately 1000 separate channels of seismic data will be transmitted via satellite in real time to the IDC in Vienna, Austria, where they will be analyzed automatically to determine routine source parameters such as location, depth, origin time, and magnitude. Although the ultimate capabilities of the monitoring system will not be known until the network is fully deployed and operational, the experience with recent nuclear tests in India and Pakistan suggests that the IDC and IMS will provide an unprecedented system for real-time global seismic monitoring with low detection thresholds.[1]

Monitoring CTBT compliance will be more challenging than past arms control treaties, because it will require high-confidence identification of any nuclear explosion, however small, carried out in remote regions of the world. The CTBT has motivated a broad program of research, focused on regional monitoring of small seismic events.[2] The results of this research are needed for two treaty monitoring goals. First, there is a need to locate all of the detected seismic events within 1000 square kilometers, because this is the largest region that can be inspected to assess a possible treaty violation. Achieving this goal will require detailed seismic calibration information (travel times, phase arrivals) for each of the IMS stations. Second, algorithms must screen out the large number of natural events that will be detected, based on location, depth, and other source characteristics. To advance these capabilities, the Department of Defense (DOD) currently supports one of the largest basic research programs in seismology in the federal government ($12 million in FY 2000). To increase the involvement of earthquake researchers, DOD plans to make all of the IMS data available for open research and hazard monitoring operations.

---

[1] B. Barker, M. Clark, P. Davis, M. Fisk, M. Hedlin, H. Israelsson, V. Khalturin, W.-Y. Kim, K. McLaughlin, C. Meade, J. Murphy, R. North, J. Orcutt, C. Powell, P.G. Richards, R. Stead, J. Stevens, F. Vernon, and T. Wallace, Monitoring nuclear tests, *Science*, **281**, 1967-1968, 1998.

[2] National Research Council, *Research Required to Support Comprehensive Nuclear Test Ban Treaty Monitoring*, National Academy Press, Washington, D.C., 137 pp., 1997.

cause it has generated the longest catalog of standardized source parameters—seismic moment, source mechanism, and centroid location—for seismicity studies worldwide.

Currently, the properties of more than 30,000 earthquakes are recorded, studied, and cataloged on an annual basis by these and other monitoring organizations. A few regional monitoring systems routinely catalog all seismicity above M 2. Broadband regional networks routinely produce moment tensor solutions for regional earthquakes greater than M 4 (e.g., Figure 4.11). In many regions, however, sensor arrays are too sparse to record events below about M 3. Further work is needed to upgrade regional networks to broadband instrumentation and digital telemetry, a task taken on by the ANSS program, and to extend cataloging procedures to include additional source parameters such as characteristic dimensions (*31*).

Volcano Seismology

Earthquake seismology plays an important role in the study of volcanoes and the prediction of volcanic eruptions (Box 4.3). Seismicity within a volcano is caused by rockfalls and avalanches, tectonic faulting, rock fracture during magma transport, and low-frequency tremors associated with the flow of melt below a volcano. Before a major eruption, earthquakes typically occur in swarms, where the rates of seismicity may be elevated by two to three orders of magnitude above background levels. Monitoring of this activity by seismic networks yields information about the shape, size, and physical state of magma reservoirs (*32*). A complete understanding of the volcanic system will require a synthesis of seismological observations into a coherent model of eruption mechanics, constrained by fluid dynamics and elastodynamics of magma flow in a porous, brittle media. Recent advances in portable instrumentation and theory for analyzing the data are stimulating important advances in this field. For example, results obtained at Redoubt volcano using nonlinear, travel-time tomography show that imaging the three-dimensional structure of a volcano is feasible down to a scale of a few hundred meters (*33*).

Beyond first-order mapping of the fluid-pathway geometry using broadband data there are many questions about the dynamics of magma transport that can be investigated using short-period seismic data. For example, two basic families of volcanic processes generate signals in the 0.1- to 1-second seismic band. The first involves volumetric sources in which the fluid plays an active role in the generation of elastic waves, and the second consists of shear or tensile sources caused by brittle rock failure. In volumetric sources, elastic radiation is generated by multiphase fluid flow through cracks and conduits; long-period events, volcanic

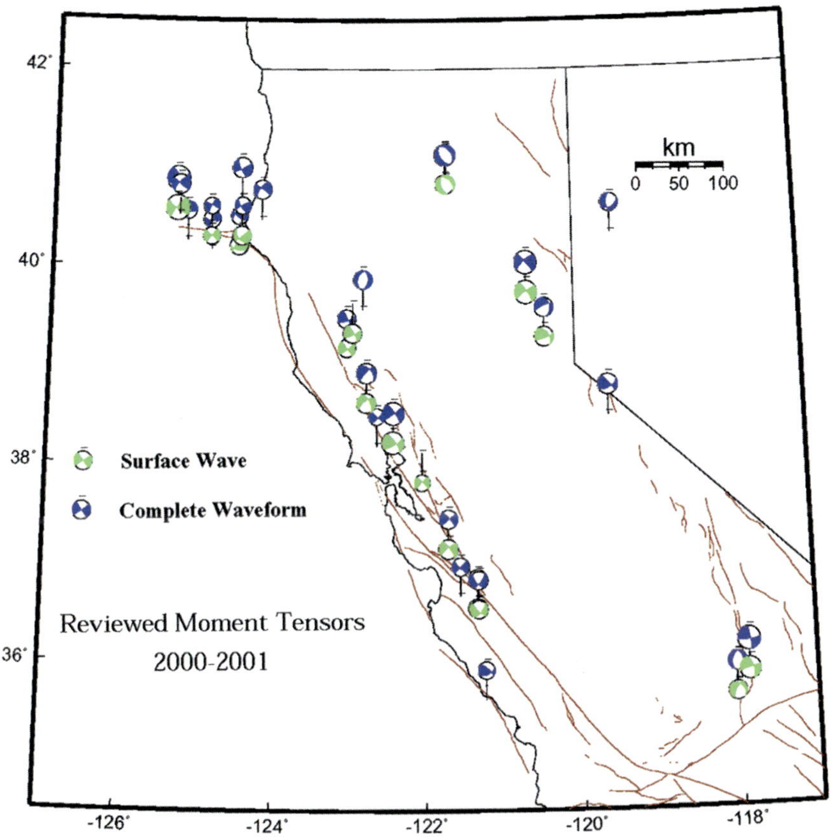

FIGURE 4.11 Moment tensor solutions for northern California obtained from the Berkeley Digital Seismic Network, illustrating advanced processing for regional earthquakes. These solutions are obtained automatically and in quasi-real time using broadband data for a subset of events of M 4 and larger. Robustness of the solution is assessed by comparing the results of two independent inversions—one in the time domain, the other in the frequency domain. SOURCE: B. Romanowicz, D. Dreger, and H. Tkalcic, University of California, Berkeley.

tremor, and seismic signals related to mechanisms of degassing in open vents are manifestations of such processes. The second family includes volcano-tectonic earthquakes, in which magmatic processes provide the source of energy for rock failure. These sources occur in the brittle rock around a magma reservoir and conduit and are associated primarily with the structural response of the volcanic edifice to the intrusion and/or

BOX 4.3 Prediction of the Mt. Pinatubo Eruption

Mt. Pinatubo in the Philippines is one of a chain of composite volcanoes known as the Luzon volcanic arc, which are being formed by the rise of magma from an eastward-dipping subduction zone along the Manila Trench. On the afternoon of April 2, 1991, villagers were surprised by a series of small explosions from a line of vents near the north flank of the summit dome. Within a few days, scientists from the Philippine Institute of Volcanology and Seismology (PHIVOLCS) installed several portable seismographs near the northwest foot of Mt. Pinatubo and began recording small earthquakes at a rate of about 40 to 140 per day. In late April, PHIVOLCS was joined by a group from the USGS, and the joint team installed a network of seven seismometers, telemetered to Clark Air Base, a major U.S. Air Force facility located just east of the volcano. Numerous small earthquakes (M lower than 2.5) continued through May, clustered in a zone 2 to 6 kilometers deep and caused by fracturing of brittle rock by rising magma. Beginning on June 1, a second cluster of earthquakes developed in the upper 5 kilometers near the fuming summit vents. A small explosion early on June 3 initiated an episode of increasing volcanic unrest characterized by intermittent minor emission of ash, increasing seismicity beneath the vents, and episodes of harmonic tremor (a prolonged rhythmic seismic signal believed to be related to sustained subsurface movement of magma or volatile material).

PHIVOLCS issued a level-3 alert on June 5, indicating the possibility of a major pyroclastic eruption within two weeks. A tiltmeter high on Mt. Pinatubo began to show a gradually increasing outward tilt early on June 6. Seismicity and the outward tilt continued to increase until late afternoon on June 7, when an explosion generated a column of steam and ash 7 to 8 kilometers high. After the explosion, seismicity decreased and the increase in outward tilt stopped. PHIVOLCS promptly announced an increase to level-4 alert (eruption possible within 24 hours) and recommended additional evacuations from the volcano's flanks. The period from June 8 through early June 12 was marked by continuing, weak ash emission and episodic harmonic tremor. On June 9, PHIVOLCS raised the alert level to 5 (eruption in progress). The radius of evacuation was extended to 20 kilometers, and the number of evacuees increased to about 25,000. The first major explosive eruption began at 0851 hours on June 12, generating a column of ash and steam that rose to 19 kilometers. Although a burst of seismic tremor had occurred several hours earlier, no specific seismic precursor immediately preceded this event; a high-amplitude seismic signal and the rise of the eruptive column seemed to begin simultaneously. Seismic records indicated that this event lasted about 35 minutes. This was the first of a series of brief explosive eruptions that occurred with increasing frequency from June 12 through 15. The climactic eruption began at 1430 hours on June 15—the world's largest in more than half a century.

The successful forecast of the Mt. Pinatubo eruption enabled Philippine civil leaders to organize massive evacuations that saved thousands of lives and greatly reduced the destruction at Clark Air Base (military aircraft worth $200 million to $275 million were also removed).[1] Nevertheless, the coincidence of the climactic eruption with a typhoon led to more than 300 deaths and extensive property damage, caused primarily by the extraordinarily broad distribution of heavy, water-saturated tephra-fall deposits.

---

[1]C. Newhall, J.W. Handley II, and P.H. Stauffer, Benefits of volcano monitoring far outweigh costs; The case of Mount Pinatubo, *U.S. Geological Survey Fact Sheet* 115-97, 3 pp., 1997.

withdrawal of fluids. Volcano-tectonic earthquakes act as stress gauges that map stress concentrations in the volcanic structure. Dense distributions of earthquake hypocenters therefore provide a signature of magma migration through volcanoes. However, gaining a better understanding of the dynamics of magma transport will require more information about the source processes for the long-period events (*34*).

4.2 TECTONIC GEODESY

The elastic strain energy unleashed in earthquakes accumulates in the Earth's crust through the imperceptibly slow motions of plate tectonics. The strain rates in tectonically active areas such as the western United States are only few parts in 10 million per year (*35*). The tools of geodesy can be used to measure these small tectonic deformations on global to local scales, furnishing data that have proven essential for estimating the long-term slip rates and seismogenic potential of lithospheric faults. In addition, geodesy provides the means to detect transient (time-localized) strains having durations from minutes to years that do not generate elastic waves and are therefore invisible to seismic monitoring. These transients comprise fault creep and stress relaxation following large earthquakes (postseismic transients), as well as the slow, localized strains that are predicted by laboratory experiments to precede dynamic faulting (deformation precursors). They also include an observed but poorly understood class of isolated events known as "silent earthquakes," which may be responsible for aseismic slip on some faults and may play a role in concentrating stress before some large earthquakes.

Tectonic geodesy includes a wide array of techniques with complementary strengths and sensitivities. Geodetic measurements vary in scale from systems that allow the recovery of three-dimensional position anywhere on the planet's surface, such as GPS, to systems that are extremely localized and sensitive, such as borehole strainmeters.

Traditional Geodetic Techniques

Many of the measurement technologies used in tectonic geodesy grew out of the needs of precise surveying. Both activities share a requirement for extremely precise measurements, and the practice of geodesy has a strong tradition of characterizing and minimizing measurement errors.

- ***Triangulation.*** This surveying technique, invented by ancient agricultural societies, can measure angles between distant points with a precision of approximately 2 arc-seconds, corresponding to a shear strain of about 5 parts per million. Triangulation requires a clear line of sight from an observing station to two or more target monuments, typically a few tens of kilometers distant (usually situated on high ground). Triangulation played an important role in estimating the strains and displacements associated with the 1906 San Francisco earthquake, providing much of the observational foundation for Reid's elastic rebound model. Triangulation is expensive, however, because observing and target sites must be occupied simultaneously by experienced personnel. Consequently, the method has been largely abandoned by the tectonic geodesy community in favor of more accurate and flexible techniques, primarily GPS (described below).

- ***Trilateration.*** In the 1970s, the ability to measure long distances with laser reflectors improved the utility of tectonic geodesy. Trilateration provided the means for repeated strain measurements over baselines of tens of kilometers with sufficient precision (about 300 parts per billion) to monitor the strain accumulation between large earthquakes (*36*). It allowed the USGS to confirm that slip rates observed over decades across major faults in California are quite similar to geological estimates, which are averaged over thousands to millions of years. Like triangulation, this method has been superceded by GPS.
- ***Spirit Leveling.*** Vertical displacements measured by spirit leveling have been used to characterize the vertical component of the deformation field associated with earthquakes (*37*). Reports of postseismic and even precursory deformation measured with leveling have been published, but the limited accuracy and the possibility of systematic error have made these reports controversial (*38*). Leveling surveys are very labor intensive and costly, although they remain the most precise way to measure relative elevations over distances of less than about 25 kilometers. Over larger distances, GPS provides a more accurate, and much more economical, alternative to leveling (*39*) and offers the tremendous advantage of continuous temporal sampling.

Space Geodetic Systems

The space program has contributed much of the new technology developed for tectonic geodesy since the 1970s. Ultraprecise methods of space-based geodesy were first pioneered in Very Long Baseline Interferometry using astronomical sources, but they reached their current state of capability by taking advantage of dedicated satellite platforms.

- ***Very Long Baseline Interferometry (VLBI) and Satellite Laser Ranging (SLR).*** The pioneering geodetic techniques of VLBI and SLR, both capable of monitoring plate motions at a global scale, were developed under the National Aeronautics and Space Administration (NASA)

Crustal Dynamics Program. VLBI uses simultaneous observations of high-frequency radio waves from extragalactic quasars to measure the baselines connecting a set of radio telescopes. Precision approaches one part per billion—millimeter changes over 1000-kilometer baselines. SLR uses laser pulses reflected from special satellites (e.g., Laser Geodynamics Satellite and Starlette) to locate optical telescopes on the ground. SLR is less precise than VLBI because of lower signal-to-noise levels and the need to solve for the motion of the satellite and ground stations. VLBI confirmed Wegener's concept of continental drift and provided a rough approximation of how the Pacific-North American plate motion is distributed across western North America (*40*). SLR also contributed data to the study of plate boundary deformation zones, particularly in the Mediterranean region. VLBI and SLR are both very cumbersome and expensive because they require sensitive instrumentation to detect very weak signals. Consequently, they have been replaced in nearly all tectonic applications with GPS measurements.

- ***Global Positioning System.*** Tectonic geodesy was rapidly transformed by the deployment of the first GPS constellation in the mid-1980s. These satellites emit strong, precisely timed radio signals easily detectable by 100-millimeter antennae and can be used to locate points anywhere on the Earth's surface. GPS instrumentation, which underlies most modern autonomous satellite navigation systems and a host of military and commercial applications, has become quite affordable since its initial development, so GPS surveying can be done by individual investigators (*41*). Because of its low cost and portability, GPS has quickly replaced other geodetic techniques for most tectonic studies, including dense sets of deformation measurements spanning plate boundaries (*42*).

GPS locations are measured relative to the orbiting satellites, whose positions are in turn estimated relative to tracking stations on the ground. Geodetic accuracy thus depends on knowing the location of the tracking stations, which are moving relative to each other because of tectonic motions. At present, the tracking station coordinates are best determined by making frequent GPS measurements at sites tied into a standard (absolute) reference frame by VLBI or SLR observations (*43*).

GPS measurements can be performed in either campaign or permanent modes. The campaign mode involves temporary occupation of geodetic benchmarks in much the same way as for earlier triangulation and trilateration surveys. A series of issues—the decreasing costs of receivers, the high labor costs of campaign measurements, the loss of precision caused by antenna setup—has motivated the installation of permanent GPS stations in configurations similar to seismic monitoring networks. Automated arrays of continuously sampled GPS receivers can measure deformation in real time as often as once per minute (*44*). Japan has the largest fixed GPS network with more than 1000 stations. In comparison, the Bay Area Regional Deformation network in northern California includes about 35 stations and the Southern California Integrated GPS Network (SCIGN) now comprises 250 stations (Figure 4.12). The arrays in Japan and southern California have tested many design elements and demonstrated what features of time-dependent strain are most significant. Permanent GPS arrays will continue to grow as receiver and data-transmission costs drop further.

Plans call for the permanent GPS networks in the western United States to be expanded and consolidated to form a major component of the Plate Boundary Observatory (PBO), proposed as part of the EarthScope program (see Chapter 6). The GPS component of the PBO would comprise more than 1000 continuously recording GPS receivers, with several hun-

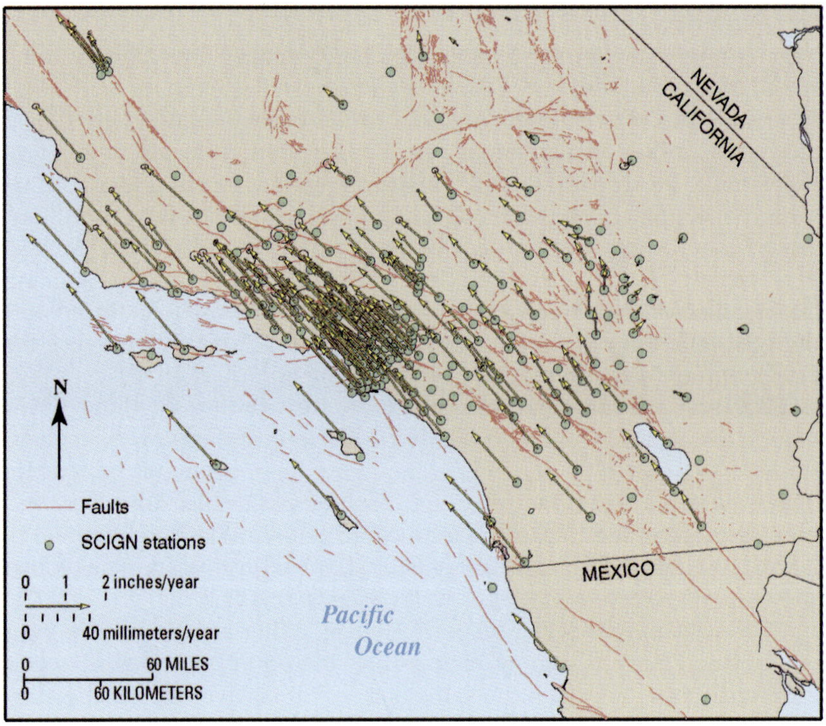

FIGURE 4.12 Velocities relative to the North American Plate of geodetic movements from 15 years of GPS measurement in Southern California, showing ongoing deformation within a complex system of faults. SOURCE: Southern California Earthquake Center and U.S. Geological Survey.

dred established as a geodetic backbone for the study of plate boundary tectonics. The remaining GPS stations would be deployed in denser clusters to provide detailed data on active faults and volcanic systems within the most active zones.

The broad coverage and high precision of GPS geodesy permit individual faults to be studied as components in strongly interacting systems rather than as isolated elements. A disadvantage of GPS is that motion can be measured only at points on the ground where receivers are located. Although the cost of GPS receivers is decreasing steadily, it is impossible to measure the deformation field densely enough to answer some key questions of earthquake science.

• ***Interferometric Synthetic Aperture Radar.*** The most recent innovation in tectonic geodesy is InSAR, which has imaged earthquake deformations at a level of detail unanticipated only 10 years ago (*45*). InSAR measures deformation by comparing reflected radar waves recorded on successive passes of a satellite from nearly identical positions. The simplest InSAR measurements are sensitive to just one component of displacement (toward the satellite), but stereoscopic measurements (pairs of images from multiple locations) allow measurement of vector displacements (Figure 4.13). InSAR is subject to errors from changes in reflective properties such as those caused by seasonal vegetation changes and snowfall, and it lacks the temporal resolution of GPS. However, the ability to map essentially continuous displacements over large swaths of active plate boundaries offers an enormous advantage.

Because they can map centimeter-level deformations with a spatial resolution on the order of 100 meters, InSAR systems are useful for determining co-seismic and interseismic slip on faults. This is particularly important in remote areas lacking GPS stations. InSAR has shown an ability to measure co-seismic slip on subsidiary faults, variations in slip distribution along strike, and slip on previously unknown faults. With its continuous coverage, InSAR systems can map surface displacements before, during, and after earthquakes or volcanic eruptions, providing time-dependent data on the mechanics of fault loading, earthquake rupture, and earthquake interaction, and they can image strain accumulation across broad tectonic zones (Figure 4.14), as well as regional subsidence induced by petroleum production and groundwater withdrawal. For example, InSAR data have been used to disentangle the latter types of motion from tectonic strains observed by GPS networks in the Los Angeles basin (Figure 4.15). Finally, InSAR has been used to detect postseismic poroelastic effects induced by fault movements (*46*). An important component of the proposed EarthScope project is a dedicated InSAR satellite, the Earth Change and Hazard Observatory (ECHO) (*47*).

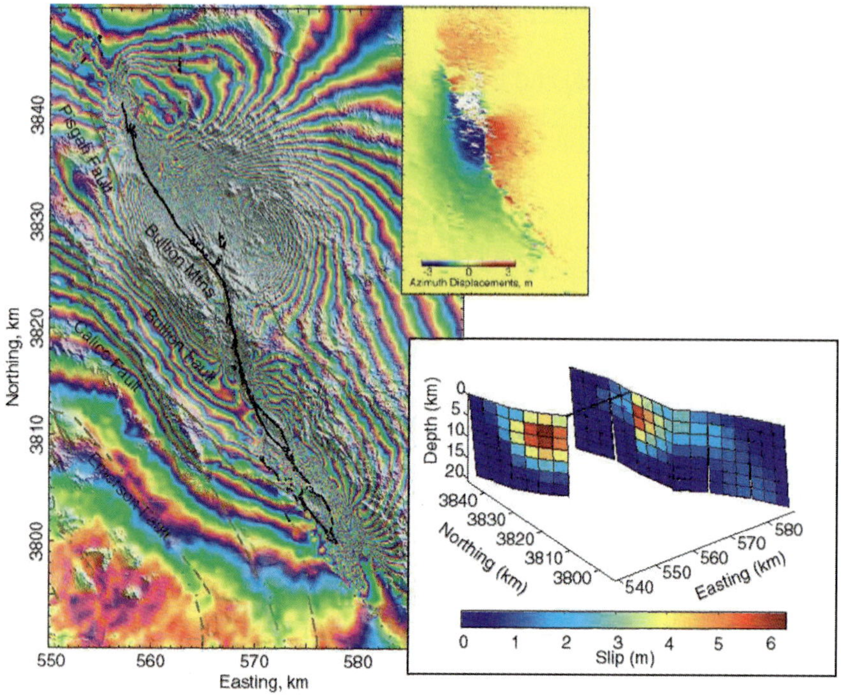

FIGURE 4.13 InSAR-derived interferogram (*left*) and azimuth offset field (*upper right*) show the surface deformation from the October 20, 1999, M 7.1 Hector Mine earthquake as determined by M. Simons et al. (2002). Colored fringes show radar phase introduced by earthquake-induced changes in distance from each point on the ground to the orbiting radar. A solution for fault slip at depth inferred from the radar data (*lower right*) as determined by Jonsson et al. (2002) indicates a maximum slip of 6 meters at 10-kilometer depth just northwest of the hypocenter. The fault has two branches in its northern part: the westernmost branch, which has greater slip than the eastern branch, has been offset in the figure for clarity. SOURCE: M. Simons, Y. Falko, and L. Rivera, Coseismic deformation from the 1999 M_W 7.1 Hector Mine, California, earthquake as inferred from InSAR and GPS observations, *Bull. Seis. Soc. Am.*, **92**, 1390-1402, 2002; S. Jonsson, H. Zebker, P. Segall, and F. Amelung, Fault slip distribution of the M_W 7.1 Hector Mine, California, earthquake estimated from satellite radar and GPS measurements, *Bull. Seis. Soc. Am.*, **92**, 1377-1389, 2002. Copyright Seismological Society of America.

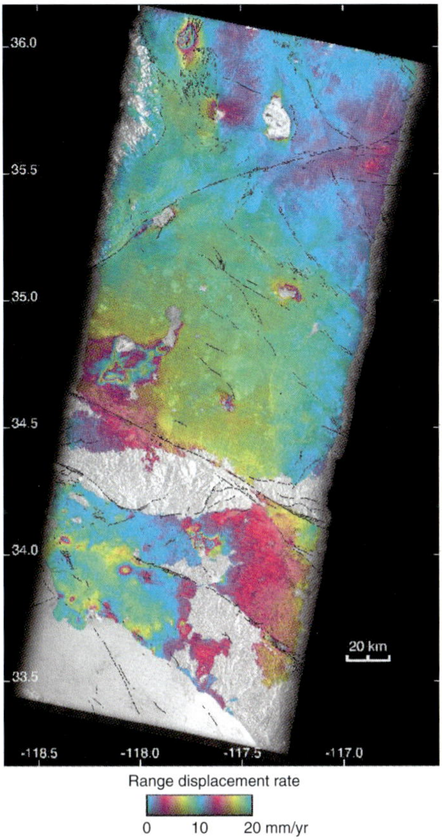

FIGURE 4.14 InSAR-observed deformation along the San Andreas fault from European Remote Sensing satellite radar. Many interferograms were combined to detect the elastic strain built up along the fault during the interval from 1992 to 2000. One color cycle shows 10 millimeters per year of ground displacement. Other local deformation signatures due to groundwater and oil withdrawal in urban areas are clearly visible. InSAR also reveals unexpected transient strain accumulation along the Blackwater-Little Lake fault system within the eastern California shear zone in a 120-kilometer-long, 20-kilometer-wide zone of concentrated shear between the southern end of the 1872 Owens Valley earthquake surface break and the northern end of the 1992 Landers earthquake surface break. The shear zone is continuous through the Garlock fault, which does not show any evidence of left-lateral slip during the same period. A dislocation model of the observed shear indicates right-lateral slip at 7 ± 3 millimeters per year on a vertical fault below 5 kilometers depth, a rate that is two to three times greater than the geologic rates estimated on northwest-trending faults in the eastern Mojave area. SOURCE: G. Peltzer, F. Cramp, S. Hensley, and P. Rosen, Transient strain accumulation and fault interaction in the eastern California shear zone, *Geology*, **29**, 975-978, 2001.

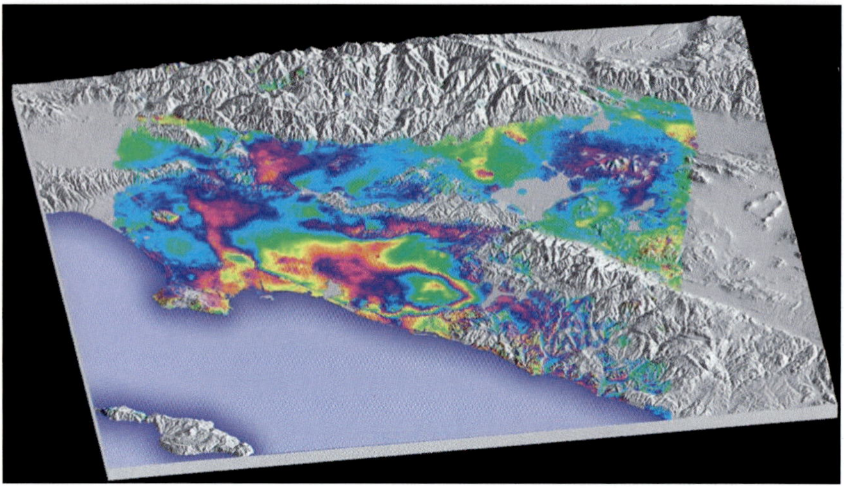

FIGURE 4.15 Perspective view of the Los Angeles region with superimposed InSAR measurements of ground motions between May and September 1999, indicating seasonal and secular variations due to groundwater withdrawal and recharge. Large regions of metropolitan Los Angeles are rising and falling by up to 11 centimeters annually, and a large portion of the city of Santa Ana is sinking at a rate of 12 millimeters per year. The repeated color banding with the large oval shows approximately 5 centimeters of subsidence. The straight line located just inside the coastline is the Newport-Inglewood fault, which controls the extent of subsidence. The small isolated bull's-eye feature north of Palos Verdes and east of downtown Los Angeles is from pumping activity in the Inglewood oil field. The motion caused by the withdrawal of water, oil, and gas from the basin contaminates GPS measurements of the deformation field. After corrections for these effects using the InSAR data, the contraction across the Los Angeles Basin is estimated to be approximately 4.5 millimeters per year. This deformation is thought to be accommodated primarily on blind thrust faults, such as those that ruptured in the 1987 Whittier Narrows and 1994 Northridge earthquakes. SOURCE: G.W. Bawden, W. Thatcher, R.S. Stein, K.W. Hudnut, and G. Peltzer, Tectonic contraction across Los Angeles after removal of groundwater pumping effects, *Nature,* **412**, 812-815, 2001. Reprinted by permission from Nature copyright 2001 Macmillan Publishers Ltd.

Strain Measurements

A separate facet of geodesy is the measurement of strain over small spatial scales using self-contained instruments called strainmeters. The discovery in 1960 of aseismic slip or "creep" on a segment of the San Andreas fault in central California (*48*) led to methods for extremely localized measurements of fault displacement using invar tapes, wire creep-meters, alignment arrays, short-baseline triangulation, and laser length surveys (*49*). Deployment of these instruments revealed both steady and episodic creep occurring at shallow depths (<4 kilometers) on some faults, often near the time of earthquakes (*50*). Aseismic creep at greater (seismogenic) depths, such as observed in central California, appears to be rare.

High-resolution laser strainmeters, borehole strainmeters, and tiltmeters are used to measure deformation very precisely in a small region. Their sensitivity approaches 10^{-12}, but long-term stability is a problem because they have small footprints susceptible to very localized, nontectonic deformations, such as ground swelling in rainstorms. The most stable instruments are the laser strainmeters and water-tube tiltmeters at Piñon Flat Observatory in California, which derive their stability from their length (>500 meters) and the "optical anchors" used to couple the end monuments to rock at about 25-meter depth (*51*). Borehole instruments suffer drift over several months, but they are very precise over shorter times and have widespread application for measuring transient deformation, including slow and silent earthquakes (*52*).

Under the best conditions, these instruments are as much as two to three orders of magnitude more sensitive than GPS at short periods. At longer periods, the relative advantage declines significantly, although long-baseline instruments may retain advantages even for time scales of years. Because they measure strain, which decays as the cube of distance from a dislocation, they must be positioned in reasonably close proximity to the source. Small numbers of borehole strainmeters have been operating for years in a few select locations. The proposed Plate Boundary Observatory would deploy several hundred borehole and perhaps several long-baseline strainmeters at strategically chosen sites along the San Andreas fault system, as well as at several volcanic systems (*53*).

Geodetic Observations of Earthquake Processes

The increasing precision and density of geodetic measurements are furnishing new constraints on how complex fault systems are loaded, how earthquakes interact, the nature of aseismic deformation transients, and how the rheological structure of the Earth's crust controls the earthquake process.

Plate Tectonics and Fault Motions Geodetic studies have shown that plate-tectonic models, based on data that average over thousands to millions of years, can accurately predict the short-term motions across plate boundary zones a few hundred kilometers wide. Denser measurements from geodetic networks place strong constraints on the slip rates on faults within these zones and are especially valuable for faults that are poorly exposed or otherwise not amenable to geological study. If proper accounting is made for fault interactions and postseismic deformations, geodetic slip rates for faults bounding small tectonic blocks with lateral dimensions of 20-50 kilometers generally agree with those estimated by geologic methods for much longer time intervals, although clear discrepancies have been documented (Figure 4.14).

By integrating the slip rate over the areas of faults, one can estimate the rate at which seismic moment is accumulating. If it can be assumed

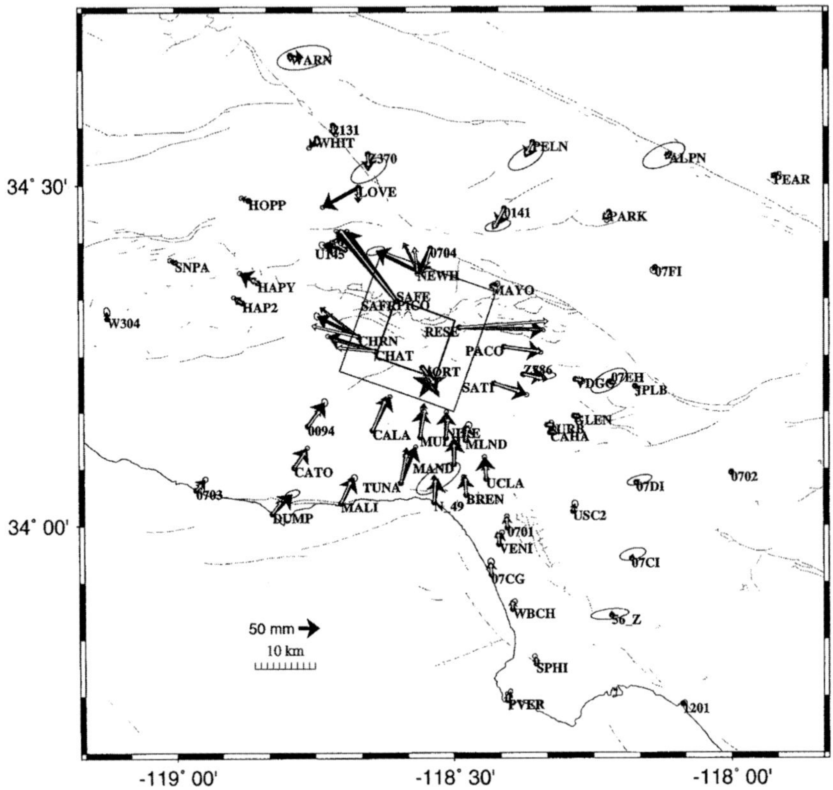

FIGURE 4.16 Horizontal displacement (heavy black vectors and the corresponding one-sigma error ellipses) associated with the Northridge earthquake measured by the GPS network in southern California. SOURCE: K.W. Hudnut, Z. Shen, M. Murray, S. McClusky, R. King, T. Herring, B. Hager, Y. Feng, A. Donnellan, and Y. Bock, Coseismic displacements of the 1994 Northridge, Calif., earthquake, *Bull. Seis. Soc. Am.*, **86**, S19-S36, 1996. Copyright Seismological Society of America.

that all of this moment is released in earthquakes (no aseismic slip) and if the relative distribution of earthquake size is known, then the long-term moment rate sets the multiplicative constant needed to infer long-term earthquake frequency (*54*). Because aseismic creep at seismogenic depths seems to account for very little of the moment budget in most continental settings, unknown fault geometry and uncertainties in the earthquake size distribution are often the limiting factors in estimating earthquake frequency (*55*).

Strain rates depend on the slip distribution and geometry of faults, so spatial variations in measured strain rate reveal significant information about the faults. For example, spatial variations of strain rate along the San Andreas fault near Parkfield, California, show details of the slip rate on the fault plane. The San Andreas is primarily locked to a depth of about 15 kilometers to the southeast of that location and is creeping to the northwest. The slip distribution is important for understanding the physical mechanism of the transition and the stress accumulation leading to future earthquakes (*56*).

The crust on either side of the creeping section of the San Andreas fault accumulates very little strain, and the geodetic displacement rate (35 millimeters per year) is virtually identical to the geologic slip rate on the San Andreas (34 millimeters per year). Elsewhere on the San Andreas and on other strike-slip faults, secular deformation is continuous across each fault. The deformation patterns can be matched with a model in which each fault is locked by friction to a depth of 10 to 20 kilometers and slips freely below that depth in a viscoelastic lower crust.

Earthquake Rupture and Subseismic Strain Events GPS data provide independent estimates of earthquake-induced fault displacement (*57*), as illustrated in Figure 4.16. Geodetically derived images of slip variations for the recent Loma Prieta and Landers earthquakes rival the spatial resolution achieved from models based on seismic data (*58*). In the case of the 1989 Loma Prieta earthquake, a geodetically derived slip model (*59*) supported seismic models that found a variation of slip direction with distance along the fault (*60*). Models of the 1992 Landers, California, earthquake based on GPS data confirm the strong spatial variability and location of slip derived from modeling the strong motion data (*61*).

The 1992 Landers, 1999 Izmit (Turkey), and 1999 Hector Mine earthquake all caused substantial deformation not detectable with existing GPS arrays. InSAR images from the Landers earthquake show co-seismic slip on the Garlock and other faults, which were not otherwise known to have slipped in the main shock (*62*). Many subsidiary faults, some previously unmapped, slipped more than 10 millimeters during the 1999 Hector Mine earthquake.

Geodetic data for the 1989 Loma Prieta and 1994 Northridge earthquakes reveal co-seismic deformation that cannot be explained by slip on the seismically determined fault plane. This deformation may have been caused by vertical variations in elastic moduli or by secondary deformation in the hanging wall of these faults (*63*).

One of the most significant discoveries in tectonic geodesy has been the detection of episodic strain transients that have earthquake-like spatial patterns but occur too slowly (over days to months) to excite seismic waves. These subseismic events, or "silent earthquakes," have been observed in volcanic areas, such as the M 7 Izu-Oshima earthquake in 1978 (*64*), and on shallow creeping faults such as the San Andreas between Parkfield and San Juan Bautista (Figure 4.17). Continuously recording stations of the Japanese GPS network detected a much larger event (M 6.5) with a duration of about a year on the subduction interface under the Bungo Channel, between the islands of Shikoku and southern Honshu (Figure 4.18). In 1999, a Canadian team used GPS networks in southwest British Columbia and northwest Washington to detect a 15-day, M 6.7 silent slip event at depths of 30-40 kilometers on the Cascadia subduction interface (*65*). Further analysis of a full decade of continuously recorded data has uncovered a quasi-periodic series of similar events with an average recurrence interval of about 14 months (*66*). It is not yet known whether such sequences are characteristic of subduction zones, nor is it understood how these subseismic transients relate in time and space to great earthquakes that occur every several hundred years along shallower segments of the subduction interface. These research problems have clear connections to the fundamental issues of earthquake predictability.

PostSeismic Deformation and Long-Term Transients The stress changes from large earthquakes cause several types of secondary deformation that can be detected by surface geodesy. These include additional slip on the fault surface (afterslip); viscous relaxation in the hotter, more ductile lower crust; and pressure-driven fluid flow. Postseismic strain transients due to one or more of these mechanisms have been documented for a number of earthquakes, such as 1906 San Francisco (*67*), 1957 Kern County (*68*), 1985 Central Chile (*69*), and 1989 Loma Prieta (*70*). The 1992 Landers earthquake, where the cumulative postseismic deformation may have equaled 10 to 20 percent of the M 7.3 mainshock, was the first to be observed by a full suite of modern geodetic methods: GPS (*71*), laser strainmeters (*72*), and InSAR (*73*). The strain transients observed by these techniques had different durations—about 6, 50, and 1000 days, respectively—either because of different instrumental sensitivities or because separate inelastic processes were operating (Figure 4.19). More recent

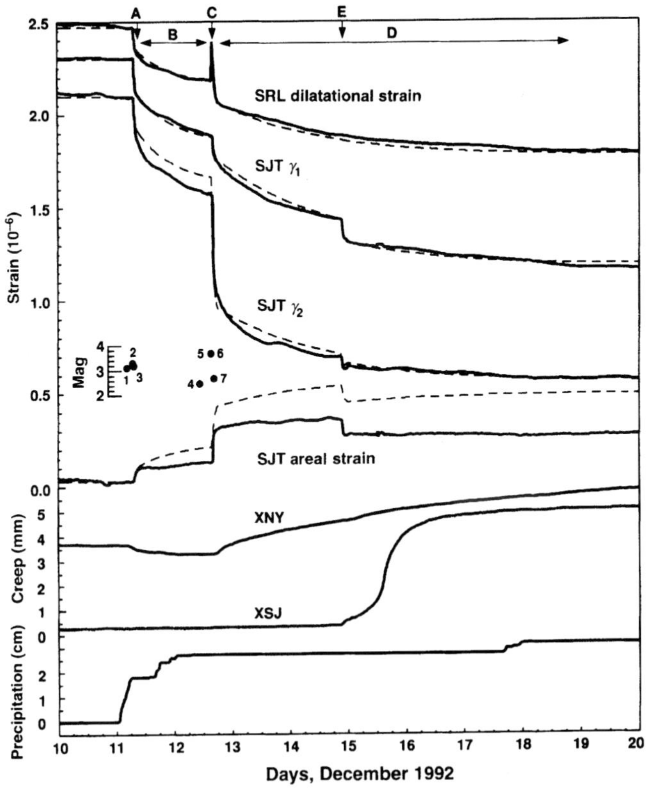

FIGURE 4.17 Borehole strain and creepmeter data from San Juan Bautista for 10 days in December 1992, showing several slow earthquakes. The observations suggests rupture speeds of 0.1 to 0.5 meter per second for these events, compared to 1 to 5 kilometers per second for ordinary earthquakes. SOURCE: A.T. Linde, M.T. Gladwin, M.J.S. Johnston, R.L. Gwyther, and R.G. Bilham, A slow earthquake sequence on the San Andreas fault, *Nature*, **383**, 65-68, 1996. Reprinted by permission from Nature. Copyright 1996 Macmillan Publishers Ltd.

earthquakes, including the 1999 Izmit and 1999 Hector Mine events, have added data sets of similar diversity that are under active analysis.

Viscoelastic diffusion following large earthquakes in California and Japan generates large strain changes, even decades later. Several authors have argued that the resulting stresses might propagate slowly, contributing to deformation and possibly earthquake triggering at very great distances from a large earthquake (*74*).

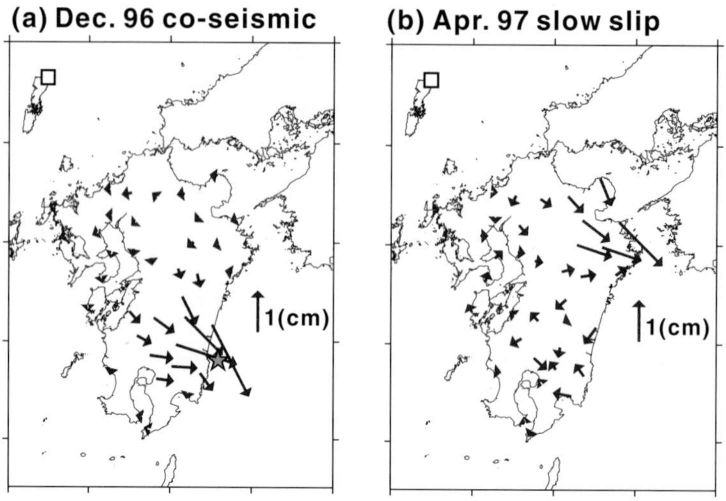

FIGURE 4.18 Observations of slow earthquake under the Bungo channel between Kyushu and Shikoku observed by the continuously recording Japanese GPS network. The GPS-derived vector motions indicated by arrows are from (a) an ordinary earthquake on the subduction interface and (b) a subsequent silent earthquake that did not radiate seismic waves. The patterns of deformation are quite similar (though displaced), indicating that the silent earthquake represents slip on a nearby patch of the subduction interface. The moment magnitude of the silent earthquake was estimated to be M 6.8. SOURCE: Modified from Y. Yagi, M. Kikuchi, and T. Sagiya, Co-seismic slip, post-seismic slip, and aftershocks associated with two large earthquakes in 1996 in Hyuga-nada, Japan, *Earth Planets Space*, **53**, 793–803, 2001.

Tectonic Deformation and Future Earthquakes Geodetic data have confirmed Reid's hypothesis that strain accumulates in the region surrounding a fault before a major earthquake, but not his conjecture that major earthquakes can be predicted from the time required to recover the strain released in the previous event. Although the latter has not yet been fully tested owing to measurement limitations and the lack of the long-term observations, Reid's notion of an earthquake "cycle" appears to be at odds with the observed complexity of the earthquake process and the inherent irregularity of stick-slip behavior.

Fault-friction models derived from laboratory data predict that observable aseismic slip in the nucleation zone of an earthquake might precede the seismic phase of fault rupture on time scales of minutes to days. Hope that such signals could be used to predict large earthquakes motivated many geodetic searches for aseismic strain precursors, primarily

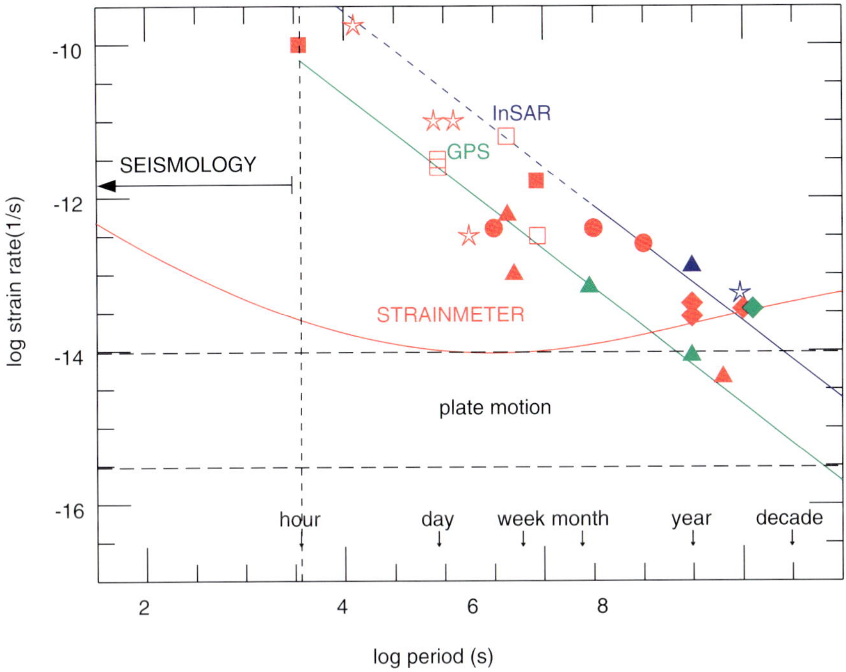

FIGURE 4.19 Thresholds of strain-rate sensitivity are shown for borehole strainmeters, GPS, and InSAR as a function of period. The diagonal lines give GPS (green) and SAR (blue) detection thresholds for 10-kilometer baselines, assuming 2-millimeter and 2-centimeter displacement resolution, respectively. GPS and InSAR strain-rate sensitivity is better at increasing periods, allowing the detection of plate motions (dashed line). Borehole strain detection threshold reaches a minimum at a period of a week and then increases with period due to Earth noise. At longer periods (months to a decade), GPS has greater sensitivity than borehole strain by one to two orders of magnitude. At intermediate periods (weeks to months) the two are comparable, and at short periods the borehole strainmeter sensitivity is one to three orders of magnitude greater. Also shown are several types of transients observed by strainmeters (red), GPS (green), and InSAR (blue). Triangles show postseismic deformation from the Landers earthquake observed by strainmeters, GPS, and InSAR. Squares show slow earthquakes; open squares show single station observations. Red and green diamonds and open circles show long-term aseismic deformation. SOURCE: PBO Steering Committee, The Plate Boundary Observatory: Creating a four-dimensional image of the deformation of western North America, White paper providing the scientific rationale and deployment strategy for a Plate Boundary Observatory based on a workshop held October 3-5, 1999. Available at <http://www.earthscope.org>.

using strainmeters and tiltmeters, which are more sensitive to short-term transients than network-based geodesy (Figure 4.19). However, no precursory strain signals have been identified reliably (*75*), presumably because the nucleation zones at depth are too small to cause measurable strains at the Earth's surface.

The role of subseismic slip events in setting the stage for future earthquakes is unknown. Such events have been observed within several plate boundary fault zones (see above), but so far no convincing relationship to seismic fault slip has been demonstrated. The deformations due to subseismic events decay much more rapidly with distance from the source than seismic waves, which makes them hard to observe. Moreover, available measurement techniques leave important parts of the space-time spectrum poorly covered (Figure 4.19). Establishing the relationship between subseismic strain and future earthquakes is a clear target for research in tectonic geodesy.

4.3 EARTHQUAKE GEOLOGY

Earthquake geology was pioneered by the postearthquake investigations of Charles Darwin, G.K. Gilbert, and B. Koto in the nineteenth century. It has since evolved into three subdisciplines: neotectonics, paleoseismology, and fault-zone geology (*76*). The methodology and research issues comprised by the first two are the subject of this section; the third is incorporated into a following section on fault and rock mechanics.

Methods and Tools

Geologists have steadily improved their acuity in reading subtle features of the geologic record, setting the stage for the process-oriented investigations that now lead the geologic study of active fault systems. For example, earthquake geologists have collaborated with paleoclimatologists to improve understanding of the youngest part of the sedimentary record—the Holocene, comprising rocks up to about 10,000 years old—which contains substantial information about prehistoric earthquakes. At the same time, technological developments have contributed new tools for geologic exploration in both space and time.

Remote Sensing Landforms are the most readily accessible expression of active tectonics because they can be viewed remotely, for example, by space-based tonal images from System Probatoire pour l'Observation de la Terre (SPOT) or Landsat satellites (Figure 4.20). The details of landform topography are particularly useful in measuring the rates of fault slip and associated deformations. For many years, stereopairs of aerial photographs, in combination with topographic and geologic maps, have been primary sources of data for earthquake geologists (Figure 4.21). Digital topographic data from more precise remote-sensing platforms have been poised for some time to substantially improve the measurement and interpretation of tectonic landforms (*77*), but progress has been frustratingly slow. In a few wealthy countries, such as the United States or Taiwan, digital elevation models (DEMs) are available at 30- to 40-meter postings, which is fine enough to be useful for neotectonic and postseismic studies (Figure 4.22); however, the resolution across most of the world is considerably poorer (1-kilometer postings are common). NASA's Shuttle Radar Topography Mission (SRTM) collected the first global, high-resolution topographic data set, sampling 80 percent of the land surface at 30-meter resolution (*78*), but national security interests have thus far prevented the release of these data.

Landforms associated with one or a few earthquakes are often so small that study requires resolution of just a few centimeters. Land-based laser-ranging "total stations" have replaced the plane table and alidade as the geologist's means of producing detailed maps. A new technology that holds great promise for the rapid mapping of the ground surface at very high resolution is light detection and ranging (LIDAR), the laser-based equivalent of radar. LIDAR systems can be mounted on light aircraft equipped with inertial and GPS guidance systems to obtain vertical resolution at the decimeter level (*79*). An example of data from the 1999 Hector Mine earthquake is shown in Figure 4.23.

Methods based on the electromagnetic spectrum cannot be used to map active tectonic structures on the seafloor, where most major plate boundaries are found. Surface ships with multibeam sonar systems can map bathymetry in swaths several times as wide as the ocean depth, yielding DEMs with a resolution comparable to those currently available for much of the land surface (*80*). This mapping capability has thus far been focused on the ridge-transform systems of the mid-ocean spreading centers. Less detailed work has been done in oceanic trench environments and on the active continental margins (Figure 4.24). Side-scan sonar systems towed in midwater use the amplitude of acoustic reflection to image small-scale features not visible by swath mapping, such as the lineations due to faulting. Sleds of instruments towed on cables within tens of meters of the seafloor can collect bathymetric data at decimeter levels, although their deployment costs are very high and they are therefore used to survey only small regions of high interest. In shallow water, swept-frequency ("chirp") sonar systems can penetrate shallow sediments to return detailed images of sedimentary layering and its disruption by earthquake faulting.

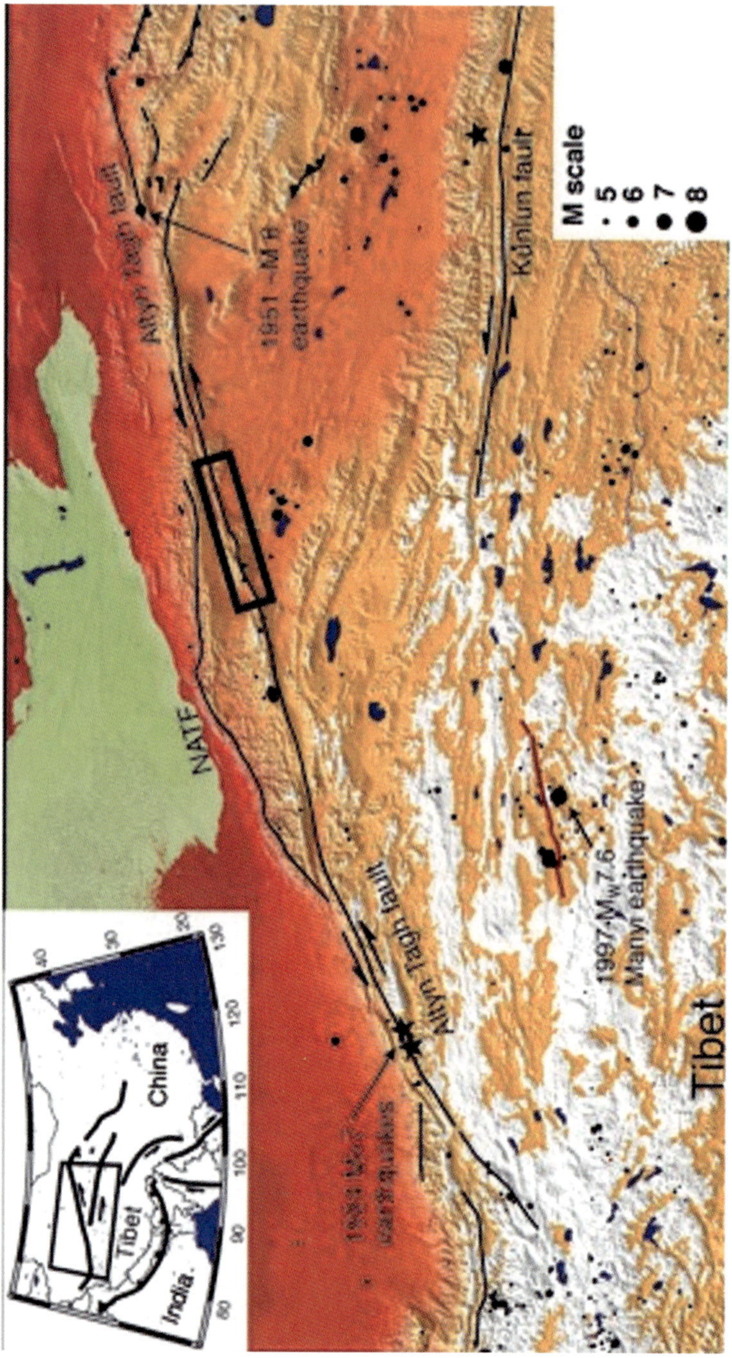
Altyn Tagh fault
NATF
Kunlun fault
M scale
5
6
7
8
Altyn Tagh fault
Manyi earthquake
Tibet
China
Tibet
India

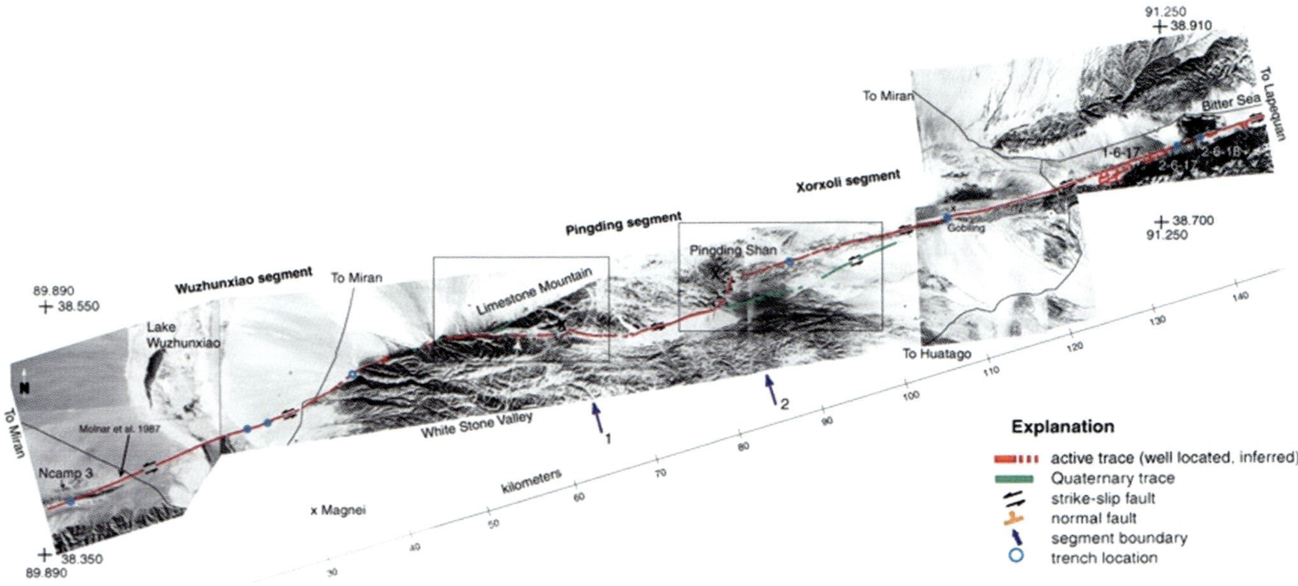

FIGURE 4.20 (*top*) Map of the Altn Tagh fault zone showing the regional tectonic setting and location in Asia. (*bottom*) Digital topography mosaic of a portion of the Altyn Tagh fault in Tibet, showing various depositional and erosional landforms that enable mapping of active faults and associated landscapes. The availability of Landsat imagery in the early 1970s allowed P. Tapponnier and P. Molnar to map the many active strike-slip faults of Tibet and surrounding regions that, at the time, were inaccessible to Western scientists. SOURCE: Z. Washburn, J.R. Arrowsmith, S.L. Forman, E. Cowgill, X.F. Wang, Y.Q. Zhang, and Z.L. Chen, Recent earthquake geology of the central Altyn Tagh fault, China, *Geology*, **29**, 1051-1054, 2001. Images are available at <http://activetectonics.la.asu.edu/atf/ATFPhotos/index.html>.

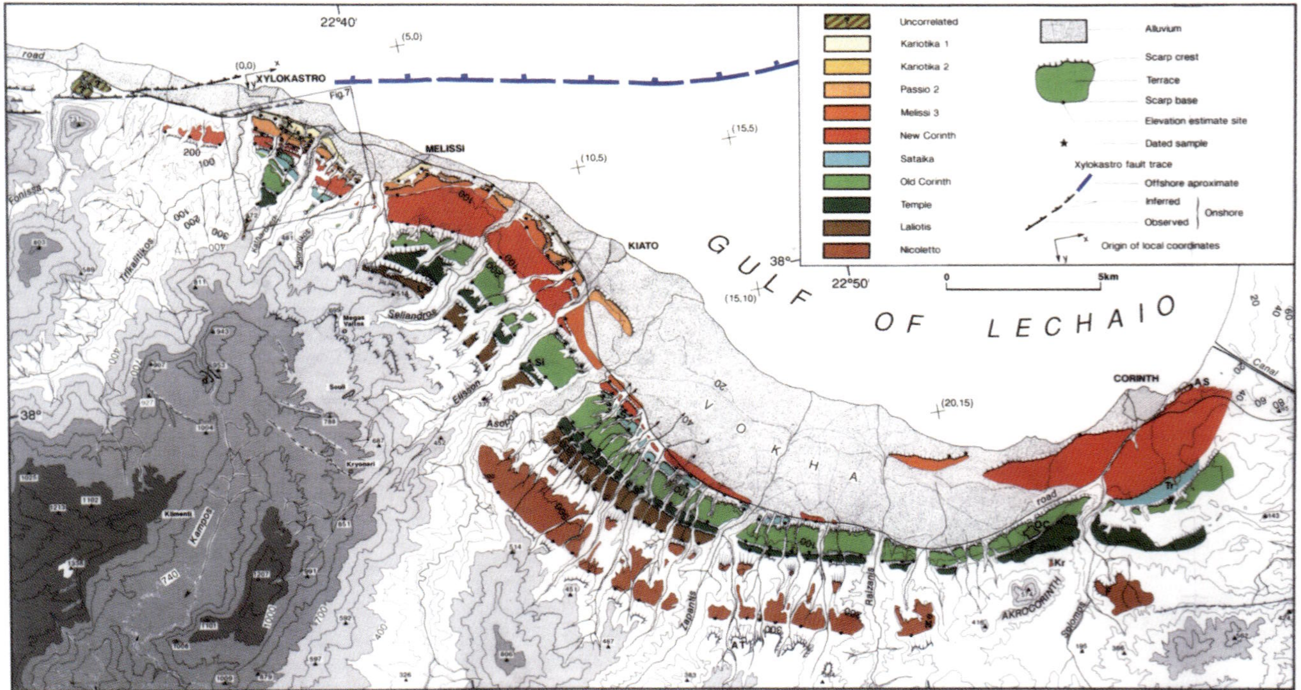

FIGURE 4.21 A flight of 10 uplifted terraces on the south flank of the Gulf of Corinth, Greece, showing progressively greater deformation with age and greater uplift toward the east. Armijo and others used these landforms to model the history and geometry of the offshore fault, as well as the response of the crust to the associated extensional deformation. SOURCE: R. Armijo, B. Meyer, G.C.P. King, A. Rigo, and D. Papanastassiou, Quaternary evolution of the Corinth Rift and its implications for the Late Cenozoic evolution of the Aegean, *Geophys. J. Int.*, **126**, 11-53, 1996. Reproduced by permission of Blackwell Publishing.

FIGURE 4.22 Shaded-relief map made from a 40-meter digital elevation model of Taiwan, used in mapping active faults and folds. Geomorphic expression of the Chelungpu fault (red line), the cause of the M 7.6 Chi-Chi earthquake in 1999, is clearly expressed. SOURCE: Charles Rubin, Central Washington University, Ellensburg, and Kerry Sieh, Caltech.

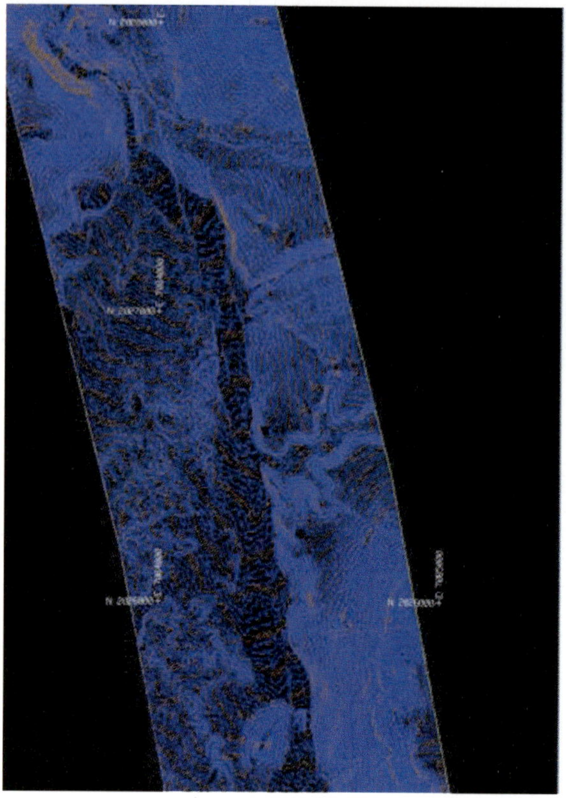

FIGURE 4.23 Topographic map of a short segment of the 1999 Hector Mine earthquake in California's Mojave desert, constructed from LIDAR data, showing small-scale features associated with the faulting. SOURCE: K. Hudnut, U.S. Geological Survey.

Geochronology Inferring crustal deformation rates and dating events requires appropriate measures of geologic time. The dates and extent of fault ruptures can be documented by historical records (*81*), but only for the last couple of millennia at most. Advances in the diversity and precision of geochronological techniques are now being applied to dating the geologic layers and erosional surfaces disturbed by prehistoric earthquakes. For example, dendrochronology (the use of annual growth rings from trees) has pinned down the dates and locations of fault ruptures in Alaska and along the San Andreas fault, and has been used to determine the dates of subduction-related submergence of coastal Washington and massive seismically induced landslides in urban Seattle (*82*). Other dating methods used in earthquake geology include tephrochronology, thermoluminescence,

FIGURE 4.24 Three-dimensional rendering of the bathymetry offshore southern California, illustrating the quality of data now achievable from swath-mapping by surface ships. The view is toward the northwest. From center ground to lower right is the trace of the right-lateral strike-slip is San Clemente Island fault. The linear ridges have formed along a restraining bend. The red hill in the background is San Clemente Island; a spectacular fault scarp forms its eastern (right) edge. SOURCE: C. Goldfinger, Oregon State University.

optically stimulated luminescence, pedology, and lichenometry (*83*). The principal isotopic techniques are radiocarbon, uranium series, helium, potassium-argon, argon-argon, and fission-track dating (*84*).

The radiocarbon technique is the workhorse for dating sediments younger than about 50,000 years, and its application has begun to supply the time series of large earthquakes for important continental fault systems (*85*). The advent of accelerator mass spectrometry (AMS) has enabled the carbon dating of samples much smaller than a gram or so by conventional techniques. Radiocarbon ages can be calibrated precisely to about 10,000 years before present using dendrochronology, yielding uncertainties as small as a decade or so. The development of AMS has also been a boon to uranium-thorium disequilibrium dating. Radiocarbon and uranium-thorium analyses of replicate samples have resulted in a calibration curve for radiocarbon dates to about 30,000 years, with uncertainties of a few hundred years (*86*).

Uranium-thorium disequilibrium dating is gaining importance in both neotectonic and paleoseismic studies. Coral heads uplifted during paleoseismic events in Vanuatu have been dated with errors of just a few years, and similarly precise timings have been made for the uplift and submergence during large earthquakes of the Sumatran subduction zone, the long-term deformation along low-latitude coastlines, and the glacial low stands and interglacial high stands of sea level in the tropics (*87*). The latter provide a basis for inferring the ages of deformed coastal deposits and surfaces at high latitudes.

Knowing the age of ground surfaces can be critical to quantifying the rate of deformation of a fold, the rate of tilt of a surface, or the rate of slip across a fault, but surfaces have been notoriously difficult to date, especially beyond the 50,000-year range of radiocarbon analysis. Surface-exposure dating by cosmogenic isotopes, especially ^{10}Be, ^{26}Al, and ^{36}Cl, has resolved some of these problems, and these techniques have been applied to faulting caused by the Indian-Asian collision, normal fault scarps in limestone in the eastern Mediterranean, and California marine terraces (*88*). The last were revealed to be tens of thousands of years younger than previously thought, implying that earthquakes such as the 1989 Loma Prieta event must be far more frequent.

Neotectonics

Although plate tectonics furnishes the first-order framework for understanding global seismicity, most tectonically active plate boundaries exhibit significant second-order complexities that are responsible for a large percentage of the destructive earthquakes of the twentieth century (see Chapter 2). Placing the resulting diversity of fault structures in a consistent kinematical framework is the program of neotectonics.

Maps of Active Faults and Folds Active faults and folds have been mapped at a scale of 1:1,000,000 to 1:10,000,000 for Japan, Turkey, the United States, New Zealand, China, and many other regions (*89*). These maps are commonly derived from interpretations of aerial photographs, satellite imagery, or bathymetry verified by field mapping and sampling. Despite this progress, no global map of active faults and folds has been compiled at even these coarse scales (*90*). Furthermore, such maps seldom are in the digital formats that allow ready access to a full range of geologic information.

More detailed maps and databases of active regions reveal the geometric and kinematic data necessary to forecast future behavior and explain past seismicity. These larger-scale maps, such as for the North Anatolian fault or the Nankai trough in Japan, often include features derived from stratigraphic data, seismic reflection profiles, seismicity, and tectonic landforms visible in topography, as well as ground-based mapping.

Slip Rates on Active Faults The earthquake production rate is a function of the rate of slip on active faults. At the San Andreas fault, where a 4000-year-old channel and a 14,000-year-old alluvial fan are offset right-laterally, the derived slip rate is about 33 millimeters per year (Figure 4.25). Here, offset during the great earthquake of 1857 was about 9 meters. If this were typical and if earthquakes were periodic, the 1857 event would repeat about every 270 years.

Such simple calculations are a starting point for determining moment-release rates, but strain relief is commonly more complex. Rates of slip along mid-ocean transform faults, for example, can be well constrained by the separation of the magnetic anomalies at the adjacent spreading centers or from globally consistent plate-motion models like NUVEL-1. However, because the rheology of rocks within transform fault zones favors aseismic rupture (*91*), the rate of production of earthquakes along oceanic transforms is almost always much lower than would be predicted from the slip rate. Cosmogenic exposure dating of surfaces in central Asia, using ^{10}Be and ^{26}Al, has begun to yield reliable slip rates for the great strike-slip faults of the Indian-Eurasian collision (*92*). Slip rates have been estimated by similar calculations for the growth of folds, blind thrusts, and other reverse faults in many regions of the globe (*93*).

Variations in sea level on ice-age time scales of 10^5 to 10^6 years have produced suites of datable landforms and strata with measurable deformations. Coastal terraces and deposits formed during sea-level highstands about 125,000, 105,000, 82,000, and 5000 years ago have been used widely to determine average rates of uplift and submergence in coastal regions ranging from less than 1 millimeter per year to about 10 millimeters per year (Figure 4.26). The rate for the Corinth fault, as an example, has been about 0.7 millimeter per year over the past several hundred thousand years (Figure 4.21).

Probing the Third Dimension Neotectonics is based on the interpretation of structures and stratigraphy at the surface; however, the extrapolation of active features to depth depends on the integration of surface data with subsurface information from drilling and seismic imaging. Seismic reflection surveys conducted for petroleum exploration and borehole data logged from oil and gas wells have furnished critical information on the three-dimensional structure of the upper crust in seismically active areas (*94*). The correlation of faults located in the upper 5 to 10 kilometers by these methods with precisely relocated earthquake hypocenters at

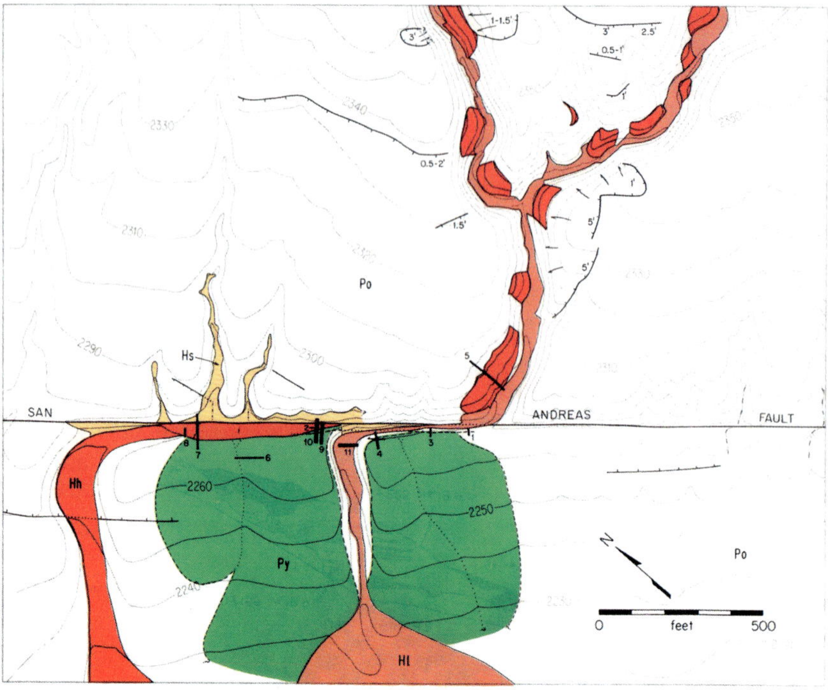

FIGURE 4.25 A detailed geologic map of Wallace Creek in the Carrizo Plain of central California. The 130- and 475-meter offsets of a stream channel (orange) and an alluvial fan apex (green) yield a Holocene slip rate for the San Andreas fault of 34 ± 3 millimeters per year. SOURCE: K. Sieh and R.H. Jahns, Holocene activity of the San Andreas fault at Wallace Creek, California, *Geol. Soc. Am. Bull.*, **95**, 883-896, 1984; see also <http://www.scec.org/wallacecreek/>.

greater depths is particularly powerful in delineating reverse faulting, and it has been used to delineate a major blind thrust beneath central Los Angeles (Figure 4.27). Future progress will likely come through "unified structural representations" that employ model-based methods and advanced visualization tools to integrate large sets of neotectonic, borehole, and seismic data, including tomographic images produced from natural earthquake sources (*95*).

Paleoseismology

Paleoseismology is the geological investigation of individual earthquakes decades, centuries, or millennia after their occurrence (*96*). Whereas neotectonics considers deformations summed over many episodes of

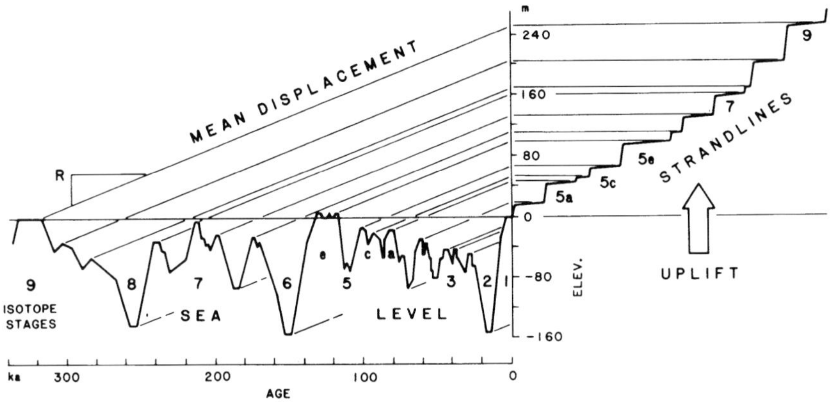

FIGURE 4.26 The rate of uplift of a series of marine terraces is commonly estimated by relating the elevation of the terraces to the history of sea-level fluctuations, which has been determined from analysis of oxygen isotope ratios in the minuscule shells of plankton accumulating on the ocean floor for hundreds of thousands of years. In this idealized drawing, the parallel, sloping lines represent the rate of uplift of the coastline. SOURCE: K.R. LaJoie, Coastal tectonics, in National Research Council, *Active Tectonics*, National Academy Press, Washington, D.C., 266 pp., 1986.

deformation, paleoseismology focuses on the geological record of specific events. Evidence reconstructed from sequences of large earthquakes spanning thousands of years is proving to be crucial to the general understanding of the size distribution of earthquakes, irregularities of the seismic process, and space-time patterns of fault slippage. By virtue of its extension through many earthquake cycles, paleoseismology provides some of the best information for long-term forecasting of major earthquakes on individual faults and for investigating the nature of the earthquake cycle. Paleoseismic features are associated with three types of processes:

1. disarticulation at a fault rupture, including fault scarps, wedges of debris at their base, fissures, and disarticulated strata of various origins;
2. changes in sea level or disruption of fluvial gradients, including ponded sediment and deformed fluvial and marine terraces; and
3. secondary effects of earthquake rupture, including manifestations of strong ground shaking such as mass wasting (landslides, rockfalls, and turbidites), liquefaction phenomena (sand blows, clastic dikes, seismites, lateral spreads), and tsunami deposits.

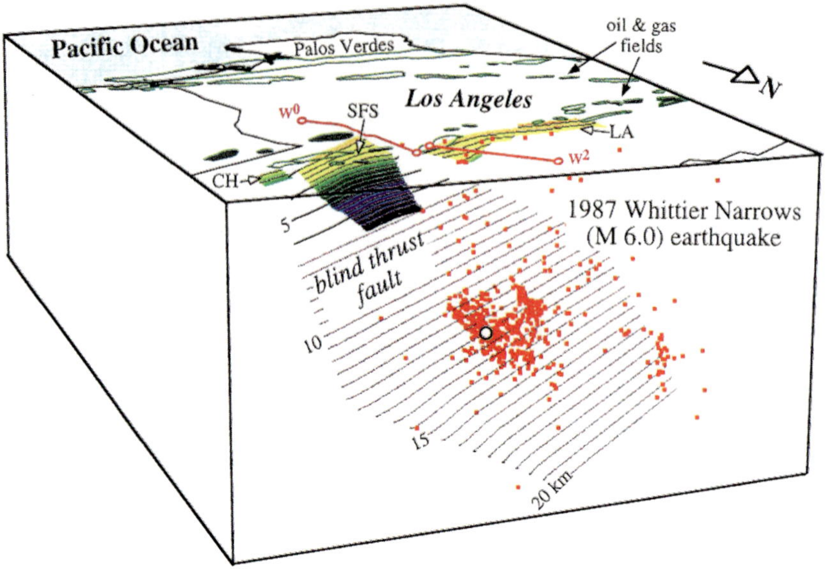

FIGURE 4.27 Southwest-looking perspective of the Puente Hills blind thrust fault (contoured in kilometers), which ruptured to produce the M 6 Whittier Narrows earthquake of 1987 (white dot). Relocated aftershocks (red dots) delineated a north-dipping plane, and seismic reflection surveys (colored contours) showed that this fault extends toward the surface beneath the active Puente Hills anticline. SOURCE: J.H. Shaw and P. Shearer, Southern California Earthquake Center.

Fault Rupture Many important active faults ruptured most recently in prehistoric or pre-instrumental time. In the case of southern California's great 1857 earthquake, the approximate length of the rupture could be deduced from written accounts of shaking, but the actual slip as a function of distance along the San Andreas fault was determined only by measurement of offset landforms more than a century after the event (*97*); these measurements implied a moment magnitude of about 7.9. Incrementally larger offsets suggested that prior events were similar in size (*98*). Other examples include ruptures along faults in New Zealand and Alaska, the North Anatolian fault, and the Fuyun fault (*99*). The erosion of dated scarps in extensional regimes, such as the Basin and Range Province of the western United States (*100*), permits determination of the sequence of ruptures along normal faults. Numerous paleoseismic investigations have used colluvial wedges as evidence of prehistoric ruptures. These features are preserved along dip-slip faults.

Paleoseismic Sea-Level and Fluvial Grade Changes Geomorphic and stratigraphic evidence for changes in sea level often indicate paleoseismic uplift or submergence of a coastline, and flights of small coastal terraces commonly indicate a series of prehistoric seismic uplifts. The extent and magnitude of uplift allow estimation of the source parameters of the underlying rupture plane. Deformed river beds, such as the Mississippi, also provide constraints on the source parameters of seismic faults and, in combination with fault-bend fold modeling, reveal the nature of the events, here one of the great 1811-1812 sequences (*101*).

Sudden submergence associated with large earthquakes of the Cascadian subduction zone and overlying faults appears in estuarine stratigraphy along the coastlines of Oregon, Washington, and British Columbia (*102*). Discovery of these records helped to change the widely held view that subduction in the Pacific Northwest was principally aseismic.

Restriction of the upward growth of coral by exposure during low tides (*103*) offers high resolution of records of submergence and uplift. This technique has been used to estimate the source parameters for the giant (M 8.9-9.2) Sumatran subduction event of 1833 (*104*).

Seismically induced landslides, rockfalls, and submarine turbidity flows are well documented, as are seismically induced liquefaction phenomena (*105*). The widespread occurrence of rockfalls and landslides, shown by lichenometry to be of similar age, may indicate paleo-earthquakes in parts of New Zealand (*106*). The age of submarine turbidite deposits off the coast of Washington and Oregon suggests that they were dislodged during large Cascadian subduction-zone earthquakes (*107*).

Ancient liquefaction features in the central United States reveal not only that the region of the 1811-1812 New Madrid earthquakes has suffered prior large earthquakes about every 600 years, but also that the Wabash Valley and other midcontinent regions are susceptible to damaging earthquakes (*108*). Paleoliquefaction gives evidence that events such as the M 7 Charleston earthquake have stricken the South Carolina region about every 1500 years (*109*). Deformed lake beds of the Dead Sea not only provide a record of strong shaking, but also show that ruptures cluster into 10,000-year sequences separated by similarly lengthy periods of quiescence (*110*).

4.4 FAULT AND ROCK MECHANICS

In the context of earthquake science, the study of fault and rock mechanics aims to describe the macroscopic phenomena of fault slip and rock deformation in terms of the microscopic transport processes that operate on crystalline and atomic scales. This discipline lies at the core of earthquake studies because it connects the phenomenology of fault-system science to the reductionist approach of condensed-matter physics. Its activities are focused in two observational environments:

1. laboratory research to characterize the properties of rocks and faults under the pressure, temperature, chemical, and strain-rate conditions that operate during the earthquake cycle—such observations are basic ingredients for the investigation of earthquake processes and the formulation of mechanistic and phenomenological models of rock friction; and
2. field research to elucidate the structure and processes of real fault zones, accounting for differences in rock types and tectonic regimes—these observations provide information on the tectonic stresses that drive lithospheric deformation and on the scaling of laboratory-based models to the parameter range of tectonic earthquakes.

The gap between laboratory scales of centimeters and field scales of kilometers has been a major stumbling block. Valuable information has come from rock-deformation and seismicity measurements in controlled environments such as boreholes and deep mines (see below), but bridging this gap relies heavily on fault-system modeling, the principal subject of Chapter 5.

Laboratory Studies of Rock Deformation

Quantifying the stress-strain response is conducted primarily in the laboratory, where rock samples are deformed in high-stress testing machines under controlled-state conditions. The sizes of the samples are small, ranging from a centimeter or less to a meter at most (*111*).

Rock Friction The modern study of fault friction began in the mid-1960s, when William Brace and his coworkers at the Massachusetts Institute of Technology first investigated stick-slip behavior as a mechanical model of earthquakes through a series of laboratory experiments. They recognized that stick-slip faulting depends on how the friction changes when slip conditions, particularly the sliding velocity, are modified. The introduction of servo-controlled testing machines in 1970, and the subsequent development of high-precision, double-direct-shear and rotary-shear devices (*112*), allowed detailed measurements of friction to be made for a wide range of materials under variable sliding conditions. This work, described in Section 2.5, led to the formulation of rate- and state-dependent friction relations (Box 4.4; Figure 4.28).

The micromechanical understanding of friction is incomplete and remains the subject of active laboratory research (*113*) as well as theoretical studies involving computer simulations of granular media (Figure 4.29).

Nevertheless, the rate-state description has enjoyed considerable success as a phenomenological theory of fault friction (*114*). It has unified the concepts of static and dynamic friction into a time-dependent theory of frictional evolution, furnished realistic constitutive equations for modeling earthquakes as stick-slip instabilities (*115*), and provided a mechanical basis for describing the key aspects of earthquake nucleation, including earthquake productivity as a function of stress (*116*) and precursory sliding. The size of the nucleation zone generating the precursory sliding is proportional to the critical slip distance D_c, so that the small laboratory values of D_c—less than a few hundred microns for gouge thicknesses up to 3 millimeters (*117*)—imply that the nucleation process will typically involves slip on a fault patch with a radius less than a few tens of meters, equivalent to only a magnitude-zero earthquake. The smallness of this aseismic moment release may explain why precursory slip has evaded detection on surface strainmeters (*118*).

Rate-dependent models of friction also furnish the framework for understanding the depth distribution of shallow earthquakes. In the notation of Box 4.4, the increase in friction with (steady-state) slipping velocity is equal to the difference between the coefficients a and b, so that the friction is *velocity strengthening* where $a - b > 0$ and *velocity weakening* where $a - b < 0$. The latter condition is necessary to nucleate dynamic faulting. How the difference $a - b$ depends on pressure, temperature, and composition can be investigated in the laboratory. The thickness of the seismogenic zone can usually be ascribed to an upper cutoff associated with velocity strengthening in a shallow layer of poorly consolidated rocks and sediments and a lower cutoff caused by a temperature-dominated increase in $a - b$ with depth (Figure 4.30).

Ductile Flow In addition to its effect on friction, raising temperature enhances the mobility of dislocations, allowing plastic behavior to occur at lower stresses, while elevating pressure suppresses the nucleation and growth of cracks by increasing the normal stresses across crack surfaces. Consequently, at some depth, called the brittle-ductile transition, rock deformation takes place entirely by ductile flow (*119*). The way in which jerky fault motions couple to this steadier, deeper flow is a complex issue, requiring detailed information about rock rheology below the seismogenic zone. For example, according to the model in Figure 4.30, crustal seismicity should stop at depths shallower than the brittle-ductile transition, which implies that the lithosphere is stronger than predicted by the early laboratory-based models, which pegged the brittle-ductile transition to the base of the seismic zone.

Detailed measurements of deformation mechanisms in rocks indicate that several deformation mechanisms contribute to the transition from

BOX 4.4 Rate- and State-Dependent Friction

In one type of rock-friction experiment, a sample is cut to introduce a fault surface and subjected to a constant normal stress σ_n across this surface. The load point in a very stiff testing apparatus is first driven at constant velocity V_0 until the friction on the fault surface obtains a steady-state value $\mu_0 = \tau_0/\sigma_n$, where τ_0 is the shear stress. As in the case of static strength, this "base friction" is observed to depend only weakly on lithology and temperature.[1] The velocity is then increased to V, and the change in μ is monitored. Many materials—rocks of various types, fault gouge, glass, paper, some metals and plastics, even wood (see Figure 4.28)—exhibit a characteristic response in which the friction first jumps quickly to a new value $a \ln (V/V_0)$ and then decays more or less exponentially by an amount $b \ln (V/V_0)$ over the distance D_c. The critical slip distance D_c varies from 2 to 100 microns and increases with surface roughness; for experiments where the fault zone contains gouge, it also increases with the particle size and thickness of the gouge. This behavior is described by a *rate-state friction equation*:

$$\mu = \mu_0 + a \ln(V/V_0) + b \ln(V_0\theta/D_c).$$

In this expression, θ is a state variable with the dimensions of time, which satisfies a first-order differential equation of the form $d\theta/dt = F(V\theta/D_c)$, where $F(1) = 0$. The latter condition implies that in steady state ($d\theta/dt = 0$), the state variable becomes $\theta_{ss} = D_c/V$ and the friction attains the value

$$\mu_{ss} = \mu_0 + (a - b) \ln(V/V_0).$$

Therefore, the sign of $a - b$ determines whether the friction associated with a velocity increase evolves to a higher value ($a - b > 0$: velocity strengthening) or a lower value ($a - b < 0$: velocity weakening). The stability of a fault to slow loading depends on this steady-state rheology and not on the details of F. Velocity weakening is required for stick-slip instabilities, and they occur in less stiff apparatus than imagined here.

The temporal behavior of the friction is determined by the details of the evolution-rate function $F = d\theta/dt$, and several forms are in common use. The version of the constitutive relation originally put forward by Dieterich[2] corresponds to

$$\frac{d\theta}{dt} = 1 - \frac{V\theta}{D_c}.$$

An alternative proposed by Ruina[3] is

$$\frac{d\theta}{dt} = \frac{V\theta}{D_c} \ln\left(\frac{V\theta}{D_c}\right).$$

Both provide adequate descriptions of the velocity-step experiments, but there are important differences in their behaviors. As the slip velocity drops to zero, the evo-

(continued on next page)

BOX 4.4 *(continued)*

lution rate vanishes in Ruina's version, whereas it goes to unity in Dieterich's. Hence, the Ruina form requires that slip occur to change friction, and is thus called a "slip law," while the Dieterich form is a "slowness law" that allows faults to strengthen during periods of stationary contact. Various analyses have been made to assess the relative merits of these evolution laws in fitting laboratory data[4] or simulating the transient behaviors of real faults. As C. Marone noted in a recent review, however, "distinguishing between them in the laboratory, even at room temperatures, has proven difficult . . . the distinction is subtle and often unresolvable owing to noise and other trends in the data." Alternate forms of the evolution function have also been investigated,[5] and the rate-state equations have been generalized to include more than one state variable[6] and changes in normal stress σ_n.[7]

---

[1]R.M. Stesky, W.F. Brace, D.K. Riley, and P.-Y.F. Robin, Friction in faulted rock at high temperature and pressure, *Tectonophysics*, **23**, 177-203, 1974; J.D. Byerlee, Friction of rocks, *Pure Appl. Geophys.*, **116**, 615-626, 1978. This is similar to the insensitivity to temperature and composition found previously in the frictional-strength experiments that established Byerlee's law.

[2] J.H. Dieterich, Modelling of rock friction 1. Experimental results and constitutive equations, *J. Geophys. Res.*, **84**, 2161-2168, 1979.

[3] A. Ruina, Slip instability and state variable friction laws, *J. Geophys. Res.*, **88**, 10,359-10,370, 1983.

[4]Among the most definitive laboratory data are the slide-hold-slide tests of N.M. Beeler, T.E. Tullis, and J.D. Weeks (The roles of time and displacement in the evolution effect in rock friction, *Geophys. Res. Lett.*, **21**, 1987-1990, 1994), which show an increase in static friction with hold time in better agreement with the Dieterich law (i.e., higher than predicted by the Ruina law).

[5]For example, G. Perrin, J.R. Rice, and G. Zheng (Self-healing slip pulse on a frictional surface, *J. Mech. Phys. Solids*, **43**, 1461-1495, 1995) consider the evolution function $d\theta/dt = 1 - (V\theta/D_c)^2$, which is an aging law (i.e., $\theta \rightarrow t$ as $V \rightarrow 0$). It yields frictional responses that are symmetric for positive and negative steps in loading rate, unlike the Dieterich form, which predicts asymmetric pulse shapes.

[6]A constitutive equation with two state variables was proposed by Ruina in his original 1983 paper; see also T.E. Tullis and J.D. Weeks (Constitutive behavior and stability of frictional sliding of granite, *Pure Appl. Geophys.*, **124**, 383-414, 1986).

[7]M.F. Linker and J.H. Dieterich, Effects of variable normal stress on rock friction: Observations and constitutive equations, *J. Geophys. Res.*, **97**, 4923-4940, 1992.

brittle behavior at shallow levels to fully plastic flow at great depths. Tensional microcracking, frictional sliding, and other nominally brittle mechanisms persist over a range of conditions where the sample shows macroscopically ductile behavior. The rheology of rocks in this "semi-brittle" field differs from the predictions of flow laws measured at higher temperatures. In particular, semibrittle behavior is characterized by dilatancy and a greater dependence of strength on pressure (*120*). Combining the laboratory data on brittle, semibrittle, and ductile behavior of rocks with temperature profiles, compositional models, and scaling laws provides estimates of the rheological structure of the lithosphere that can be tested with field observations of deformation (*121*).

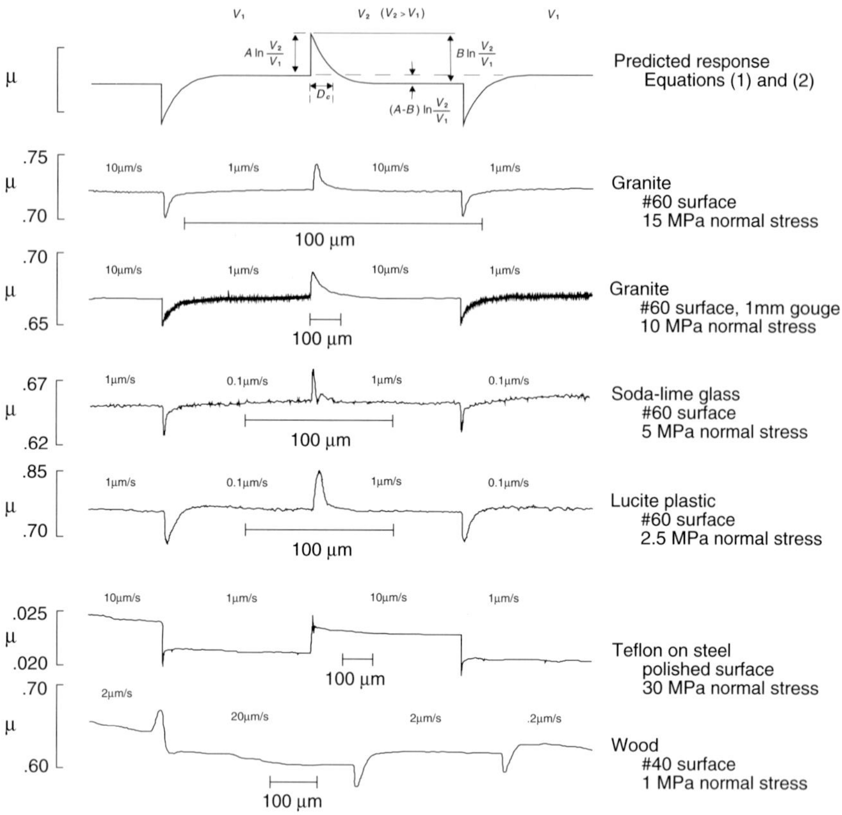

FIGURE 4.28 Results of laboratory experiments showing a similar frictional response for a variety of surfaces, including rock, glass, plastic, steel, and wood. Coefficient of friction μ is plotted for stepwise changes in slipping velocity V. The upper panel is the predicted response for the slowness version of the rate- and state-dependent friction law, described in Box 4.4. The similarity indicates the wide applicability of this formalism. SOURCE: J.H. Dieterich and B. Kilgore, Direct observation of frictional contacts: New insights for state-dependent properties, *Pure Appl. Geophys.*, **143**, 283-302, 1994.

Field Studies of Faulting

Rock masses involved in faulting have large-scale structural features (joints, gouge zones, and compositional boundaries) that make their behavior different from the rock materials tested in the laboratory. Data gathered in the field are therefore essential in establishing the applicability of laboratory models and determining scaling relationships for the key constitutive parameters.

(a)

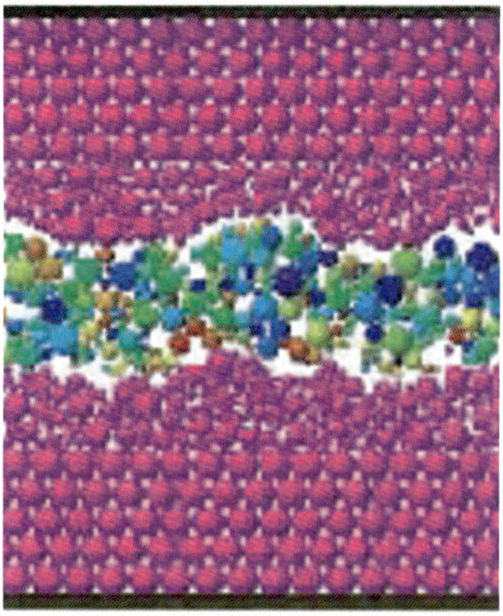

(b)

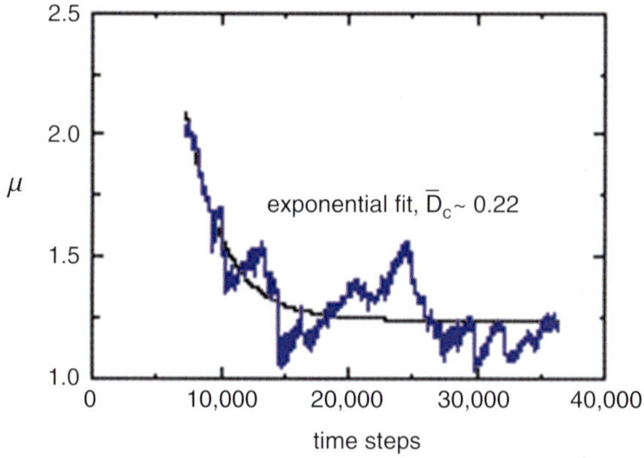

FIGURE 4.29 (a) Two-dimensional lattice solid model of fault gouge between two rough surfaces used in numerical simulations of frictional behavior. (b) Decay of macroscopic friction averaged over several slide-hold-slide cycles computed for this model (blue curve), and an exponential fit to obtain the critical slip distance D_c (black curve). SOURCE: S. Abe, J.H. Dieterich, P. Mora, and D. Place, Simulation of the influence of rate- and state-dependent friction on the macroscopic behavior of complex fault zones with the lattice solid model, *Pure Appl. Geophys.*, **159**, 1967-1983, 2002.

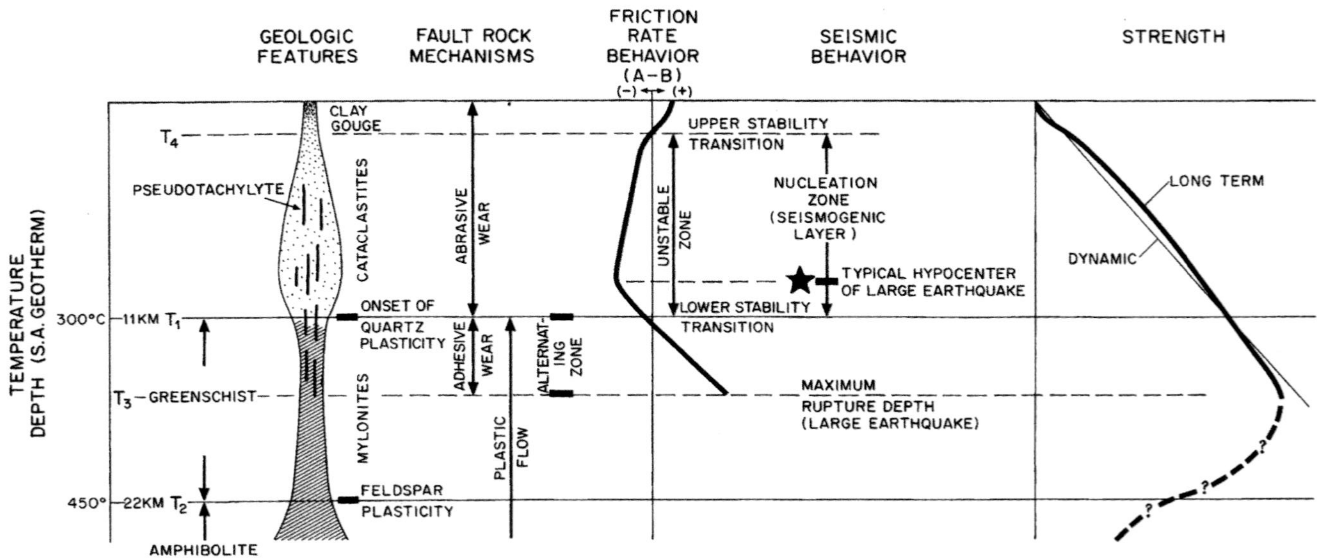

FIGURE 4.30 Hypothetical model of a major continental strike-slip fault such as the San Andreas, showing (from left to right) the depth variation of geologic features, deformation mechanisms, friction rate parameter $a - b$, seismic behavior, and strength (fault yield stress). Earthquakes can nucleate in the unstable zone where $a - b$ is negative, implying friction that weakens with increasing sliding velocity. SOURCE: C.H. Scholz, *The Mechanics of Earthquakes and Faulting*, Cambridge University Press, Cambridge, U.K., p. 129, 1990. Reprinted with the permission of Cambridge University Press.

Fault Mechanics Cracks occur in three primary modes: mode-I tensile cracks where the displacement is normal to the plane of the crack, mode-II cracks where displacements are parallel to the crack plane and normal to the crack edge, and mode-III cracks where displacements are parallel to the crack plane and the crack edge. A fundamental problem, identified in the 1970s, is that simple mode-II shear cracks, which resemble faults, cannot propagate in their own plane owing to the stress concentrations at the crack tip (*122*). Detailed studies over the next decade demonstrated that shear cracks propagate by forming tensile cracks at their tips to relieve the local stress concentrations (*123*). The extension of these tensile cracks then plays a fundamental role in linking up parallel mode-II cracks in an en echelon stair-step pattern, concentrating shear stresses on the plane of the mode-II crack. Geologic observations indicate that fault formation processes are broadly similar. For example, there is a well-defined correlation between the total width of a fault zone (i.e., the width of the en echelon shear faults) and the total displacement on a fault (*124*). For individual faults, there is similar correlation due to the formation of gouge along the fault plane (*125*). Observations of the structures, displacements, and mineralogy within large granitic bodies have provided strong evidence for the coupled growth of mode-I and mode-II cracks in the nucleation and formation of large-scale shear faults (*126*).

On a larger scale, the field analysis of fault structures furnishes information on the apparent coefficient of friction governing fault strength. As deformation progresses within a fault system, crustal blocks undergo rotations and the faults bounding these blocks become misoriented with respect to the regional stress system, eventually causing them to "lock up" at a critical angle to the maximum compressive stress (*127*). Under the assumption that the sliding surfaces have zero cohesive strength (i.e., their strength is proportional to the normal stress), the critical angles can be used to estimate the coefficient of friction. Low-displacement faults in a variety of tectonic regimes typically give values consistent with Byerlee's law: $\mu = 0.6\text{-}0.85$ (*128*). The notable exceptions are shallow-dipping extensional detachment faults and major transform faults like the San Andreas, which are anomalously weak and may require some mechanism to maintain fluid overpressures.

Fault-Zone Petrology Field-scale observations provide important constraints on the genesis of fine-grained rocks in fault zones (*129*). C. Lapworth first described highly deformed rocks, which he called mylonites (literally, "milled rock"), in the Moine thrust of northwestern Scotland, but it was not until the latter part of the 20th century that geologists recognized that recrystallization and plastic flow can cause the fine grain sizes found in these high-temperature rocks (*130*). Fine-grained fault rocks are now generally classified

8. U.S. Geological Survey, *Requirements for an Advanced National Seismic System*, U.S. Geological Survey Circular 1188, Denver, Colo., 56 pp., 1999. This report lists 41 regional networks comprising 3095 earthquake-monitoring stations that are operated by the USGS, research universities, state geological surveys, private companies, and other organizations throughout the United States.

9. A major program to densify regional networks in seismically active regions was launched by the USGS in the late 1960s (J.P. Eaton, W.H.K. Lee, and L.C. Pakiser, Microearthquakes and mechanics of earthquake generation, San Andreas fault, *Tectonophysics*, **9**, 259-282, 1970). The principal goal was the accurate location of small earthquakes to delineate fault structures and measure changes in low-level seismicity that might be precursory to larger earthquakes; therefore, emphasis was placed on recording the arrival times of initial *P* waves using very sensitive, high-frequency instruments with precise timing control.

10. National Research Council, *Assessing the Nation's Earthquakes: The Health and Future of Regional Seismic Networks*, National Academy Press, Washington, D.C., 67 pp., 1990. That report emphatically recommended that "the federal government should establish a more rational, coordinated, and stable means of support for the seismic networks of the United States." Several important steps have been taken toward improving and coordinating seismic monitoring efforts at the regional level. Two coordinating bodies have been formed: the Council of the National Seismic System, comprising primarily the operators of conventional weak-motion monitoring networks (<http://www.cnss.org>), and the Consortium of Organizations for Strong Motion Observational Systems, whose membership mostly includes strong-motion network operators.

11. The piezoelectric transducers used in mine monitoring have usable response to 7 kilohertz and are recorded at rates of up to 10,000 samples per second. Local networks comprising these sensors can locate events of magnitudes below –2 in rock volumes with linear dimensions of 1 kilometer.

12. E. Richardson and T.H. Jordan, Seismicity in deep gold mines of South Africa: Implications for tectonic earthquakes, *Bull. Seis. Soc. Am.*, **92**, 1766-1782, 2002.

13. W.H. Bakun and A.G. Lindh, The Parkfield, California prediction experiment, *Earthq. Predict. Res.* **3**, 285-304, 1985. In addition to the USGS, participating institutions include the University of California, Berkeley; Lawrence Berkeley National Laboratory; University of California, Riverside; University of Wisconsin; Rensselaer Polytechnic Institute; University of California, San Diego; and Australia's Commonwealth Scientific and Industrial Research Organization (CSIRO) Exploration and Mining.

14. A. Michelini and T.V. McEvilly, Seismological studies at Parkfield: 1. Simultaneous inversion for velocity structure and hypocenters using B-splines parameterization, *Bull. Seis. Soc. Am.*, **81**, 524-552, 1991.

15. For example, C.G. Sammis, R.M. Nadeau, and L.R. Johnson, How strong is an asperity? *J. Geophys. Res.*, **104**, 10,609-10,619, 1999.

16. For example, E.D. Karageorgi, T.V. McEvilly, and R. Clymer, Seismological studies at Parkfield, IV. Variations in controlled-source waveform parameters and their correlation with seismicity, 1987-1994, *Bull. Seis. Soc. Am.*, **87**, 39-49, 1997.

17. R.M. Nadeau and T.V. McEvilly, Fault slip rates at depth from recurrence intervals of repeating microearthquakes, *Science*, **285**, 718-1138, 1999.

18. R.M. Nadeau and L.R. Johnson, Seismological studies at Parkfield, VI. Moment release rates and estimates of source parameters for small repeating earthquakes, *Bull. Seis. Soc. Am.*, **88**, 790-814, 1998.

19. W.L. Ellsworth, M.V. Matthews, R.M. Nadeau, S.P. Nishenko, P.A. Reasenberg, and R.W. Simpson, A physically-based earthquake recurrence model for estimation of long-term earthquake probabilities, Workshop on Earthquake Recurrence: State of the Art and Directions for the Future, Instituto Nazionale de Geofisica, Rome, February 22-25, 1999.

20. A national network was recommended, for example, in National Research Council, *U.S. Earthquake Observatories: Recommendations for a New National Network*, National Academy Press, Washington, D.C., 122 pp., 1980.

21. The National Strong-Motion Network operated by the USGS currently involves 900 accelerographs at 628 stations in 32 states and the Caribbean (see <http://nsmp.wr.usgs.gov>).

22. The PEER database is unique in distributing strong-motion data that have been processed to remove instrument response and noise and integrated for velocity and displacement using a standard algorithm, which makes the data especially suitable for engineering applications (see <http://peer.berkeley.edu/smcat>).

23. National Research Council, *Recommendations for the Strong-Motion Program in the United States*, National Academy Press, Washington, D.C., 59 pp., 1987.

24. See <http://www.k-net.bosai.go.jp/>.

25. The PASSCAL pool comprises a variety of instrument types and configurations, including more than 400 portable, seismic instruments. IRIS manages a PASSCAL instrument center at New Mexico Tech in Socorro, New Mexico, which provides software and logistical support to scientists in the design and execution of their experiments. RAMP instruments can be shipped anywhere in the world in less than 24 hours. See <http://www.iris.edu/passcal/passcal.htm>.

26. The USGS maintains a pool of about 100 portable digital seismographs in Menlo Park, California, and Golden, Colorado, for aftershock studies, and it also operates portable seismographs in urban arrays (e.g., San Jose, California; Seattle, Washington) to record small and moderate earthquakes that can be analyzed to determine site response and basin effects at frequencies of 0.1 to 20 hertz. These deployments have proven that even weak ground motions from nearby M 2 earthquakes can be recorded successfully in high-noise urban environments.

27. Simplifications are often made in parameterizing Earth structure. For example, high-frequency seismic waves are relatively insensitive to independent variations in the mass density ρ except at reflecting interfaces; attenuation in pure compression can usually be ignored relative to attenuation in shear, and low-frequency waves are relatively insensitive to small-scale variations in the shear attenuation factor. On the other hand, the wave velocities in some regions of the Earth can be moderately anisotropic (i.e., they depend on the direction of wave propagation, as well as position), which requires the introduction of additional elastic parameters.

28. The NEIS telemeters data from the GSN, USNSN, and other networks, locating approximately 15,000 earthquakes annually, and publishes data on these events in a variety of formats. The Quick Epicenter Determinations, a very preliminary list of earthquakes, is computed daily and is available via the Internet. The Preliminary Determination of Epicenters (PDE) is published and distributed weekly to those contributing data to the NEIS. The PDE, Monthly Listing is published monthly and is also available over the Internet. The Earthquake Data Report, also a monthly publication, provides additional and more detailed information for the use of seismologists on a data exchange basis. Other publications include CD-ROMs, maps, and an annual book *United States Earthquakes*.

29. The ISC was formed in Edinburgh in 1964 to continue the work of the British Geological Survey in producing the International Seismological Summary (described in Section 2.3). In 1970, with the help of the United Nations Educational, Scientific and Cultural Organization and other international scientific bodies, the center was reconstituted as an international nongovernmental body, funded by interested institutions in about 7 countries; today nearly 50 countries fund the ISC. The ISC analysis of earthquake data is undertaken in monthly batches and begins after 22 months to allow the information used to be as complete as possible; the final product, the *Bulletin of the International Seismological Centre*, is published routinely about two years after the data are collected.

30. Harvard CMT solutions are published quarterly in the *Physics of the Earth and Planetary Interiors* and are accessible on-line. Quick CMTs are produced in near-real time and distributed to interested parties via e-mail. The CMT methodology is described by A.M. Dziewonski, T.-A. Chou, and J.H. Woodhouse, Determination of earthquake source parameters from waveform data for studies of global and regional seismicity, *J. Geophys. Res.*, **86**, 2825-2852, 1981; and A.M. Dziewonski and J.H. Woodhouse, An experiment in systematic study of global seismicity: Centroid-moment tensor solutions for 201 moderate and large earthquakes of 1981, *J. Geophys. Res.*, **88**, 3247-3271, 1983. The Quick CMT algorithm is discussed by G. Ekström, Rapid earthquake analysis utilizes the Internet, *Computers in Physics*, **8**, 632-638, 1994.

31. Second moments of the space-time distribution of faulting, which include characteristic dimensions, duration, and directivity, can be recovered from low-frequency seismic waves (J. McGuire, L. Zhao, and T.H. Jordan, Teleseismic inversion for the second-degree moments of earthquake space-time distributions, *Geophys. J. Int.*, **145**, 661-678, 2001).

32. H.M. Iyer, Seismological detection and delineation of magma chambers: Present status with emphasis on the western USA, in *Volcanic Seismology. IAVCEI Proceedings in Volcanology* 3, P. Gasparini, R. Scarpa, and K. Aki, eds., Springer, Berlin, pp. 299-338, 1992.

33. H.M. Benz, B.A. Chouet, P.B. Dawson, J.C. Lahr, R.A. Page, and J.A. Hole, Three-dimensional P and S wave velocity structure of Redoubt Volcano, Alaska, *J. Geophys. Res.*, **101**, 8111-8128, 1996.

34. B.A. Chouet, R.Y. Koyanagi, and K. Aki, The origin of volcanic tremor in Hawaii. Part II: Theory and discussion, in *Volcanism in Hawaii*, R.W. Decker, T.L. Wright, and P.H. Stauffer, eds., U.S. Geological Survey Professional Paper 1350, U.S. Government Printing Office, Washington, D.C., pp. 1259-1280, 1987; B.A. Chouet, R.A. Page, C.D. Stephens, J.C. Lahr, and J.A. Power, Precursory swarms of long-period events at Redoubt Volcano (1989-1990), Alaska: Their origin and use as a forecasting tool, in *The 1989-1990 Eruptions of Redoubt Volcano, Alaska*, T.P. Miller and B.A. Chouet, eds., *J. Volcan. Geotherm. Res.*, **62**, 95-135, 1994; B.R. Julian, Volcanic tremor: Nonlinear excitation by fluid flow, *J. Geophys. Res.*, **99**, 11,859-11,877, 1994.

35. Strain can be derived from the fractional changes in the lengths of baselines connecting points of geodetic observation. In California, 100-kilometer baselines that cross the San Andreas fault are observed to change at rates of 20-35 millimeters per year, corresponding to strain rates of $2.0\text{-}3.5 \times 10^{-7}$ per year.

36. J.C. Savage and W.H. Prescott, Precision of geodolite distance measurements for determining fault movements, *J. Geophys. Res.*, **78**, 6001-6008, 1973. The high precision presupposes corrections for atmospheric refraction using temperature and humidity measurements made by helicopter or light aircraft along the line of sight. Thus, precise trilateration measurements are expensive and limited to locations where aircraft can fly safely.

37. Spirit leveling measurements are made by sighting graduated leveling rods through a telescope whose axis is precisely horizontal. Its precision is about one part per million under the best conditions, but larger systematic errors exist. Relative elevations are measured by adding many increments along the survey track, so the elevation error accumulates following a random walk (i.e., in proportion to the square root of the survey length).

38. For example, scientists hypothesized that the "Palmdale bulge" was a possible precursor to a major earthquake. Leveling data were responsible for the reported apparent uplift of about 350 millimeters centered on the San Andreas fault near Palmdale, California (R.O. Castle, J.P. Church, and M.R. Elliott, Aseismic uplift in southern California, *Science*, **192**, 251-254, 1976). The Palmdale bulge is thought to have been an artifact caused by atmospheric refraction and poor calibration of the leveling rods (see C.C. Chi, R.E. Reilinger, L.D. Brown, and J.E. Oliver, Leveling circuits and crustal movements, *J. Geophys. Res.*, **85**, 1469-1474, 1980; D.D. Jackson, W.B. Lee, and C.C. Liu, Aseismic uplift in southern Califor-

nia: An alternative interpretation, *Science*, **210**, 534-536, 1981; W.E. Strange, The impact of refraction correction on leveling interpretations in southern California, *J. Geophys. Res.*, **86**, 2809-2834, 1981; R.S. Stein, Role of elevation-dependent errors on the accuracy of geodetic leveling in the southern California uplift, 1953-1979, in *Earthquake Prediction—An International Review*, D.W. Simpson and P.G. Richards, eds., American Geophysical Union, Maurice Ewing Series, 4, Washington, D.C., pp. 441-456, 1981).

39. Errors in leveling lines accumulate as a constant times the square root of the line length L. The constant of proportionality is typically 3 millimeters when L is expressed in kilometers, such that for $L = 25$ kilometers, the standard error is about 15 millimeters. Measurement precision with GPS is about 1 to 2 millimeters in horizontal position and 5 to 10 millimeters in vertical position, which can be achieved over short length scales for observation periods of a few hours and at length scales of hundreds of kilometers with day-long observations. Thus, several years of continuous GPS recording can yield steady-state velocity estimates with a precision of under 1 millimeter per year. The precision of such long-term measurements depends critically on stable, deeply anchored monuments.

40. T.A. Herring, I.I. Shapiro, T.A. Clark, C. Ma, J.W. Ryan, B.R. Schupler, C.A. Knight, G. Lundqvist, D.B. Shaffer, N.R. Vandenberg, B.E. Corey, H.F. Hinteregger, A.E.E. Rogers, J.C. Webber, A.R. Whitney, G. Elgered, B.O. Ronnang, and J.L. Davis, Geodesy by radio interferometry: Evidence for contemporary plate motion, *J. Geophys. Res.*, **91**, 8341-8347, 1986; T.H. Jordan and J.B. Minster, Measuring crustal deformation in the American West, *Sci. Am.*, **256**, 48-58, 1988.

41. The utility of GPS for a variety of military, commercial, and recreational purposes has reduced the price of navigation-quality receivers with resolutions of about 10 meters to a few hundred dollars. However, geodetic-quality receivers that operate off the GPS carrier phase are an order of magnitude more expensive, partly because they require better electronics and must process two frequencies to correct for signal delays caused by charged particles in the Earth's ionosphere. Errors in GPS data come from errors in the reference frame, drift of the clocks onboard the satellites, refraction in the ionosphere and troposphere, multipath reflection of radio waves from the satellites, and so forth. These error sources can be included as terms in the basic equations to model GPS signals, and with sufficiently redundant data, the errors can be reduced dramatically. Large continuous networks are especially valuable for this purpose. In tectonic geodesy an additional error arises from nontectonic motions of the survey points caused by soil motions and fluid withdrawal in the region of the survey point. To reduce site instability, the Southern California Continuous GPS Network has developed an effective type of monument fixed at four points, each more than 10 meters below ground, by stainless steel rods welded to the surface monument.

42. K. Feigl, D. Agnew, Y. Bock, D. Dong, A. Donnellan, B. Hager, T. Herring, D. Jackson, T. Jordan, R. King, S. Larsen, K Larson, M. Murray, Z. Shen, and F. Webb, Space geodetic measurement of crustal deformation in central and Southern California, 1984-1992, *J. Geophys. Res.*, **98**, 21,677-21,712, 1993.

43. The International GPS Service for Geodynamics (IGS) integrates several forms of geodetic measurement to provide precise estimates of the satellite orbital parameters, the locations of selected tracking stations, and other important information needed in GPS processing. For more information, see <http://igscb.jpl.nasa.gov/>.

44. Because of the need to integrate information into a global network, GPS data processing and interpretation depend critically on data sharing. The International GPS Services for Geodynamics, the Universities NAVSTAR Consortium, and the Southern California Earthquake Center, among many organizations, have made great progress in publishing GPS data freely over the Internet, contributing greatly to scientific progress in tectonic geodesy.

45. The first InSAR image of an earthquake displacement field was published by D. Massonnet, M. Rossi, C. Carmona, F. Adragna, G. Peltzer, K. Feigl, and T. Rabaute (The displacement field of the Landers earthquake mapped by radar interferometry, *Nature*, **364**, 138-142, 1993), who used a series of radar images acquired by the European Remote Sensing (ERS) satellites to construct an interferogram of the 1992 Landers earthquake (M 7.3).

46. G. Peltzer, P. Rosen, F. Rogez, and K. Hudnut, Postseismic rebound in fault stepovers caused by pore fluid flow, *Science*, **273**, 1202-1204, 1996.

47. The proposed ECHO mission would be carried out jointly between NASA, NSF, and the USGS to provide spatially continuous strain measurements over wide geographic areas. The design goals of the proposed InSAR mission are dense spatial (100 meters) and temporal (every eight days) coverage of the entire North American-Pacific plate boundary with vector solutions accurate to 2 millimeters on spatial scales of 50 kilometers over all terrain types, which exceeds the capabilities of existing and planned international SAR missions. Spatially continuous, but intermittent, InSAR images complement continuous GPS point measurements and will therefore contribute to the EarthScope science objectives.

48. Stable aseismic slip was discovered at the Cienega Winery, which straddles the San Andreas fault south of Hollister, California (K.V. Steinbrugge, E.G. Zacher, D. Tocher, C.A. Whitten, and C.N. Clair, Creep on the San Andreas fault [California]—Analysis of geodetic measurements along the San Andreas fault, *Bull. Seis. Soc. Am.*, **50**, 396-404, 1960). The walls of the winery building have been progressively offset at a rate of 11 millimeters per year since it was built in 1948. This "creeping section" of the San Andreas extends 160 kilometers from San Juan Bautista to Parkfield, California.

49. Near-fault tectonic geodesy is reviewed by A.G. Sylvester in National Research Council, *Active Tectonics*, National Academy Press, Washington, D.C., pp. 164-180, 1986.

50. C.R. Allen, M. Wyss, J.N. Brune, A. Grantz, and R.E. Wallace, Displacements on the Imperial, Superstition Hills, and San Andreas faults triggered by the Borrego Mountain earthquake, U.S. Geological Survey Professional Paper 787, Reston, Va., pp. 87-104, 1972; S.S. Schulz, G. Mavco, R.O. Burford, and W.D. Smith, Long-term fault creep observations in central California, *J. Geophys. Res.*, **87**, 6977-6982, 1982; R.O. Burford, The response of creeping parts of the San Andreas fault to earthquakes on nearby faults; Two examples, *Pure Appl. Geophys.*, **126**, 499-529, 1988; C.H. Thurber, Creep events preceding small to moderate earthquakes on the San Andreas fault, *Nature*, **380**, 425-428, 1996.

51. D.C. Agnew, Strainmeters and tiltmeters, *Rev. Geophys.*, **24**, 579-624, 1986.

52. I.S. Sacks, S. Suyehiro, A.T. Linde, and J.A. Snoke, Slow earthquakes and stress redistribution, *Nature*, **275**, 599-602, 1978; A.T. Linde, S. Suyehiro, I. Miura, I.S. Sacks, and A. Takagi, Episodic aseismic earthquake precursors, *Nature*, **334**, 513-515, 1988; M.T. Gladwin, High-precision multicomponent borehole deformation monitoring, *Rev. Sci. Instr.*, **55**, 2011-2016, 1984.

53. PBO Steering Committee, The Plate Boundary Observatory: Creating a four-dimensional image of the deformation of western North America, White paper providing the scientific rationale and deployment strategy for a Plate Boundary Observatory based on a workshop held October 3-5, 1999. Available at <http://www.earthscope.org>.

54. The Southern California Earthquake Center employed this strategy in its 1995 earthquake hazard estimate. See Working Group on California Earthquake Probabilities, Seismic hazards in southern California: Probable earthquake 1994 to 2024, *Bull. Seis. Soc. Am.*, **85**, 379-439, 1995. In combination with geologic estimates of fault slip rate, they used GPS estimates from K. Feigl, D. Agnew, Y. Bock, D. Dong, A. Donnellan, B. Hager, T. Herring, D. Jackson, T. Jordan, R. King, S. Larsen, K Larson, M. Murray, Z. Shen, and F. Webb, Space geodetic measurement of crustal deformation in Central and Southern California, 1984-1992, *J. Geophys. Res.*, **98**, 21,677-21,712, 1993.

55. See, for example, E. Field, D. Jackson, and J.F. Dolan, A new look at earthquake occurrence in southern California: No deficit or huge earthquake required, *Bull. Seis. Soc. Am.*, **89**, 559-578, 1999.

56. Harris and Segall (R. Harris and P. Segall, Detection of a locked zone at depth in the Parkfield, California segment of the San Andreas fault, *J. Geophys. Res.*, **92**, 7945-7962, 1987) identified a locked patch apparently surrounded by a creeping region. They identified the locked patch with the rupture zone of the 1966 Parkfield earthquake, suggesting that the geodetic data provided evidence for a future earthquake on the same rupture surface. Sung and Jackson (L. Sung and D. Jackson, Geodetic evidence of the seismic potential at Parkfield, California, *Geophys. Res. Lett.*, **15**, 820-823, 1988) showed that a smooth transition from creeping to locked, without the asperity, also fit the geodetic data within their accuracy.

57. For example, models of the 1992 Landers, California, earthquake (K.W. Hudnut, Y. Bock, M. Cline, P. Fang, Y. Feng, J. Freymueller, X. Ge, W.K. Gross, D. Jackson, M. Kim, N.E. King, J. Langbein, S.C. Larsen, M. Lisowski, Z.K. Shen, J. Svarc, and J. Zhang, Coseismic displacements of the 1992 Landers earthquake sequence, *Bull. Seis. Soc. Am.*, **84**, 625-645, 1994) confirm the strong spatial variability and location of slip derived from modeling of the strong-motion data. In the case of the 1989 Loma Prieta earthquake, a geodetically derived slip model (T. Arnadottir and P. Segall, The 1989 Loma Prieta earthquake imaged from inversion of geodetic data, *J. Geophys. Res.*, **99**, 21,835-21,855, 1994) supported seismic models that found a variation of rate with distance along the fault in that earthquake (G.C. Beroza and H. Krawinkler, Near-source modeling of the Loma Prieta earthquake; Evidence for heterogeneous slip and implications for earthquake hazard, *Bull. Seis. Soc. Am.*, **81**, 1603-1621, 1991).

58. P. Segall and M. Matthews, Time dependent inversion of geodetic data, *J. Geophys. Res.*, **102**, 22,391-22,409, 1997.

59. T. Arnadottir and P. Segall, The 1989 Loma Prieta earthquake imaged from inversion of geodetic data, *J. Geophys. Res.*, **99**, 21,835-21,855, 1994.

60. G.C. Beroza and H. Krawinkler, Near-source modeling of the Loma Prieta earthquake; Evidence for heterogeneous slip and implications for earthquake hazard, *Bull. Seis. Soc. Am.*, **81**, 1603-1621, 1991.

61. J. Freymueller, N.E. King, and P. Segall, The co-seismic slip distribution of the Landers earthquake, *Bull. Seis. Soc. Am.*, **84**, 646-659, 1994.

62. D. Massonnet, M. Rossi, C. Carmona, F. Adragna, G. Peltzer, K. Feigl, and T. Rabaute, The displacement field of the Landers earthquake mapped by radar interferometry, *Nature*, **364**, 138-142, 1993; G. Peltzer, P. Rosen, F. Rogez, and K. Hudnut, Postseismic rebound in fault step-overs caused by pore fluid flow, *Science*, **273**, 1202-1204, 1996.

63. T. Arnadottir and P. Segall, The 1989 Loma Prieta earthquake imaged from inversion of geodetic data, *J. Geophys. Res.*, **99**, 21,835-21,855, 1994; K.W. Hudnut, Y. Bock, M. Cline, P. Fang, Y. Feng, J. Freymueller, X. Ge, W.K. Gross, D. Jackson, M. Kim, N.E. King, J. Langbein, S.C. Larsen, M. Lisowski, Z.K. Shen, J. Svarc, and J. Zhang, Co-seismic displacements of the 1992 Landers earthquake sequence, *Bull. Seis. Soc. Am.*, **84**, 625-645, 1994; Z.-K. Shen, B.X. Ge, D.D. Jackson, D. Potter, M. Cline, and L.Y. Sung, Northridge earthquake rupture models based on the Global Positioning System measurements, *Bull. Seis. Soc. Am.*, **86**, S37-S48, 1996.

64. I.S. Sacks, A.T. Linde, J.A. Snoke, and S. Suyehiro, A slow earthquake sequence following the Izu-Oshima earthquake of 1978, in *Earthquake Prediction—An International Review*, D.W. Simpson and P.G. Richards, eds., American Geophysical Union, Maurice Ewing Series, 4, Washington, D.C., pp. 617-628, 1981.

65. H. Dragert, K. Wang, and T.S. James, A silent slip event on the deeper Cascadia subduction interface, *Science*, **292**, 1525-1528, 2001.

66. M.M. Miller, T. Melbourne, D.J. Johnson, and W.Q. Sumner, Periodic slow earthquakes from the Cascadia subduction zone, *Science*, **295**, 2432, 2002.

67. W. Thatcher, Present-day crustal movements and the mechanics of cyclic deformation, in *San Andreas Fault System, California*, R.E. Wallace, ed., U.S. Geological Survey Professional Paper 1515, Washington, D.C., pp. 189-205, 1990.

68. D.D. Jackson, Z.-K. Shen, D. Potter, X.-B. Ge, and L. Sung, Southern California deformation, *Science*, **277**, 1621-1622, 1997.

69. Two limnographs on different sides of Rapel Lake showed water-level differences, implying tilting, in response to the 1985 earthquake in Chile. The tilting had a characteristic time of 10 months. A similar signal was also seen on tide gauge records of sea level recorded at a distance of 100 kilometers. There was no co-seismic change in lake level, so the postseismic and co-seismic deformations were generated differently. Based on these observations Barrientos concluded that the postseismic slip occurred updip from the co-seismic slip. See S.E. Barrientos, Dual seismogenic behavior; The 1985 central Chile earthquake, *Geophys. Res. Lett.*, **22**, 3541-3544, 1995.

70. For the Loma Prieta earthquake, fault-normal compression after the earthquake exceeded that from the earthquake itself. Savage and others explain this observation with a model of postseismic fault-zone collapse with a time constant of 1.3 years (J.C. Savage, M. Lisowski, and J.L. Svarc, Postseismic deformation following the 1989 (M = 7.1) Loma Prieta, California, earthquake, *J. Geophys. Res.*, **99**, 13,757-13,765, 1994). Bürgmann and others model the same observations with a combination of continuing slip on the mainshock plane and aseismic slip on reverse faults located to the northeast of the mainshock fault plane (R. Burgmann, P. Segall, M. Lisowski, and J.P. Svarc, Postseismic strain following the 1989 Loma Prieta earthquake from GPS and leveling measurements, *J. Geophys. Res.*, **102**, 4933-4955, 1997).

71. Z.-K. Shen, D.D. Jackson, Y. Feng, M. Cline, M. Kim, P. Fang, and Y. Bock, Postseismic deformation following the Landers earthquake, California, 28 June 1992, *Bull. Seis. Soc. Am.*, **84**, 780-791, 1994. These authors found a postseismic signal with a decay time of about 1 month following the Landers earthquake, with slip in their model concentrated on the southern half of the mainshock rupture zone. They also inferred slip at depth below the Banning segment of the San Andreas fault, which could have serious consequences for possible earthquakes on the San Andreas. Savage and Svarc inferred that postseismic slip was concentrated on the northern half of the mainshock rupture zone and that postseismic deformation is still ongoing (J.C. Savage and J.L. Svarc, Postseismic deformation associated with the 1992 Mw = 7.3 Landers earthquake, Southern California, *J. Geophys. Res.*, **102**, 7565-7577, 1997); they also found evidence for fault-zone collapse in the Landers postseismic zone.

72. F.K. Wyatt, D.C. Agnew, and M.T. Gladwin, Continuous measurements of crustal deformation for the 1992 Landers earthquake sequence, *Bull. Seis. Soc. Am.*, **84**, 768-779, 1994.

73. D. Massonnet, W. Thatcher, and H. Vadon, Detection of postseismic fault-zone collapse following the Landers earthquake, *Nature*, **382**, 612-616, 1996. Interferograms spanning intervals after the Landers mainshock show clear postseismic signals consistent with ongoing slip in the rupture zone as well as motions concentrated within fault offsets. The latter have been attributed to pressure-driven fluid flow into dilatant regions (G. Peltzer, P. Rosen, F. Rogez, and K. Hudnut, Postseismic rebound in fault step-overs caused by pore fluid flow, *Science*, **273**, 1202-1204, 1996).

74. P.A. Rydelek and I.S. Sacks, Asthenospheric viscosity and stress diffusion; A mechanism to explain correlated earthquakes and surface deformations in NE Japan, *Geophys. J. Int.*, **100**, 39-58, 1990; I.S. Sacks and F.F. Pollitz, *Analysis of Postseismic Crustal Motions in California*, Carnegie Institution, Washington, D.C., 5 pp. + 7 plates, 1992; F.F. Pollitz and I.S.

Sacks, Modeling of postseismic relaxation following the great 1857 earthquake, Southern California, *Bull. Seis. Soc. Am.*, **82**, 454-480, 1992; F.F. Pollitz and I.S. Sacks, Consequences of stress changes following the 1891 Nobi earthquake, Japan, *Bull. Seis. Soc. Am.*, **85**, 796-807, 1995; F. Press and C. Allen, Patterns of seismic release in the southern California region, *J. Geophys. Res.*, **100**, 6421-6430, 1995.

75. The best bounds on the sizes of strain precursors are from recent earthquakes in California and Japan. See M.L.S. Johnson, A.T. Linde, and M.T. Gladwin (Near-field high resolution strain measurements prior to the October 18, 1989 Loma Prieta M_S 7.1 earthquake, *Geophys. Res. Lett.*, **17**, 1777-1780, 1990) for the Loma Prieta earthquake, and F.K. Wyatt, D.C. Agnew, and M. Gladwin (Continuous measurements of crustal deformation for the 1992 Landers earthquake sequence, *Bull. Seis. Soc. Am.*, **84**, 768-779, 1994) for the Landers earthquake.

76. Developments in this field are summarized in the textbook *The Geology of Earthquakes*, by R.S. Yeats, K. Sieh, and C.R. Allen (Oxford University Press, Oxford, U.K., 568 pp., 1997).

77. National Research Council, *Active Tectonics*, National Academy Press, Washington, D.C., 266 pp., 1986.

78. The SRTM mission and results are described at <http://jpl.nasa.gov/srtm>.

79. An example is the Airborne Topographic Mapper, mounted on an Otter aircraft and used in the U.S. Topographic Change Mapping Project, a joint venture among NOAA, NASA, and USGS; see <http://www.csc.noaa.gov/crs/tcm/>.

80. In deep water (3 kilometers), oceanographic swath-mapping systems yield bathymetric maps with a horizontal resolution of about 60 meters and a vertical precision of a few meters. The resolution and precision improve more or less linearly with decreasing water depth.

81. For example, N.N. Ambraseys and C.P. Melville, *A History of Persian Earthquakes*, Cambridge University Press, Cambridge, U.K., 219 pp., 1982; Y. Sugiyama, Neotectonics of southwest Japan due to the right-oblique subduction of the Philippine Sea plate, *Geof. Int.*, **33**, 53-76, 1994; D.C. Agnew and K. Sieh, A documentary study of the felt effects of the great California earthquake of 1857, *Bull. Seis. Soc. Am.*, **68**, 1717-1729, 1978; K. Satake, K. Shimazaki, Y. Tsuji, and K. Ueda, Time and size of a giant earthquake in Cascadia inferred from Japanese tsunami records of January 1700, *Nature*, **379**, 246-249, 1996.

82. See R. Page, Dating episodes of faulting from tree rings: Effects of the 1958 rupture of the Fairweather fault on tree growth, *Geol. Soc. Am. Bull.*, **81**, 3085-3094, 1970; V.C. LaMarche, Jr. and R.E. Wallace, Evaluation of effects of trees on past movements on the San Andreas fault, northern California, *Geol. Soc. Am. Bull.*, **83**, 2665-2676, 1972; G.C. Jacoby, P.R. Sheppard, and K.E. Sieh, Irregular recurrence of large earthquakes along the San Andreas fault; Evidence from trees, *Science*, **241**, 196-199, 1988; D.K. Yamaguchi, B.F. Atwater, D.E. Bunker, B.E. Benson, and M.S. Reid, Tree-ring dating the 1700 Cascadia earthquake, *Nature*, **389**, 922-923, 1997.

83. The age range, resolution, and applications of different techniques for dating surficial materials are described in J.S. Noller, J.M. Sowers, and W.R. Lettis, *Quaternary Geochronology: Methods and Applications*, American Geophysical Union, Washington, D.C., 582 pp., 2000.

84. J.A. Spotila, K.A. Farley, and K. Sieh, Uplift and erosion of the San Bernardino Mountains associated with transpression along the San Andreas fault, California, as constrained by radiogenic helium thermochronometry, *Tectonics*, **17**, 360-378, 1998; C.R. Bacon, M.A. Lanphere, and D.E. Champion, Late Quaternary slip rate and seismic hazards of the West Klamath Lake fault zone near Crater Lake, Oregon Cascades, *Geology*, **27**, 43-46, 1999; J. Lee, C.M. Rubin, and A. Calvert, Quaternary faulting history along the Deep Springs fault, California, *Geol. Soc. Am. Bull.*, **113**, 855-869, 2001; A.K. Jain, D. Kumar; S. Singh, A.

Kumar, and N. Lal, Timing, quantification and tectonic modelling of Pliocene-Quaternary movements in the NW Himalaya; Evidence from fission track dating, *Earth Planet. Sci. Lett.*, **179**, 437-451, 2000; B.J. Szabo and J.N. Rosholt, Uranium-series nuclides in the Golden fault, Colorado, U.S.A.: Dating latest fault displacement and measuring recent uptake of radionuclides by fault-zone materials, *Appl. Geochem.*, **4**, 177-182, 1989.

85. For example, K.E. Sieh, Prehistoric large earthquakes produced by slip on the San Andreas fault at Pallet Creek, California, *J. Geophys. Res.*, **83**, 3907-3939, 1978; K. Sieh, M. Stuiver, and D. Brillinger, A more precise chronology of earthquakes produced by the San Andreas fault in southern California, *J. Geophys. Res.*, **94**, 603-623, 1989. See Box 4.5.

86. E. Bard, B. Hamelin, R.G. Fairbanks, and A. Zindler, Calibration of the ^{14}C timescale over the past 30,000 years using mass spectrometric U-Th ages from Barbados corals, *Nature*, **345**, 405-410, 1990.

87. F.W. Taylor, C. Frohlich, J. Lecolle, and M. Strecker, Analysis of partially emerged corals and reef terraces in the central Vanuatu arc: Comparison of contemporary coseismic and nonseismic with Quaternary vertical movements, *J. Geophys. Res.*, **92**, 4905-4933, 1987; J. Zachariasen, K. Sieh, F. Taylor, R.L. Edwards, and W.S. Hantoro, Submergence and uplift associated with the giant 1833 Sumatran subduction earthquake: Evidence from coral microatolls, *J. Geophys. Res.*, **104**, 895-919, 1999.

88. L.A. Perg, R.S. Anderson, and R.C. Finkel, Use of a new ^{10}Be and ^{26}Al inventory method to date marine terraces, Santa Cruz, California, USA, *Geology*, **29**, 879-882, 2001; J. Van der Woerd, F.J. Ryerson, P. Tapponnier, Y. Gaudemer, R. Finkel, A.S. Meriaux, M. Caffee, G. Zhao, and Q. He, Holocene left-slip rate determined by cosmogenic surface dating on the Xidatan segment of the Kunlun fault (Qinghai, China), *Geology*, **26**, 695-698, 1998; L.C. Benedetti, R.C. Finkel, G.C.P. King, R. Armijo, D. Papanastassiou, F.J. Ryerson, F. Flerit, and D. Farber, Earthquake time-slip history of the Kaparelli fault (Greece) from in situ chlorine-36 cosmogenic dating, *EOS Trans. Am. Geophys. Union*, **82**, F931, 2001.

89. Examples include neotectonic maps of Japan (Y. Kinugasa, E. Tsukada, and H. Yamazaki, Neotectonic map of Japan, *Geological Atlas of Japan*, 2nd ed., Asakura Publishing Company, Ltd., Tokyo, sheet 5, 1992), Turkey (F. Saroglu, O. Emre, and I. Kuscu, The east Anatolian fault zone of Turkey, *Annales Tectonicae*, Suppl. 6 (Special Issue), 99-125, 1992), Alaska (G. Plafker, L.M. Gilpin, and J.C. Lahr, Neotectonic map of Alaska, in *The Geology of Alaska*, G.B. Plafker and H. Berg, eds., Decade of North American Geology, **G-1**, Geological Society of America, Boulder, Colo., pl. 12 (map), 1994), southern Tibet (R. Armijo, P. Tapponnier, J.L. Mercier, and T.-L. Han, Quaternary extension in southern Tibet: Field observations and tectonic implications, *J. Geophys. Res.*, **91**, 13,803-13,872, 1986), and Sumatra (K. Sieh and D. Natawidjaja, Neotectonics of the Sumatran fault, *J. Geophys. Res.*, **105**, 28,295-28,336, 2000).

90. The World Map of Major Active Faults being compiled under Project II-2 of the International Lithosphere Program is a step in this direction; see<http://www.gfz-potsdam.de/pb4/ilp96/projects.htm>.

91. L.A. Reinen, J.D. Weeks, and T.E. Tullis, The frictional behavior of serpentinite: Implications for aseismic creep on shallow crustal faults, *Geophys. Res. Lett.*, **18**, 1921-1924, 1991; The frictional behavior of lizardite and antigorite serpentinites: Experiments, constitutive models, and implications for natural faults, *Pure Appl. Geophys.*, **143**, 317-358, 1994.

92. J. Van der Woerd, F.J. Ryerson, P. Tapponnier, A.-S. Meriaux, Y. Gaudemer, B. Meyer, R.C. Finkel, M.W. Caffee, Z. Guoguang, and X. Zhiqin, Uniform slip-rate across the Kunlun fault: Implications for seismic behavior and large-scale tectonics, *Geophys. Res. Lett.*, **27**, 2353-2356, 2000.

93. J.-C. Lee, Y.-G. Chen, K. Sieh, K. Mueller, W.-S. Chen, H.-T. Chu, Y.-C. Chan, C. Rubin, and R. Yeats, A vertical exposure of the 1999 surface rupture of the Chelungpu fault at Wufeng, western Taiwan: Structural and paleoseismic implications for an active thrust fault, *Bull. Seis. Soc. Am.*, **91**, 914-929, 2001.

94. R.S. Yeats, Large-scale Quaternary detachments in Ventura basin, southern California, *J. Geophys. Res.*, **88**, 569-583, 1983.

95. Developing a unified structural representation for southern California has been set as a high-priority goal of the Southern California Earthquake Center; see *Southern California Earthquake Center, Science Plan for 2002-2007*, University of Southern Calif., 9 pp., 2001, available at <http://www.scec.org/aboutSCEC/documents/science.plan.2002/>.

96. This definition of paleoseismology is offered in the historical overview by R.S. Yeats and C.S. Prentice, Introduction to special session: Paleoseismology, *J. Geophys. Res.*, **101**, 5847-5853, 1996. A survey of the subject is given by J.P. McCalpin, ed., *Paleoseismology*, International Geophysics Series 62, Academic Press, San Diego, Calif., 588 pp., 1996.

97. D.C. Agnew and K. Sieh, A documentary study of the felt effects of the great California earthquake of 1857, *Bull. Seis. Soc. Am.*, **68**, 1717-1729, 1978. Geologic features corresponding to individual earthquake offsets on the San Andreas, including the 1857 event, were first recognized by R.E. Wallace (Notes on stream channels offset by the San Andreas fault, southern Coast Ranges, California, in *Proceedings of a Conference on Geological Problems of the San Andreas Fault System*, W.R. Dickinson and A. Grantz, eds., Stanford University Publications in Geological Science **11**, Stanford, Calif., pp. 6-21, 1968).

98. K. Sieh, M. Stuvier, and D. Brillinger, A more precise chronology of earthquakes produced by the San Andreas fault in southern California, *J. Geophys. Res.*, **94**, 603-623, 1989.

99. See K.R. Berryman, S. Beanland, A. Cooper, H. Cutten, R. Norris, and P. Wood, The Alpine fault, New Zealand: Variation in Quaternary structural style and geomorphic expression, *Annales Tectonicae*, Suppl. 6 (Special Issue), 126-163, 1992; K. Sieh, A review of geological evidence for recurrence times of large earthquakes, in *Earthquake Prediction—An International Review*, D. Simpson and P. Richards, eds., American Geophysical Union, Maurice Ewing Series 4, Washington, D.C., pp. 181-207, 1981; A.A. Barka, Slip distribution along the North Anatolian fault associated with the large earthquakes of 1939-1967, *Bull. Seis. Soc. Am.*, **86**, 1238-1254, 1996; Q.-D. Deng and P.-Z. Zhang, Research on the geometry of shear fracture zones, *J. Geophys. Res.*, **89**, 5699-5710, 1984.

100. R.E. Wallace, Profiles and ages of young fault scarps, north-central Nevada, *Geol. Soc. Am. Bull.*, **88**, 1267-1281, 1977.

101. K. Mueller, J. Champion, M. Guccione, and K. Kelson, Fault slip rates in the modern New Madrid seismic zone, *Science*, **286**, 1135-1138, 1999.

102. J. Clague, Evidence for large earthquakes at the Cascadia subduction zone, *Rev. Geophys.*, **35**, 439-460, 1997.

103. F.W. Taylor, C. Frohlich, J. Lecolle, and M. Strecker, Analysis of partially emerged corals and reef terraces in the central Vanuatu arc: Comparison of contemporary coseismic and nonseismic with Quaternary vertical movements, *J. Geophys. Res.*, **92**, 4905-4933, 1987; R.L. Edwards, F.W. Taylor, and G.J. Wasserburg, Dating earthquakes with high-precision thorium-230 ages of very young corals, *Earth Planet. Sci. Lett.*, **90**, 371-381, 1988.

104. J. Zachariasen, K. Sieh, F. Taylor, R.L. Edwards, and W.S. Hantoro, Submergence and uplift associated with the giant 1833 Sumatran subduction earthquake: Evidence from coral microatolls, *J. Geophys. Res.*, **104**, 895-919, 1999; K. Sieh, S. Ward, D. Natawidjaja, and B. Suwargadi, Crustal deformation at the Sumatran subduction zone revealed by coral rings, *Geophys. Res. Lett.*, **26**, 3141-3144, 1999.

105. See R.S. Yeats, K. Sieh, and C.R. Allen, *The Geology of Earthquakes*, Oxford University Press, Oxford, U.K., 568 pp., 1997, for a more extensive enumeration and discussion.

106. W.B. Bull, J. King, F. Kong, T. Moutoux, and W.M. Phillips, Lichen dating of coseismic landslide hazards in alpine mountains, *Geomorph.*, **10**, 253-264, 1994.

107. J. Adams, Paleoseismicity of the Cascadia subduction zone: Evidence from turbidites off the Oregon-Washington margin, *Tectonics*, **9**, 569-583, 1990.

108. P.J. Munson, S.F. Obermeier, C. Munson, and E.R. Hajic, Liquefaction evidence for Holocene and latest Pleistocene seismicity in the southern halves of Indiana and Illinois: A preliminary overview, *Seis. Res. Lett.*, **68**, 521-536, 1997.

109. S.F. Obermeier, G.S. Gohn, R.F. Weems, R.L. Gelinas, and M. Rubin, Geologic evidence for recurrent moderate to large earthquakes near Charleston, South Carolina, *Science*, **227**, 408-411, 1985.

110. S. Marco, M. Stein, A. Agnon, and H. Ron, Long-term earthquake clustering: A 50,000-year paleoseismic record in the Dead Sea Graben, *J. Geophys. Res.*, **101**, 6179-6191, 1996.

111. Most laboratory experiments in triaxial, direct-shear, and rotary-shear machines that can attain high pressures, temperatures, and fluid pressures involve rock samples with maximum dimensions of a few centimeters. The USGS has conducted stick-slip experiments at room conditions on a biaxially loaded granite sample of dimension $1.5 \times 1.5 \times 0.4$ cubic meters; the sample was saw-cut on the diagonal and loaded by jacks at the edges, allowing both shear and normal stresses to be varied on a precut fault area of 2×0.4 square meters (J.H. Dieterich, N.G.W. Cook, and H.C. Heard, Potential for geophysical experiments in large scale tests, *Geophys. Res. Lett.*, **8**, 653-656, 1981; D.A. Lockner and P.G. Okubo, Measurements of frictional heating in granite, *J. Geophys. Res.*, **88**, 4313-4320, 1983). Slip events were observed with seismic moments as large as 3×10^6 newton-meters; this corresponds to a moment magnitude of about –1.7, which overlaps with the sizes of microearthquakes that have been recorded by seismic sensors in deep mines.

112. M.S. Paterson, *Experimental Rock Deformation—The Brittle Field*, Springer-Verlag, Berlin, 254 pp., 1978. The development of double-direct-shear, rotary-shear, and other types of testing machines was discussed by T.E. Tullis and J.D. Weeks (Constitutive behavior and stability of frictional sliding of granite, *Pure Appl. Geophys.*, **124**, 383-414, 1986).

113. The micromechanics of rate-state friction has been discussed by T.E. Tullis (Rock friction constitutive behavior from laboratory experiments and its implications for an earthquake prediction field monitoring program, *Pure Appl. Geophys.*, **126**, 555-588, 1988) and C.G. Sammis and S.J. Place (The micromechanics of friction in a granular layer, *Pure Appl. Geophys.*, **142**, 777-794, 1994). Recent advances in atomic-scale tribology (the study of friction, lubrication, and wear) are described by G. Hähner and N. Spencer (Rubbing and scrubbing, *Physics Today*, **51**, 22-27, 1998).

114. See the review by C. Marone, Laboratory-derived friction laws and their application to seismic faulting, *Ann. Revs. Earth Planet. Sci.*, **26**, 643-696, 1998.

115. P. Okubo, Dynamic rupture modeling with laboratory-derived constitutive relations, *J. Geophys. Res.*, **94**, 12,321-12,335, 1989; A. Cochard and R. Madariaga, Dynamic faulting under rate-dependent friction, *Pure Appl. Geophys.*, **142**, 419-445, 1994; N. Lapusta, J.R. Rice, Y. Ben-Zion, and G. Zheng, Elastodynamic analysis for slow tectonic loading with spontaneous rupture episodes on faults with rate and state dependent friction, *J. Geophys. Res.*, **105**, 23,765-23,789, 2000.

116. J.H. Dieterich, A constitutive law for rate of earthquake production and its application to earthquake clustering, *J. Geophys. Res.*, **99**, 2601-2618, 1994; J.H. Dieterich, V. Cayol, and P. Okubo, The use of earthquake rate changes as a stress meter at Kilauea volcano, *Nature*, **408**, 457-460, 2000.

117. K. Mair and C. Marone, Friction from simulated fault gouge at a wide range of velocities, *J. Geophys. Res.*, **104**, 28,888-28,894, 1999. As first hypothesized by D.J. Andrews (Rupture propagation with finite stress in antiplane strain, *J. Geophys. Res.*, **81**, 3579-3587, 1976), the critical slip distance for fault zones of finite width W may obey $D_c = \gamma_c W$, where γ_c is a "critical strain." Experiments by C. Marone and B. Kilgore (Scaling of the critical slip distance for seismic faulting with shear strain in fault zones, *Nature*, **362**, 618-621, 1993) gave $\gamma_c = 10^{-2}$, provided W was interpreted to be the width of the "active gouge" containing shear bands actually involved in the slip.

118. J.H. Dieterich and B. Kilgore, Implications of fault constitutive properties for earthquake prediction, *Proc. Natl. Acad. Sci.*, **93**, 3787-3794, 1996.

119. B. Evans, J.T. Frederich, and T.F. Wong, The brittle-ductile transition in rocks: Recent experimental and theoretical progress, in *The Brittle-Ductile Transition in Rocks: The Heard Volume*, A.G. Duba, W. Durham, J. Handin, and H. Wang, eds., American Geophysical Union, Washington, D.C., pp. 1-20, 1990.

120. N.L. Carter and S.H. Kirby, Transient creep and semibrittle behavior of crystalline rocks, *Pure Appl. Geophys.*, **116**, 807-839, 1978; C.H. Scholz, *The Mechanics of Earthquakes and Faulting*, Cambridge University Press, Cambridge, U.K., 439 pp., 1990.

121. D.L. Kohlstedt, B. Evans, and S.J. Mackwell, Strength of the lithosphere: Constraints imposed by laboratory experiments, *J. Geophys. Res.*, **100**, 17,587-17,602, 1995.

122. B.R. Lawn and T.R. Wilshaw, *Fracture of Brittle Solids*, Cambridge University Press, Cambridge, U.K., 204 pp., 1975.

123. C.H. Scholz, Mechanics of faults, *Ann. Rev. Earth Planet Sci.*, **17**, 309-334, 1989.

124. J.J. Walsh and J Waterson, Analysis of the relation between displacements and dimensions of faults, *J. Struct. Geol.*, **10**, 238-247, 1988.

125. W.L. Power, T.E. Tullis, and J.D. Weeks, Roughness and wear during brittle faulting, *J. Geophys. Res.*, **93**, 15,268-15,278, 1988.

126. P. Segall and D.D. Pollard, Nucleation and growth of strike slip faults in granite, *J. Geophys. Res.*, **88**, 555-568, 1983.

127. A. Nur, H. Ron, and O. Scotti, Kinematics and mechanics of tectonic block rotations, in *Slow Deformation and Transmission of Stress in the Earth*, S.C. Cohen and P. Vanicek, eds., American Geophysical Union, Geophysics Monograph 49, Washington, D.C., pp. 31-46, 1989.

128. R.H. Sibson, An assessment of field evidence for "Byerlee" friction, *Pure Appl. Geophys.*, **142**, 645-662, 1994.

129. See R.H. Sibson, Earthquakes and rock deformation in crustal fault zones, *Ann. Rev. Earth Planet. Sci.*, **14**, 149-175, 1986.

130. C. Lapworth, The Highland controversy in British geology, *Nature*, **32**, 558-559, 1885; J.M. Christie, Mylonitic rocks of the Moine thrust-zone in the Assynt region, north-west Scotland, *Trans. Geol. Soc. Edinburgh*, **18**, 79-93, 1960; T.H. Bell and M.A. Ethridge, Microstructure of mylonites and their descriptive terminology, *Lithos.*, **6**, 337-348, 1973.

131. R.H. Sibson, Fault rocks and fault mechanisms, *J. Geol. Soc. London*, **133**, 191-213, 1977; J. Magloughlin, F.M. Chester, and J. Spray, Penrose conference report: Fine-grained fault rocks, *GSA Today*, **6**, 33-37, 1996.

132. M.A. Etheridge and J.C. Wilkie, Grain size reduction, grain boundary sliding, and the flow strength of mylonites, *Tectonophysics*, **58**, 159-178, 1979.

133. R.H. Sibson, Fault zone models, heat flow, and the depth distribution of earthquakes in the continental crust of the United States, *Bull. Seis. Soc. Am.*, **72**, 151-163, 1982; R.H. Sibson, Roughness at the base of the seismogenic zone: Contributing factors, *J. Geophys. Res.*, **89**, 5791-5800, 1984.

134. J.P. Evans and F.M. Chester, Fluid-rock interaction in faults of the San Andreas system: Inferences from San Gabriel fault rock geochemistry and microstructures, *J. Geophys. Res.*, **100**, 13,007-13,020, 1995; F.M. Chester and J.S. Chester, Ultracataclasite structure and friction processes of the Punchbowl fault, San Andreas system, California, *Tectonophysics*, **295**, 199-221, 1998.

135. Y.-G. Li, K. Aki, D. Adams, A. Hasemi, and W.H.K. Lee, Seismic guided waves trapped in the fault zone of the Landers, California earthquake of 1992, *J. Geophys. Res.*, **99**, 11,705-11,722, 1994.

136. S.E. Schulz and J.P. Evans, Mesoscopic structure of the Punchbowl fault, southern

California, and the geological and geophysical structure of active strike-slip faults, *J. Struct. Geol.*, **22**, 913-930, 2000.

137. H. Kanamori and T.H. Heaton, Microscopic and macroscopic physics of earthquakes, in *Physics of Earthquakes*, J.B. Rundle, D.L. Turcotte, and W. Klein, eds., American Geophysical Union Monograph, Washington, D.C., pp. 147-163, 2000.

138. The other energy terms include seismic-wave radiation, gravitational energy, and the energy in the formation of new surfaces, all of which are thought to be small relative to heat dissipation (C.H. Scholtz, *The Mechanics of Earthquakes and Faulting*, Cambridge University Press, Cambridge, U.K., 439 pp., 1990).

139. R.H. Sibson, Generation of pseudotachylyte by ancient seismic faulting, *Geophys. J. R. Astron. Soc.*, **43**, 775-794, 1975; R.H. Maddock, Melt origin of fault-generated pseudotachylytes demonstrated by textures, *Geology*, **11**, 105-108, 1983.

140. C.H. Scholz, Shear heating and the state of stress on faults, *J. Geophys. Res.*, **85**, 6174-6184, 1980.

141. C.H. Scholz, J. Beavan, and T.C. Hanks, Frictional metamorphism, argon depletion, and tectonic stress on the Alpine fault, New Zealand, *J. Geophys. Res.*, **84**, 6770-6782, 1979.

142. P. Molnar, W.-P. Chen, and E. Padovani, Calculated temperatures in overthrust terrains and possible combinations of heat sources responsible for the Tertiary granites in the Greater Himalaya, *J. Geophys. Res.*, **88**, 6415-6429, 1983.

143. The fault-slip directions determined from historic offsets or from grooves and slickensides on exposed fault surfaces can be used as stress indicators in both extensional and compressional regimes. The horizontal component of normal-fault slip is taken as the direction of least principal stress, while the horizontal component of reverse-fault slip determines the direction of greatest principal stress. As in the case of seismologically derived focal mechanisms, these geological interpretations can be biased if the faulting occurs on misoriented planes of weakness; however, the bias associated with normal and reverse faults tends to be small, because strength anisotropy tends to be in the σ_1-σ_3 plane and not in planes containing the σ_2 axis.

144. Dike intrusions, which are common in extensional regimes, are particularly reliable estimators of stress orientation (±5-10 degrees); even in regions with abundant geologic fabric, the orientation of the intrusions follows planes perpendicular to the axis of least principal stress (R.L. Christiansen and E.H. McKee, Late Cenozoic volcanic and tectonic evolution of the Great Basin and Columbia intermountain regions, in *Cenozoic Tectonics and Regional Geophysics of the Western Cordillera*, R.B. Smith and G.P. Eaton, eds., Geological Society of America Memoir 152, Boulder, Colo., pp. 283-312, 1978).

145. This problem was first recognized by D.P. McKenzie (Relation between fault plane solutions and direction of principal stresses, *Bull. Seis. Soc. Am.*, **59**, 591-601, 1969). In his analysis, he assumed that the deviatoric stresses in the crust were much smaller than those needed to fracture intact rock and therefore faulting must occur on weak, preexisting fractures; he showed that for the general case, σ_1 must lie in the same quadrant as the *P* axis, but could differ by as much as 90 degrees. Using more realistic constraints on rock strength, C.B. Raleigh, J.H. Healy, and J.D. Bredehoeft (Faulting and crustal stress at Rangely, Colorado, in *Flow and Fracture of Rocks*, H.C. Heard, I.Y. Borg, and N.L. Carter, eds., American Geophysical Union, Geophysical Monograph Series 16, Washington, D.C., pp. 275-284, 1972) concluded that the bias was not likely to be more than 35-40 degrees.

146. M.L. Zoback and M.D. Zoback (Faulting patterns in north-central Nevada and strength of the crust, *J. Geophys. Res.*, **85**, 275-284, 1980) found that composite fault-plane solutions for a large number of earthquakes agreed to within 3 degrees with the observed fault orientations from the northern Basin and Range, even though the standard deviation in the *T*-axis azimuth for an individual mechanism was substantial (25 degrees).

147. J. Angelier, Tectonic analysis of fault slip data sets, *J. Geophys. Res.*, **89**, 5835-5848,

1984; J.W. Gephart and D.W. Forsyth, An improved method for determining the regional stress tensor using earthquake focal mechanism data: Application to the San Fernando earthquake sequence, *J. Geophys. Res.*, **89**, 9305-9320, 1984. This method can also be used to constrain the relative magnitude of the maximum deviatoric stress, $\sigma_1 - \sigma_3$, which governs the amount of scatter.

148. B.C. Haimson and C. Fairhurst, In-situ stress determination at great depth by means of hydraulic fracturing, in *Rock Mechanics—Theory and Practice, Proceedings, 11th Symposium on Rock Mechanics, Berkeley, 1969*, W.H. Somerton, ed., Society of Mining Engineers of AIME, New York, pp. 559-584, 1970; A. McGarr and N.C. Gay, State of stress in the Earth's crust, *Ann. Rev. Earth Planet. Sci.*, **6**, 405-436, 1978. Stress orientations can also be estimated from the strain observed inside boreholes after overcoring relieves a component of the in situ stress, although this method cannot be used far from the free surface and is more sensitive to local inhomogeneities.

149. See summary comments in the special volume on mechanical involvement of fluids in faulting by S. Hickman, R. Sibson, and R. Bruhn (Introduction to special session: Mechanical involvement of fluids in faulting, *J. Geophys. Res.*, **100**, 12,831-12,840, 1995).

150. R. Emmermann and J. Lauterjung, The German Continental Deep Drilling Program KTB: Overview and major results, *J. Geophys. Res.*, **102**, 18,179-18,201, 1997.

151. See M.D. Zoback and R. Emmermann, eds., *Scientific Rationale for Establishment of an International Program of Continental Scientific Drilling*, Report of the International Conference on Scientific Drilling, Potsdam, Germany, 126 pp., August 20-September 1, 1993; H.C. Larsen and I. Kushiro, eds., *Report of the Conference of Cooperative Ocean Riser Drilling (CONCORD)*, Tokyo, 116 pp., July 22-24, 1997.

5

Earthquake Physics and Fault-System Science

Earthquake research focuses on two primary problems. Basic earthquake science seeks to understand how earthquake complexity arises from the brittle response of the lithosphere to forces generated within the Earth's interior. Applied earthquake science seeks to predict seismic hazards by forecasting earthquakes and their site-specific effects. Research on the first problem began with attempts to place earthquake occurrence in a global framework, and it contributed to the discovery of plate tectonics; research on the second was driven by the needs of earthquake engineering, and it led to the development of seismic hazard analysis. The historical separation between these two problems, reviewed in Chapter 2, has been narrowed by an increasing emphasis on dynamical explanations of earthquake phenomena. In this context, the term dynamics implies a consideration of the forces (stresses) within the Earth that act to cause fault ruptures and ground displacements during earthquakes. The stress fields responsible for deep-seated earthquake sources cannot be measured directly, but they can be inferred from models of earthquake systems that obey the laws of physics and conform to the relationships between stress and deformation (rheology) observed in the laboratory.

This chapter describes how this physics-based approach has transformed the field into an interdisciplinary, system-level science—one in which dynamical system models become the means to explain and integrate the discipline-based observations discussed in Chapter 4. The chapter begins with an essay on the central problems of dynamics and prediction, which is followed by five sections on areas of intense interdisciplinary research: fault systems, fault-zone processes, rupture dynamics, wave propagation, and seismic hazard analysis. Each of the latter summarizes the current understanding and articulates major goals and key questions for future research.

5.1 EARTHQUAKE DYNAMICS

For present purposes, the term "dynamical system" can be understood to mean any set of coupled objects that obeys Newton's laws of motion—rocks or tectonic plates, for example (*1*). If one can specify the positions and velocities of each of these objects at any given time and also know exactly what forces act on them, then the state of the system can be determined at a future time, at least in principle. With the advent of large computers, the numerical simulation of system behavior has become an effective method for predicting the behavior of many natural systems, especially in the Earth's fluid envelopes (e.g., weather, ocean currents, and long-term climate change) (*2*). However, many difficulties face the application of dynamical systems theory to the analysis of earthquake behavior in the solid Earth. Forces must be represented as tensor-valued stresses (*3*), and the response of rocks to imposed stresses can be highly nonlinear. The dynamics of the continental lithosphere involves not only the sudden fault slips that cause earthquakes, but also the folding of sedimentary layers near the surface and the ductile motions of the hotter rocks in the lower crust and upper mantle. Moreover, because earthquake source regions are inaccessible and opaque, the state of the lithosphere at seismogenic depths simply cannot be observed by any direct means, despite the conceptual and technological breakthroughs described in Chapter 4.

From a geologic perspective, it is entirely plausible that earthquake behavior should be contingent on a myriad of mechanical details, most unobservable, that might arise in different tectonic environments. Yet earthquakes around the world share the common scaling relations, such as those noted by Gutenberg and Richter (Equation 2.5) and Omori (Equation 2.8). The intriguing similarities among the diverse regimes of active faulting make earthquake science an interesting testing ground for concepts emerging from the physics of complex dynamical systems. One consequence of recent interactions between these fields is that theoretical physicists have adopted a family of idealized models of earthquake faults as one of their favorite paradigms for a broad class of nonequilibrium phenomena (*4*). At the same time, earthquake scientists have become aware that earthquake faults may be intrinsically chaotic, geometrically fractal, and perhaps even self-organizing in some sense. As a result, an entirely new subdiscipline has emerged that is focused around the development and analysis of large-scale numerical simulations of deformation dynamics. Combined with insightful physical reasoning and intriguing new laboratory and field data, these investigations promise a better understanding of seismic complexity and predictability.

Complexity and the Search for Universality

Earthquakes are clearly complex in both the commonsense and the technical meanings of the word. At the largest scales, complexity is manifested by features such as the aperiodic intervals between ruptures, the power-law distribution of event frequency across a wide range of magnitudes, the variable patterns of slip for earthquakes occurring at different times on a single fault, and the richness of aftershock sequences. Individual events are also complex in the disordered propagation of their rupture fronts and the heterogeneous distributions of residual stress that they leave in their wake. At the smallest scales, earthquake initiation appears to be complex, with a slowly evolving nucleation zone preceding a rapid dynamic breakout that sometimes cascades into a big rupture. Among the many open issues in this field are the questions of whether these different kinds of complexity might be related to one another and, if so, how.

The most ambitious and optimistic reason for considering the ideas of dynamical systems theory is the hope that one might discover universal features of earthquake-like phenomena. Such features would, of course, be extremely interesting from a fundamental scientific point of view. They might also have great practical value, for example, as a basis for interpreting seismic records or for making long-term hazard assessments. Two thought-provoking, complementary concepts that look as if they might bring some element of universality to earthquake science are fractality and self-organized criticality. The first describes the geometry of fault systems; the second is an intrinsically dynamic hypothesis that pertains to the complex motions of these systems. Although each has provoked its own point of view among earthquake scientists—that seismic complexity is, on the one hand, primarily geometric in origin or, on the other hand, primarily dynamic—it seems likely that both concepts contain some elements of the truth and that neither is a complete description of the behavior of the Earth.

There is substantial evidence that fault geometry is fractal, at least in some cases and over some ranges of length scales. Fractality is a special kind of geometric complexity that is characterized by scale invariance (*5*). That is, images of the same system made with different magnifications are visually similar to one another; there is no intrinsic length scale such as a correlation length or a feature of recognizable size that would enable an observer to determine the magnification simply by looking at the image.

One result of such a property in the case of fault zones is that there would be a broad, power-law distribution of the lengths of the constituent fault segments (*6*). If, in the simplest conceivable scenario, the seismic moment of the characteristic earthquake on each segment were proportional to its length, and each segment slipped at random, then the moment distribution would also be a power law. This picture is too simplistic to be a plausible explanation of the Gutenberg-Richter relation, but it may contain some element of the truth.

Self-organized criticality refers to the conjecture that a large class of physical systems, when driven persistently away from mechanical equilibrium, will operate naturally near some threshold of instability, and will therefore exhibit something like thermodynamic critical fluctuations (*7*). Earthquake faults, or arrays of coupled faults, seem to be natural candidates for this kind of behavior; such systems are constantly being driven by tectonic forces toward slipping thresholds (*8*). If the thermodynamic analogy were valid, then the fluctuations—the slipping events—would be self-similar and scale invariant, and their sizes would obey power-law distributions. More important, systems with this self-organizing property would always be at or near their critical points. Critical behavior, with strong sensitivity to small perturbations and intrinsic unpredictability, would be a universal characteristic of such systems.

Elementary Models of Earthquake Dynamics

The ideas of fractality and dynamic self-organization have inspired a wide range of theoretical models of seismic systems. These models are almost invariably numerical: that is, they are studied primarily by means of large-scale computation. One class is cellular automata in which highly simplified rules for the behavior of large numbers of coupled components attempt to capture the essential features of complex seismic systems (*9*). Almost all cellular automata are related in some ways to the original one-dimensional slider-block model of Burridge and Knopoff (*10*), illustrated in Figure 5.1. Perhaps the most important result to emerge from such studies so far is the discovery that some of the simplest of these models, even the completely uniform Burridge-Knopoff model with a plausible, velocity-weakening dynamic friction law, are deterministically chaotic (*11*).

A chaotic system, by definition, is one in which the accuracy needed to determine its motion over some interval of time grows rapidly, in fact exponentially, with the length of the interval. Two identical systems that are set in motion with almost but not quite the same initial conditions may move in nearly the same way for a while. If these systems are chaotic, however, their motions eventually will differ from each other and, after a

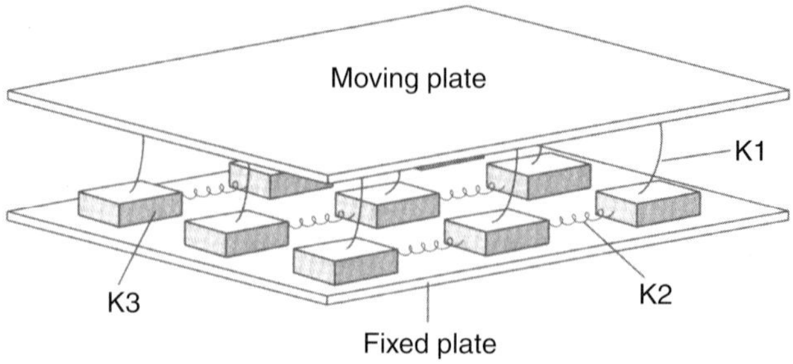

FIGURE 5.1 Two-dimensional version of the Burridge-Knopoff spring-slider model. Leaf springs (K1) connect a moving plate to an array of smaller sliding blocks. These blocks are in turn connected to their nearest neighbors via coil springs (K2, K3). Sliding blocks also have a frictional contact with a fixed plate. Simulated earthquakes from models of this type display a wide variety of complexity. SOURCE: P. Bak, *How Nature Works: The Science of Self-Organized Criticality*, Springer-Verlag, New York, 226 pp., 1996. Copyright permission granted by Springer-Verlag.

sufficiently long time, will appear to be entirely uncorrelated. The correlation time depends sensitively on the difference in the initial conditions. In the context of predictability, this means that any uncertainty in one's knowledge of the present state of a deterministically chaotic system produces a theoretical limit on how far into the future one can determine its behavior reliably, a topic explored further below.

One theoretical issue that has attracted a lot of attention has come to be known as the question of smooth versus heterogeneous fault models. This issue arose initially as a result of the unexpected success of the uniform Burridge-Knopoff slider-block models in producing very rough but interesting caricatures of complex earthquake-like behavior, which fueled speculation that some of the slip complexity of natural earthquakes might be generated by the nonlinear dynamics of stressing and rupture on essentially smooth and uniform faults. The more conventional and perhaps obvious assumption is that the heterogeneity of fault zones—their geometric disorder and strong variations of lithological properties—plays the dominant role. It appears that earthquake faults, when modeled in any detail, have relevant length and time scales that invalidate simple scaling assumptions. For example, the tectonic loading speed (meters per century) combined with known friction thresholds and elastic moduli of rocks suggests natural characteristic intervals (hundreds of years) between large slipping events. Models that incorporate these features produce event distributions in which the large events fail to be self-similar (*12*).

Another example is the thickness of the seismogenic layer, which is less than the rupture scale for larger earthquakes. It, too, seems likely to produce scaling violations both in dynamic behavior and in the geometry of fault systems (see Section 5.2).

The existence of relevant length and time scales does not, per se, invalidate dynamical scaling theories; it may merely limit their ranges of validity. In some smooth-fault models, for example, it appears that the small, localized seismic events are self-similar over broad ranges of sizes; however, the large, delocalized events look quite different and are substantially more frequent than would be predicted by extrapolating the scaling distribution for the small events (*13*), as in the "characteristic earthquake" model discussed in Section 2.6. The picture may change appreciably if one considers large arrays of coupled faults and, especially, if one includes the mechanism for creation of new faults as a part of the dynamical system. It is possible that this global system, in some as yet poorly understood average sense, may come closer to a pure form of self-organized criticality.

Chaos and Predictability

The theoretical issue of earthquake predictability (as distinct from the practical issue of how to predict specific earthquakes) remains a central, unresolved issue. The wide range of event sizes described by the Gutenberg-Richter law, the obvious irregularities in intervals between large events, the fact that chaotic behavior occurs commonly in very simple earthquake-like models, and many other clues, all argue in favor of chaos and thus for an intrinsic limit to predictability. The interesting question is what bearing this theoretical limit might have on the kinds of earthquake prediction that are discussed elsewhere in this report. If one could measure all the stresses and strains in the neighborhood of a fault with great accuracy, and if one knew with confidence the physical laws that govern the motion of such systems, then the intrinsic time limit for predictability might be some small multiple of the average interval between characteristic large events on the fault. Most of the seismic energy is released in the large events; thus, it seems reasonable to suppose that the system suffers most of its memory loss during those events as well. If this supposition were correct, earthquake prediction on a time scale of months or years—intermediate-term prediction of the sort described in Section 2.6—would, in principle, be possible.

The difficulty, of course, is that one cannot measure the state of a fault and its surroundings with great accuracy, and one still knows very little about the underlying physical laws. If these gaps in knowledge could be filled, then predicting earthquakes a few years into the future might be no

that it must be crucial because, without a coupling between the top and bottom of the fault, there is no restoring force to limit indefinitely large slip or, equivalently, to couple kinetic energy of slip back into stored elastic energy. It is hard to see how the dynamics of large events, especially rupture arrest and pulse propagation, can be studied sensibly without full three-dimensional analyses.

The issues of how to make progress toward realism are theoretical as well as computational. There is an emerging realization among theorists working on earthquake dynamics, and in solid mechanics more generally, that the problems with which they are dealing are far more difficult mathematically than they had originally supposed. One of the reasons that small-scale features can control large-scale behavior, as mentioned above, is that these features enter the mathematical statement of the problem as singular perturbations. For example, the surface tension in the solidification problem and the viscosity in certain shock-front problems enter the equations of motion as coefficients of the highest derivative of the dependent variable. As such, they completely change the answer to questions as basic as whether or not physically acceptable solutions exist and how many parameters or boundary conditions are needed to determine them. A related difficulty that is emerging, especially in problems involving elasticity, is that the equations of motion are often expressed most accurately as singular integral equations. Except for a few famous cases due largely to Muskhelishvili (*17*), such equations are not analytically solvable. There are not even good methods for determining the existence of solutions, nor are there reliable numerical algorithms for finding solutions when they do exist. In general, the ability to resolve the uncertainties regarding connections between model ingredients and physical phenomena will depend on advances in both mathematics and computer science. These problems are solvable, but they are indeed difficult.

5.2 FAULT SYSTEMS

Most theories of earthquake dynamics presume that essentially all major earthquakes occur on thin, preexisting zones of weakness, so that the behavior of the biggest events derives from the slip dynamics of a fault network. There are strongly different conceptions of fault systems, all of which may have merit for some purposes (*18*). Faults can be modeled as smooth Euclidean surfaces of displacement discontinuity in an otherwise continuous medium; fault systems can be represented as fractal arrays of surfaces; fault segments can be regarded as merely the deforming borders between blocks of a large-scale granular material transmitting stress in a force-chain mode. Representing the crust as a fault system is especially useful on the interseismic time scales relevant to fault interactions, seismicity distributions, and the long-term aspects of the postseismic response.

Fault-system dynamics involves highly nonlinear interactions among a number of mechanical, thermal, and chemical processes—fault friction and rupture, poroelasticity and fluid flow, viscous coupling, et cetera—and sorting out how these different processes govern the cycle of stress accumulation, transfer, and release is a major research goal. Moreover, progress on the problem of seismicity as a cooperative behavior within a network of active faults has the potential to deliver huge practical benefits in the form of improved earthquake forecasting. The latter consideration sets a direction for the long-term research program in earthquake science.

Architecture of Fault Systems

Thermal convection and chemical differentiation are driving mass motions throughout the planetary interior, but the slip instabilities that cause earthquakes appear to be confined to the relatively cold, brittle boundary layers that constitute the Earth's lithosphere. With sufficient knowledge of the rheologic properties of the lithosphere and the necessary computational resources, it should be possible to set up simulations of mantle convection that reproduce plate tectonics from first principles, including the localization of deformation into plate boundary zones. However, the nonlinearity of the rheology and its sensitivity to pressure, temperature, and composition (especially the minor but critical constituent of water) make this a difficult problem (*19*). Tough computational issues are also posed by the wide range of spatial scales that must be represented in numerical models. Strain localization is most intense on plate boundaries that involve the relatively thin oceanic crust, although there are exceptions. One is a region of diffuse though strong seismicity (up to moment magnitude [M] 7.8) in the central Indian Ocean that may represent an incipient plate boundary (*20*). The study of these juvenile features may shed light on the localization problem.

In continents, earthquakes are typically distributed across broad zones in which active faults form geometrically and mechanically complicated networks that accommodate the large-scale plate motions. This diffuse nature is clearly related to the greater thickness and quartz-rich composition of the continental crust, as described in Section 2.4. The structure of continental fault zones is thought to be complicated by variations in frictional behavior with depth, changes in wear mechanisms, and a brittle-ductile transition (Figure 4.30), although the details remain highly uncertain.

Interesting issues also arise from attempts to understand how the complexities are related to the long geological history of the continents. In the southwestern United States, for example, the fault systems that produce high earthquake hazards have developed over tens of millions of years by tectonic interactions among the heterogeneous ensemble of accreted terrains that constitute the North American continental lithosphere and the oceanic lithosphere of the Farallon and Pacific plates. These interactions have created a zone of deformation a thousand kilometers wide that extends from the continental coastline to the Rocky Mountains. The "master fault" of this plate-boundary zone is the strike-slip San Andreas system, but other types of faults participate in the deformation, from extension in the Basin and Range to contraction in the Transverse Ranges. Likewise, the great thrust faults that mark the subduction zones of the northwestern United States and Alaska are accompanied by secondary faulting distributed for considerable distances landward of the subduction boundary. Within the continental interior far from the present-day plate boundaries, deformation is localized on reactivated, older faults, and some of these structures are capable of generating large earthquakes (see Section 3.2).

The geometric complexity of fault systems is fractal in nature, with approximately self-similar roughness, segmentation, and branching over length scales ranging from meters to hundreds of kilometers (Figure 3.2). Fault systems also have mechanical heterogeneities due to lithologic contrasts, uneven damage, and possibly pressurized compartments within fault zones (*21*). The understanding of fault system architecture and earthquake generation in such systems is at a rudimentary stage of development.

Fault Kinematics and Earthquake Recurrence

The subject of fault kinematics pertains to descriptions of earthquake occurrence and slip of individual faults at different time scales, and the partitioning of slip among faults to accommodate regional deformation. An important goal of this characterization is to address the fundamental question of how slow and smoothly distributed regional deformations across fault systems, as seen in geodetic observations, are eventually transformed, principally at the time of earthquakes, into localized slip on particular faults. To build a comprehensive picture of this process requires synthesis of detailed geologic, geophysical, and seismic observations. At present, some regions—particularly portions of California and Japan—have sufficient information to describe the recent history of large earthquakes, to make estimates of the long-term average of slip rates of the principal faults, and to map the surface strain field across fault systems. Though comprehensive descriptions of fault-system kinematics are not yet possible in any region, some generalizations have emerged on fault-system behavior at different time scales.

Across periods of perhaps a million years, fault systems evolve as slip brings different geologic formations into juxtaposition, new faults become activated, and previously existing faults go dormant. Processes on these time scales are undoubtedly important for understanding the origins and evolution of fault-system architecture. However, for estimations of earthquake probabilities and simulations of seismic activity on shorter times scales, an assumption of fixed fault-system geometry appears to be a reasonable approximation.

On time scales of a thousand years and less, there is clear evidence that earthquake activity is not stationary in time or space. That is, some regions show episodes of high earthquake activity followed by long periods of relative inactivity. Perhaps the best known example of episodic earthquake activity on a regional scale is from the north Anatolian fault in Turkey (Figure 3.21). Similarly in China, which has a long historical record of major earthquakes, it is evident that large regions have been episodically activated for many decades followed by long interludes of low earthquake activity (*22*). In the United States, geologic studies in Nevada, the eastern California shear zone, and elsewhere have found evidence for periods of high seismic activity across broad regions followed by long intervals with little or no geologic evidence of faulting activity (*23*).

Questions relating to the repeatability and recurrence intervals of large earthquakes on shorter time scales are of particular importance for the evaluation of earthquake probabilities used in seismic hazard analysis. Current approaches to estimating earthquake probabilities assume either that earthquakes occur randomly in time, but at some fixed rate, or that major earthquakes have sufficient periodicity to permit estimates of probability to be made based on elapsed time from the previous earthquake on a fault segment. Few large faults have ruptured more than once during the instrumental or historical period, and only in rare cases have the ruptures been documented well enough to enable unambiguous comparisons of the sequential ruptures. Hence, discussions of the periodicity (or aperiodicity) of large earthquakes, and the degree to which earthquake source parameters vary through several slip events, are dominated by conjecture. One approach to evaluating repeatability and periodicity of earthquakes employs seismic data from smaller earthquakes. Along the creeping portion of the San Andreas fault in central California, M 4 to M 5 earthquakes have been frequent enough to enable studies of their similarity. Waveforms from these moderate events can be sorted into nearly identical groups, establishing the existence of small, active fault patches, each generating nearly identical characteristic earthquakes with well-defined periodicities (*24*). These characteristic patches appear to be driven by aseismic slip of the surrounding regions of the fault plane (*25*).

Another approach to characterizing the repeatability and degree of periodicity of large earthquakes is based on paleoseismic investigations, which seek to reconstruct fault slip and earthquake histories over periods of thousands of years. Available paleoseismic data suggest that major earthquakes often involve distinct fault segments that tend to slip persistently in a similar manner from earthquake to earthquake. Some examples include portions of the Wasatch fault in Utah, the Superstition Hills fault in California, and the Lost River Range fault, a normal fault in Idaho (*26*). However, in other cases, more varied behavior among the segments appears to be the norm. For example, both the Imperial fault in southern California and the North Anatolian fault in Turkey have failed in a different manner in historic time (*27*). In some cases, paleoseismic data support the concept of periodicity, while in other situations, earthquake occurrence appears to have been aperiodic. These observations, together with episodic regional activation at long time scales, imply that simple characterizations of earthquake repeatability and periodicity may not be possible.

Seismicity and Scaling

Earthquake scaling laws, and the circumstances under which they break down, furnish insights on fault interactions that carry important ramifications for seismic hazard analysis and earthquake prediction. For instance, the seismicity of individual faults does not follow the Gutenberg-Richter relation (*28*), indicating that the frequency-magnitude power law is a property of the fault system, perhaps related to the fractal distributions of fault sizes. The Gutenberg-Richter relation also appears to break down for large earthquakes, where the earthquake rupture width is constrained by the depth extent of the seismogenic zone (*29*). The scaling laws for earthquake parameters at larger magnitudes also seem to be bounded by the thickness of the seismogenic zone. Although this topic has created a great deal of controversy, recent results suggest that the scaling of slip with rupture length in earthquakes is consistent with scale-independent rupture physics (*30*).

Uncertainty also exists on the breakdown of self-similarity and the Gutenberg-Richter relation at small magnitudes. Theoretical studies, which employ laboratory-derived fault friction laws, indicate there should be some minimum fault length for earthquake fault slip as defined by the nucleation zone for earthquake initiation (see Section 5.3). This dimension is of fundamental importance for two reasons. First, it sets a scale length that must be respected for realistic simulations of the earthquake initiation and rupture propagation processes. Second, it defines the dimensions of the region of precursory strains related to the earthquake nucleation process. Small scaling lengths impose severe restrictions on numerical calculations and could also mean that precursory phenomena related to earthquake nucleation may be difficult or impossible to detect.

Stress Interactions and Short-Term Clustering

Although major earthquakes generally tend to be associated with large faults easily recognized at the surface, instrumentally recorded seismicity indicates that smaller earthquakes become more diffusely distributed as their size decreases. The smallest earthquakes often arise on faults with no known surface expression. Stress-mediated interactions among these fractal fault systems can be explored by using the scaling behavior of the seismicity to monitor system organization as a function of time. This type of regional seismicity analysis offers the most promising approach to intermediate-term prediction.

A widely studied type of fault interaction arises from the permanent change of the stress field following an earthquake. According to the Coulomb stress condition for frictional failure (Equation 2.1), an increase in the magnitude of the shear stress acting across a fault should push it closer to failure, while an increase in normal stress should increase the effective frictional strength, thus retarding failure. An important recent discovery is that regional seismicity appears to be correlated with the relatively small Coulomb stress increments calculated from static dislocation models of large earthquakes (*31*). This interpretation of seismicity has been largely successful in explaining the patterns of aftershocks as well as regions of reduced seismicity ("stress shadows") following large events along the San Andreas fault system (*32*), the 1999 Izmit earthquake in Turkey (*33*), and various earthquakes in Japan, Italy, and elsewhere (*34*).

The Coulomb stress calculations usually assume purely elastic interactions at the time of the mainshock. This is a reasonable approximation in the outer layers of the brittle crust, but it does not describe known postseismic processes, which include ductile flow below the seismogenic zone, fault creep (earthquake afterslip), and poroelastic effects (due to fluid flow) that all result in extended intervals of stressing in the region of a large earthquake (Section 4.2). The role these postseismic effects have in controlling, or altering, aftershocks sequences is presently not well understood, but the stress changes due to these processes are usually rather small compared to the immediate stress change caused by the mainshock.

Aftershocks are thought to be primarily a response of the surrounding fault system to stress changes caused by the mainshock fault slip. That is, the Coulomb stress changes drive the aftershock fault planes to failure.

Aftershocks are an extreme example of short-term earthquake clustering that appears to be quite distinct from the long-term regional clustering of large earthquakes discussed above. Aftershocks can temporarily increase the local seismicity rates to more than 10,000 times the pre-mainshock level. Although Coulomb stress interactions provide an explanation for many aftershock patterns, those models alone do not account for either the rates of seismicity that occur in response to the stress changes or the subsequent decay of rates inversely proportional to time, as expressed in Omori's aftershock decay law (Equation 2.8). The most fully developed explanation for these and other properties of aftershocks is based on the rate- and state-dependent fault frictional properties observed in laboratory experiments (see Section 4.4). These frictional properties require that the initiation of earthquake slip (earthquake nucleation) be a delayed instability process in which the time of an earthquake is nonlinearly dependent on stress changes (*35*). This approach has resulted in a state-dependent model for earthquake rates that provides quantitative explanations for observed aftershock rates in response to a stress change, the Omori decay law, and various other features of aftershocks (Box 5.1).

Aftershocks can also be generated by dynamic stresses during the passage of seismic waves. At large epicentral distances, these transients are much greater than the static Coulomb stresses, although they act only over short intervals. Short-term dynamic loading was responsible for triggering seismicity across the western United States after the 1992 Landers, California earthquake (*36*). Immediately following the Landers earthquake, bursts of seismicity were observed at locations more than 1000 kilometers from the mainshock (Figure 5.2). The mechanisms for after-

BOX 5.1 State-Dependent Seismicity

A physically based method for quantitative modeling of the relationships between stress changes and earthquake rates is provided by the rate- and state-dependent representation of fault friction. This approach treats seismicity as a sequence of earthquake nucleation events and specifically includes the time and stress dependence of the earthquake nucleation process as required by rate- and state-dependent friction. The result is a general state-dependent formulation for earthquake rates:[1]

$$R = \frac{r}{\gamma \dot{S}_r}, \quad d\gamma = \frac{1}{A\sigma}\left[dt - \gamma dS\right], \tag{1}$$

where R is earthquake rate (in some magnitude interval), γ is a state variable, t is time, and S is Coulomb stress. The normalizing constant r is defined as the steady-state earthquake rate at the reference stressing rate $\dot{S}_r$. A is a dimensionless fault

(continued on following page)

BOX 5.1 *(continued)*

constitutive parameter with values in the range 0.005 to 0.015. For this model, the Coulomb stress function is defined as

$$S = \tau - [\mu - \alpha]\sigma , \tag{2}$$

where τ and σ are shear and effective normal stress, respectively, acting across the fault planes that generate earthquakes; μ is the coefficient of fault friction; and α is a constant with values in the range $0 < \alpha < \mu$. In equation (1) the term $A\sigma$ is treated as constant (i.e., the changes in stress are negligible relative to total normal stress). A property of seismicity predicted by this model is that stress perturbations drive seismicity rates away from a steady-state condition set by the Coulomb stressing rate S^Y and seismicity seeks to return to steady state over the characteristic time $t_a = A\sigma/S^Y$.

The effects of a nearby earthquake on seismicity rates are given by the solution of (1) for a stress step ΔS,

$$R = \frac{rS^Y / S_r^Y}{\left[\frac{\dot{S}}{\dot{S}_r}\exp\left(\frac{-\Delta S}{A\sigma}\right) - 1\right]\left[\exp\left(\frac{-t}{t_a}\right)\right] + 1}, \dot{S} \neq 0 \tag{3}$$

where $t = 0$ at the time of the step. At $t > t_a$, earthquake rates approach a constant background rate, and at $t < t_a$ this solution acquires the form of Omori's aftershock decay law (Equation 2.8):

$$R = \left(\frac{a}{c + t}\right)^p , \tag{4}$$

with $p = 1$, $a = rt_a$, and $c = t_a \exp(-\Delta S/A\sigma)$. Model predictions of aftershock rates, time-dependent expansion of the aftershock zone, proportionality of aftershock duration t_a to the inverse of stressing rates, and spatially averaged aftershock decay by $t^{-0.8}$ all appear to be consistent with the data.[2] Variations of this approach have been used to model foreshocks[3] and the statistics of earthquake pairs. Recently the state-dependent seismicity formulation has been used to invert earthquake rate for the changes in stress that drive earthquakes at Kilauea, Hawaii.[4]

---

[1] J.H. Dieterich, A constitutive law for rate of earthquake production and its application to earthquake clustering, *J. Geophys. Res.*, **99**, 2601-2618, 1994; J.H. Dieterich, V. Cayol, and P. Okubo, The use of earthquake rate changes as a stress meter at Kilauea volcano, *Nature*, **408**, 457-460, 2000.

[2] Equation 3 predicts aftershock decay by t^{-1} in regions where stress change is homogeneous. However, stress change caused by earthquake faulting varies greatly by position and distance from the rupture. Integration of Equation 3 over a finite region using the stress distribution around a circular rupture gives a net aftershock decay by approximately $t^{-0.8}$. Similar average decay rates are found for aftershocks, frequency of earthquake pairs, and frequency of foreshock-mainshock pairs (L.M. Jones, Foreshocks and time-dependent earthquake hazard assessment in southern California, *Bull. Seis. Soc. Am.*, **75**, 1669-1679, 1985; Y.Y. Kagan and D.D. Jackson, Long term earthquake clustering, *Geophys. J. Int.*, **104**, 117-133, 1991).

[3] J.H. Dieterich and B. Kilgore, Implications of fault constitutive properties for earthquake prediction, *Proc. Natl. Acad. Sci.*, **93**, 3787-3794, 1996.

[4] J.H. Dieterich, V. Cayol, and P. Okubo, The use of earthquake rate changes as a stress meter at Kilauea volcano, *Nature*, **408**, 457-460, 2000.

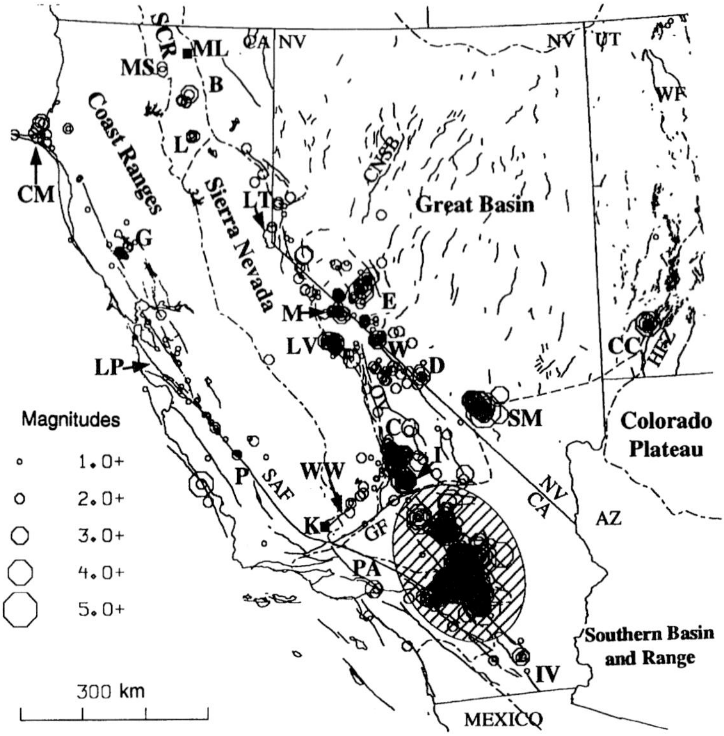

FIGURE 5.2 Map of the western United States showing areas of increased seismicity dynamically triggered by seismic waves from the 1992 Landers earthquake. SOURCE: D.P. Hill, P.A. Reasenberg, A. Michael, W. Arabaz, G.C. Beroza, J.N. Brune, D. Brumbaugh, S. Davis, D. DePolo, W.L. Ellsworth, J. Gomberg, S. Harmsen, L. House, S.M. Jackson, M. Johnston, L. Jones, R. Keller, S. Malone, S. Nava, J.C. Pechmann, A. Sanford, R.W. Simpson, R.S. Smith, M. Stark, M. Stickney, S. Walter, and J. Zollweg, Seismicity remotely triggered by the magnitude 7.3 Landers, California, earthquake, *Science*, **260**, 1617-1623, 1993. Copyright 1993 American Association for the Advancement of Science.

shock triggering by seismic waves are poorly understood but may involve fluid-rock interactions or triggering of local deformations that produce permanent stress changes after the waves have passed through a region.

Foreshocks Foreshocks are generally thought to arise by one of two mechanisms. The first proposes that a mainshock following a foreshock has an identical origin to that of aftershocks. In this case, earthquake frequency-magnitude statistics predict that occasionally an aftershock will be larger than the prior event, which by definition makes the prior event a foreshock (*37*). The other proposed mechanism for foreshocks is that premonitory processes, perhaps the fault creep related to mainshock nucleation, result in stress changes that drive the foreshock process in surrounding areas. Models based on state-dependent earthquake rates indicate that both mechanisms are in general agreement with time and distance statistics of foreshock-mainshock pairs (*38*).

Short-term clustering, as manifest in foreshock-mainshock pairs and aftershocks, attests to large but transient changes in the probabilities of additional earthquakes that occur whenever an earthquake takes place. The concepts of stress interaction and state-dependent seismicity permit physically based calculations of earthquake probability following large earthquakes (*39*). This approach has been used to evaluate the changes in earthquake probability that arose as a consequence of stress interactions along the Anatolian fault in Turkey (*40*) and following the M 6.9 earthquake that struck Kobe, Japan, in 1995 (*41*).

Accelerating Seismicity and Intermediate-Term Prediction A central issue for earthquake prediction is the degree to which the seismicity clustering can be used to monitor the stress changes leading to large earthquakes. Various studies have shown that large earthquakes tend to be preceded by clusters of intermediate-sized events (*42*). This increase in seismicity can be fit to a time-to-failure equation in the form of a power law, which is commonly used by engineers to describe progressive failures that result from the accumulation of structural damage (*43*). The power-law time-to-failure equation is also expected if large earthquakes represent critical points for regional seismicity (*44*).

As described in Section 5.1, regional seismicity has many of the characteristics of a self-organized critical system, including power-law (Gutenberg-Richter) frequency-size statistics and fractal spatial distributions of hypocenters. However, the near-critical behavior of fault systems is the subject of some debate. If the crust continuously maintains itself in a critical state, as originally proposed by Bak and Tang, then all small earthquakes will have the same probability of growing into a big event. This hypothesis has been used as the physical basis for assertions that earthquake prediction is inherently impossible (*45*). Alternatively, the crust could repeatedly approach and retreat from a critical state. The working hypotheses for this latter view are (1) large regional earthquakes become more probable when the stress field becomes correlated over increasingly larger distances, (2) this approach to a critical state is reflected in an acceleration of regional seismicity, and (3) a system-spanning event destroys criticality on its network, creating a period of relative quiescence after which the process repeats by rebuilding correlation lengths toward criticality and the next large event (*46*). It is the decay of the post-event stress shadows by continuing tectonic deformation that introduces predictability into the system.

The seismic cycle implied by these hypotheses agrees with some important aspects of the data on seismic stress shadows and accelerating seismicity (*47*). Many issues remain to be resolved, however. Quantitative testing will require precisely formulated numerical models adapted to specific fault networks (i.e., computer simulations with realistic representations of fault and block geometries, rheologies, and tectonic loadings). Such "system-level" models are in the early stages of development. The long-term clustering statistics generated by the models must be understood in terms of the underlying dynamics (*48*), and these behaviors will have to be evaluated against the extended earthquake records now being provided by paleoseismology (see Section 4.3). The key step is to deploy the models in regulated prediction environments to rigorously test their predictive skill.

Key Questions

- What are the limits of earthquake predictability, and how are they set by fault-system dynamics?
- Which aspects of the seismicity are scale invariant, and which are scale dependent? How do these scaling properties relate to the underlying dynamics of the fault system? Under what circumstances is it valid to extrapolate results based on low-magnitude seismicity to large-earthquake behavior?
- Are there patterns in the regional seismicity that are related to the past or future occurrence of large earthquakes? For example, are major ruptures preceded by enhanced activity on secondary faults, temporal changes in *b* values, or local quiescence? Can the seismicity cycles associated with large earthquakes be described in terms of repeated approaches to, and retreats from, a regional critical point of the fault system?
- On what scales, if any, is the seismic response to tectonic loading stationary? What are the statistics that describe seismic clustering in time and space, and what underlying dynamics (e.g., mode-switching) control this episodic behavior? Is clustering observed in some fault systems due to repeated ruptures on an individual fault segment or to rupture overlap from multiple segments? Is clustering on an individual fault related to regional clustering encompassing many faults?
- What systematic differences in fault strength and behavior are attributable to the age and maturity of the fault zone, lithology of the wall rock, sense of slip, heat flow, and variation of physical properties with depth? Are mature faults such as the San Andreas weak? If so, why?

- To what extent do fault-zone complexities, such as bends, step-overs, changes in strength, and other "quenched heterogeneities," control seismicity? How applicable are the characteristic earthquake and slip-patch models in describing the frequency of large events? How important are dynamic cascades in determining this frequency? Do these cascades depend on the state of stress, as well as the configuration of fault segments?
- How does the fault system respond to the abrupt stress changes caused by earthquakes? To what extent do the stress changes from a large earthquake change nearby seismicity rates and advance or retard large earthquakes on adjacent faults? How does stress transfer vary with time (*49*)?
- What controls the amplitude and time constants of the postseismic response, including aftershock sequences and transient aseismic deformations? In particular, how important are the induction of self-driven accelerating creep, fault-healing effects, poroelastic effects (which involve the hydrostatic response of porous rocks to stress changes), and coupling of the seismogenic layer to viscoelastic flow at depth?
- What special processes occur at borders or transition regions between creeping zones, whether localized on faults or distributed, and fault zones that are locked between seismic events? Do lineations of microseismicity provide evidence for processes along such borders?
- What part of aseismic deformation on and near faults occurs as episodes of slip or strain versus steady creep?

5.3 FAULT-ZONE PROCESSES

The move toward physics-based modeling of earthquakes dictates that research be focused on relating small-scale processes within fault zones to the large-scale dynamics of earthquakes and fault systems. Earthquakes have many scale-invariant and self-similar features, yet numerical simulations must assume some smallest length scale in a grid or mesh, as well as a shortest time step, in order to discretize the computational problem. The issue then becomes how to refine the discretization adequately so that principal phenomena are represented qualitatively, if not at the quantitatively correct small size scale. There is also the question of whether it is possible to capture the wealth of processes that occur on sub-grid scales through judicious parameterizations. For example, rate- and state-dependent friction laws suggest that processes at a scale smaller than the coherent slip patch size can be swept into the macroscopic constitutive description. This characteristic dimension appears to be a very small, however—on the order of 0.1 to 10 meters (see Section 5.4). Numerical resolution of processes at that size scale is well beyond the capability of current three-dimensional earthquake simulations (*50*).

Damage Mechanics

The question of how well earthquakes can be approximated as propagating dislocations on idealized friction-bound fault planes is also tied to the degree of rheological breakdown and damage in regions of significant lateral extent away from the rupture surface. Such damage zones can be investigated on large scales by seismological field experiments using fault-zone trapped waves (*51*) as well as by gravity and electromagnetic methods (*52*). On smaller scales, processes of rock failure can be studied in the laboratory and their effects observed by field work on exhumed faults.

Recent years have seen strong focus on the possibilities that fractal and granular aspects might be major parts of the observed complexity of fault systems and of fault-zone response. Nevertheless, over the same time, close geological investigations of exhumed fault zones (*53*) have strengthened the viewpoint that much of the observed complexity of damage zone and secondary fault structures bordering large-slip faults could indeed be a relatively inactive relic of evolution and that, with ongoing slip accumulation, faults become more like Euclidean surfaces (*54*). For example, studies at the Punchbowl and North Branch San Gabriel faults (*55*) show abundant complexity of structure, with damaged and faulted rock that extends on order of 100 meters from the fault core. Yet a severely granulated ultracataclastic core on the order of only 100 millimeters wide seems to have accumulated all significant slip, summing to several kilometers of motion. Also, a principal fracture surface that may be only a few millimeters wide seems to have hosted large amounts of slip, presumably corresponding to the last several earthquakes, whereas there is little evidence of significant slip accumulation on secondary faults in the damaged border zone.

This does not at all imply that the damaged zone is irrelevant to fault dynamics. First, it is a storage site for pore fluids. Second it provides a heterogeneity of elastic properties that may allow slip on the main fault, if not well centered within the damaged zone, to induce changes in normal stress, with consequences for frictional instability (*56*). Third, as a zone of low strength, it may react inelastically to the high stresses associated with a propagating rupture front. Stresses acting off the main fault plane become much larger than those along it as the rupture approaches what, in elastic-brittle dynamic crack theory, would be its limiting speed (*57*). It is likely that faulted rock within that border region acts as a macroscale plastic zone when rupture speed approaches the limit speed, so that much of the inferred fracture energy of earthquake faulting may emanate from energy dissipation in the damage zone rather than exclusively from the main fault surface itself (as often assumed in relating seismic observations to parameters of slip-weakening rupture description). Also, the high off-fault stresses may activate rupture along fortuitously oriented, branch fault structures that intersect the main fault. Such a process is a possible source of spontaneous arrest of rupture and of intermittence of rupture propagation speed (enriching the radiated seismic spectrum at high frequencies), and it can be correlated to natural examples of macroscopic branching of the rupture path (*58*).

Friction of Fault Materials

Experimentally determined constitutive laws, such as those presented in Box 4.4, have been validated for slip rates between about 10^{-10} and 10^{-3} meter per second. As such, they cover the range from plate rates to rates at which incipient dynamic instabilities are well under way, so they probably provide an appropriate description of frictional processes during earthquake nucleation and postseismic response. In the common form of these laws, the logarithmic dependence of stress on sudden changes in sliding velocity, introduced empirically, is now generally assumed to descend from an Arrhenius activated rate process governing creep at asperity contacts (*59*). That is, the slip rate V for each active mechanism at the contacting asperities is proportional to $e^{-Q/RT}$, where the activation energy Q is diminished linearly by stress over the narrow range sampled in experiments. This leads at once to the instantaneous ln V dependence of friction coefficient in the range for which forward-activated jumps are vastly more frequent than backward ones. Considering the backward jumps regularizes the ln V dependence at $V = 0$ (*60*). Experiments on optically transparent materials, including quartz, have linked the state evolution slip distance D_c to the sliding necessary to wipe out the original contact population and replace it with a new one (*61*). These experiments also showed time-dependent growth of contact junctions, which is a mechanism by which strength depends on the maturity of the contact population (measured by the state variable). Further, models have proposed thermally activated creep as a mechanism for contact growth that delivers a steady-state friction coefficient proportional to ln V (*62*), which is often observed, at least over limited ranges, in experiments.

The above description outlines the simplest physical understanding of the empirically derived friction laws. To confidently extend these relations to situations not directly studied in the laboratory, it will be important to put them on a firmer basis, in a way that deals more completely with contact statistics and the actual granular structure of fault-zone cores and that recognizes the possibility of multiple deformation mechanisms with different dependencies on temperature, stress, and the chemical environment. A simple version is to assume that deformation in the fault zone can include both slip on frictional surfaces and more distributed creep deformation, with both processes taking place under the same stress (*63*). Based on earlier hydrothermal studies of granite and quartz gouge (*64*), F. Chester suggests that response can be modeled by three mechanisms: solution transfer, cataclastic flow, and localized slip (*65*). Each is assumed to follow a rate- and state-dependent law, but with additional terms to represent effects of changing temperature. Studies of this kind, firmly rooted in materials physics, are needed to extrapolate laboratory data confidently over a range of hydrothermal conditions to very long times at temperature on natural faults, to infer in situ stress conditions and the conditions of local stress and slip rate necessary to nucleate a frictional instability.

Earthquake Mechanics in Real Fault Zones

It may be conjectured that different physical mechanisms prevail at contacts during the most violent seismic instabilities, when average slip rates reach 1 meter per second and maximum slip rates near the rupture front might be as great as 10^2 meters per second. In that range, the dynamics of rapid stress fluctuations from sliding on a rough surface, openings of the rupture surfaces, microcracking, and fluidization of finely comminuted fault materials may result in a different velocity dependence, possibly with a dramatic weakening. Most significantly, very large temperatures will be generated in the rapid, large slips of large earthquakes. These are expected to lead to thermal weakening, but there is presently very limited laboratory study of the process (*66*). When two surfaces slide rapidly, compared to heat diffusion times at the scale of the asperity contacts, a first thermal weakening is due to flash heating and thermal softening of the contacts (*67*). With poor conductors such as rocks, continued shear—especially along narrow surfaces as inferred for the Punchbowl fault (*68*)—would necessarily lead to local melting. The amount of melt generated in actual faulting events is not well constrained. Pseudotachylytes (amorphous rocks, rapidly cooled from the melt) are sometimes seen as fillings of faults and of veins that run off them and at dilatational jogs (*69*). An open question is, how much of the finest-grain gouge components are also the result of rapid cooling of a melt that has been squeezed into narrow pore spaces where it solidified and thermally cracked to small fragments upon cooling.

Although there are presently few experimental constraints on response in the high-slip-rate range, experiments and coordinated theory for this range are essential to understanding the overall stress levels at which faults operate, the heat outflow from faults (and whether its lowness is paradoxical or not), and the mode of rupture along them. For the latter, it is now understood (*70*) that strong velocity weakening together with low shear stress levels over the region through which a rupture propagates promote self-healing of the rupture behind the front, a phenomenon found in numerical simulations (*71*) and observed in real events (*72*). Yet whether it is velocity weakening or some other process or fault-zone property that controls the observed mode of rupture remain to be clarified. Good experiments and observations are a must, and velocity-weakening constitutive response is not the only route to short slip duration. They can also be induced by strong fault heterogeneity (*73*) and as a consequence of even fairly modest dissimilarity of elastic properties between the two elastic blocks bordering a fault zone (*74*).

Provided that typical laboratory friction coefficients for rocks (0.5 to 0.7) apply and that pore pressure is hydrostatic, the shear strength that must be overcome to initiate slip at, say, 10-kilometer depth is estimated to be about 100 megapascals. This is much larger than seismic stress drops, typically on the order of 1 to 10 megapascals. Thus, one option is that faults slide during large earthquake slips at stresses on the order of 100 megapascals. This is, however, in conflict with the well-known lack of a sharply peaked heat outflow over the San Andreas fault (see Section 2.5). It is also difficult to reconcile with observations (*75*) of a steep inclination (60 to 80 degrees) with the San Andreas fault of the principal compression direction in the adjoining crust. The possible ways around this problem are the subject of much discussion. It has been argued (*76*) that the heat flow data are unreliable, being influenced by shallow topographically driven groundwater flows, and that stress directions are dictated by bordering tectonics and are a misinterpreted signal of tectonics in the bordering regions. However, many workers have not been as ready to dismiss these considerations and have sought other modes of explanation. Pore pressure that is greatly elevated over hydrostatic, and nearly lithostatic, at seismogenic depth has been invoked. Also, the possibility has been raised that fault-zone material within well-slipped faults has anomalously low friction, due either to its mineralogical or its morphological evolution (e.g., possibly stabilizing hydrophilic phases with low friction comparable to that of montmorillonite clay (*77*)) or to the inclusion of weak lithologies, possibly serpentine, in the fault. In contrast to these propositions for zones of active tectonics such as the San Andreas, faults intersected by the few deep drill holes in stable continental crust seem to be at hydrostatic pore pressure and to carry maximum shear stresses consistent with friction coefficients in the range 0.5 to 0.7 (*78*). Thus, it is important to better constrain these possibilities by drilling, such as that planned in the San Andreas Fault Observatory at Depth (SAFOD) component of the EarthScope Program, as well as by examinations of exhumed faults, to establish if and why major plate-bounding faults are different in composition or fluid pressurization.

Yet another possibility is that dynamic weakening may be responsible for the low-stress observations along the San Andreas fault. Sources could include severe thermal weakening, including melt formation, in rapid, large slips, as above, or the formation of gouge structures that accommodate slip by rolling with little frictional dissipation (*79*). In the case of sliding between elastically dissimilar materials, there is coupling between spatially inhomogeneous sliding and alteration of normal (clamping) stress. Mathematical solutions have been constructed that allow a pulse of slip to occur in a region of locally diminished clamping stress and hence diminished frictional dissipation (*80*). Experiments on foam rubber blocks (*81*) show a similar effect, even leading to surface separation. Analogous effects have not been found in laboratory rock experiments in the large sawcut apparatus at the U.S. Geological Survey (USGS)-Menlo Park, and the mechanism in the foam rubber remains obscure (nonlinearities in the surrounding continuum-like field could contribute); however, something similar to this could be found for natural faults, possibly as a result of the interaction of the fault core with the damaged zone adjoining it.

These considerations highlight the importance of determining the composition, structure, and physical state of fault-zone materials; of determining their rheology, especially in rapidly imposed large slips; and of understanding the dynamical processes within the core and their interaction with the heterogeneity and possible localized failure processes in the damaged border zones. At larger scales, there is a need for better characterization of fault junctions and of the structure and mechanical properties of fault-jog materials, over or through which rupture jumps in transferring slip from one fault segment to another.

Key Questions

- Which small-scale processes—pore-water pressurization and flow, thermal effects and melt generation, geochemical alteration of minerals, solution transport effects, contact creep, microcracking and rock damage, gouge comminution and wear, gouge rolling—are important in describing the earthquake cycle of nucleation, dynamic rupture, and postseismic healing?
- What fault-zone properties determine velocity-weakening versus velocity-strengthening behavior? How do these properties vary with temperature, pressure, and composition?

• What rheologies govern the shallow deformation of fault zones? When does fault creep occur near the surface? Do lightly consolidated sediments allow distributed inelastic deformation?

• How does fault strength drop as slip increases immediately prior to and just after the initiation of dynamic fault rupture? Are dilatancy and fluid-flow effects important during nucleation?

• What is the nature of near-fault damage and how can its effect on fault-zone rheology be parameterized? Can damage during large earthquake ruptures explain the discrepancy between the small values of the critical slip distance found in the laboratory (less than 100 microns) and the large values inferred from the fracture energies of earthquakes and assumptions about the drop from peak strength for slip initiation to dynamic friction strength (5 to 50 millimeters if the strength drop is 100 megapascals, but an order of magnitude higher for 10 megapascals)?

• Are the broad damage zones observed for some faults relics of the evolution of a through-going fault system on what was a misoriented array of poorly connected fault segments that were reactivated or originated as joints? Do the damage zones result from misfit stresses generated by the sliding of surfaces with larger-scale fractal irregularities? Are they just passive relics or do they also play a significant role in the dynamics of individual events?

• How does fault-zone rheology depend on microscale roughness, mesoscale offsets and bends, variations in the thickness and rheology of the gouge zone, and variations in porosity and fluid pressures? How can the effects of these or other physical heterogeneities on fault friction be parameterized in phenomenological laws based on rate and state variables?

• How does fault strength vary as the slip velocities increase to values as great as 1 meter per second or more? How much is frictional weakening enhanced during high-speed slip by thermal softening at asperity contacts and by local melting?

• How do faults heal? Is the dependence of large-scale fault healing on time logarithmic, as observed over much shorter times in the laboratory? What small-scale processes govern the healing rate, and how do they depend on temperature, stress, mineralogy, and pore-fluid chemistry?

• How does rupture on a major fault interact with faults in the bordering regions? Is this interaction a source of intermittent rupture propagation and resulting enriched high-frequency radiated energy or of the spontaneous arrest of ruptures? Are the high seismically inferred fracture energies (on the order of 100 times laboratory values for initially intact rock under high confining stress) actually due to induction of extensive frictional inelasticity in that border zone? Is fracture energy misinterpreted as being due to slip weakening on a single major fault versus a network of dynamically stressed secondary faults?

- When does the rupture path follow a fault that branches off from the major failure surface? What is the role of pre-stress magnitudes and orientations and of the dynamically altered stress distribution near the rupture front? How do ruptures surmount stepovers? Are elastic descriptions adequate for the stepped-over material, or is there an essential role for damaged rock and smaller fault structures within the stepover region?

5.4 RUPTURE DYNAMICS

Earthquake rupture entails nonlinear and geometrically complex processes near the fault surface, generating stress waves that evolve into linear (anelastic) waves at some distance from the fault. Better knowledge of the physics of rupture propagation and frictional sliding on faults is therefore critical to understanding and predicting earthquake ground motion. Research on rupture processes may also contribute to improvements in earthquake forecasting because of the dynamical connection between the evolution of the stress field on interseismic time scales and the stress heterogeneities created and destroyed during earthquakes.

Rupture Initiation

The process leading to the localized initiation of unstable stick-slip in laboratory (*82*) and theoretical (*83*) models of the earthquake process is referred to as earthquake nucleation. In frictional fault models, stick-slip instabilities can begin only in regions where the progression of slip causes the fault friction to decrease. For the rate-state model, this situation corresponds to velocity weakening—when the steady-state friction μ_{ss} decreases with velocity V:

$$\frac{d\mu_{ss}}{d \ln V} = a - b < 0. \tag{5.1}$$

The dimensionless rate dependence $a - b$ can vary with rock composition, temperature, and pressure. Equation 5.1 defines the condition at which earthquake nucleation can occur (Figure 4.30). However, a correspondence between the depth range at which earthquakes occur and the region where $a - b$ is negative has not been confirmed by independent observations of velocity weakening, and there is no micromechanical theory that can be used to extrapolate laboratory data to crustal conditions. Nevertheless, the available lab information on the effect of temperature on the constitutive parameters, combined with inferred geotherms, suggests a reasonable degree of agreement between the depth at which $a - b$ is expected to become positive and the depth at which earthquakes stop.

As a fault is loaded, stress will fluctuate about the quasi-steady value $\tau_{ss} = \mu_{ss}\sigma_n$. Where stress is a bit higher than τ_{ss}, the slip rate increases slightly and occasionally a fluctuation will occur over a large enough area to initiate an instability. The criterion for instability is that the patch size be larger than a critical value L_c:

$$L_c \approx \frac{GD_c}{(b-a)\sigma_n}, \tag{5.2}$$

where G is the shear modulus. As nucleation begins, slip concentrates within a region of characteristic dimension L_c, and slip rate increases inversely with the time to instability (Figure 5.3). To what extent this type of behavior occurs in the Earth and what the size of L_c might be are two of the key questions in the science of earthquakes.

Earthquake nucleation is difficult to observe on faults in the Earth for two reasons. First, it is predicted to occur only over a spatially limited nucleation zone. If this zone is small, it will be difficult to detect. Second, nucleation may be a largely aseismic process such that it will not generate seismic waves. There are, however, observations that constrain possible models of earthquake nucleation, and these can be grouped into two classes: those that suggest the nucleation zone is small and those that suggest the nucleation zone is large.

Several types of observations point to a small nucleation zone (L_c less than 100 meters). Borehole strainmeter data provide the most sensitive measurements of small strain signals in the near field. These data show no evidence of strain precursors at levels that correspond to about 1 percent of the mainshock seismic moment (*84*). This suggests that the nucleation zone and the amount of slip within it must be small (Figure 5.4). A second line of evidence comes from rupture dimensions of the smallest earthquakes, which place an upper bound on the size of the nucleation zone since slip over an area less than L_c must be stable. Microearthquakes on the San Andreas fault recorded on the downhole instruments of the deep Cajon Pass borehole have source dimensions of about 10 meters (*85*). This places an upper bound on the size of the nucleation zone, at least locally, though fault roughness, gouge thickness, and apparent normal stress all affect L_c and will vary spatially. If the laboratory parameters for smooth faults applied to faults in nature, the minimum earthquake size would be on the order of 1 to 10 meters. Direct evidence for a lower-magnitude cutoff at the upper end of this range (near M 0) comes from the microseismicity observed by sensitive networks in the deep gold mines of South Africa (*86*).

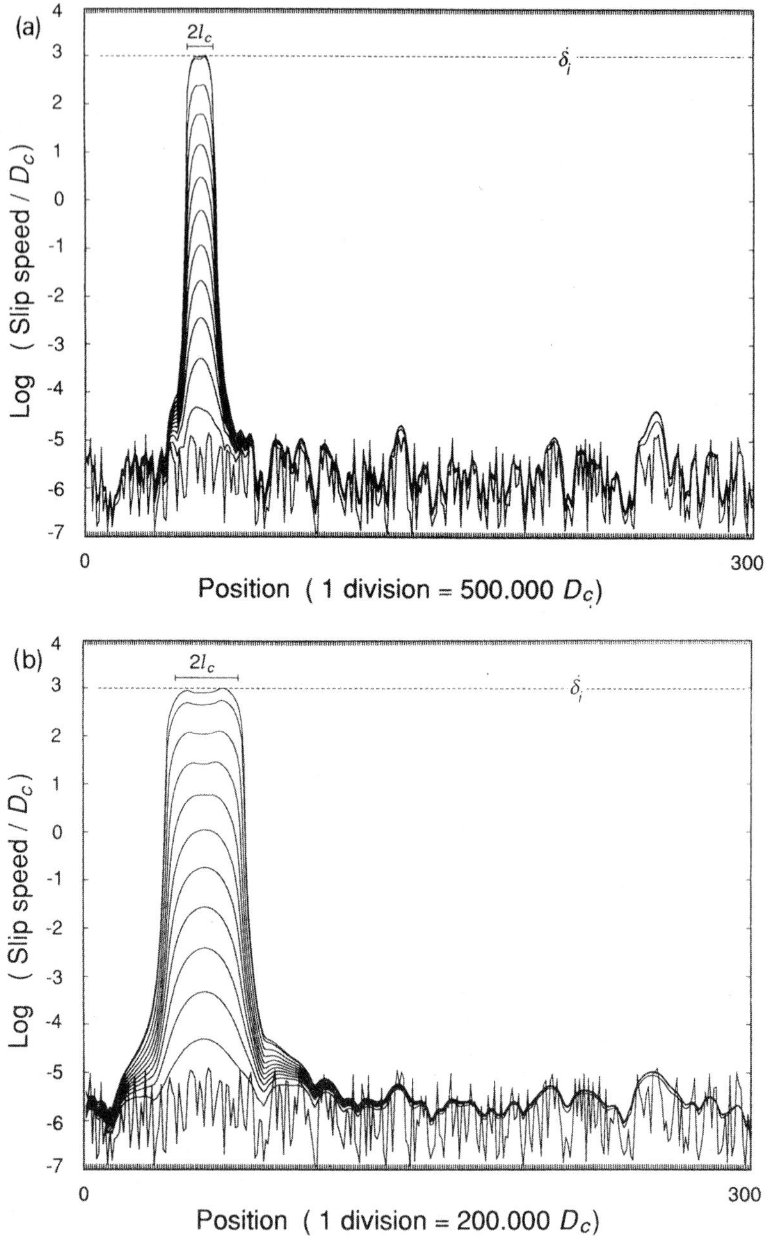

FIGURE 5.3 Numerical simulation of earthquake nucleation on a fault governed by rate-state friction with $A = 0.004$, $B = 0.006$, $\theta = 10^9$ s, $G = 10^4\sigma$, $\tau = 0$, and initial shear stress from $(\mu'_0 + 0.06)\sigma$ to $(\mu'_0 + 0.08)\sigma$. The fault is discretized with 300 elements and initialized with random stress fluctuations. Lines show the slip speeds at suc-

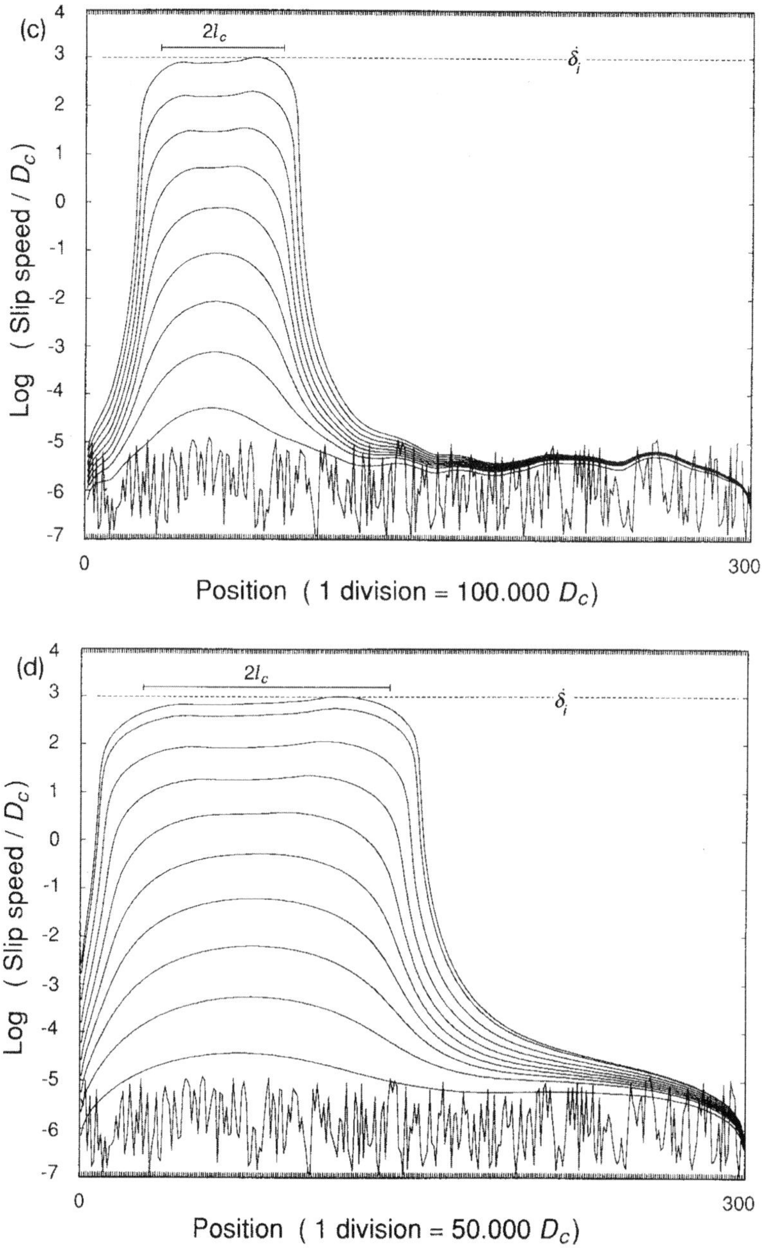

cessive times in the calculation; the time intervals are decreased by one order of magnitude for every order-of-magnitude increase in slip velocity. SOURCE: J.H. Dieterich, Earthquake nucleation on faults with rate- and state-dependent strength, *Tectonophysics*, **211**, 115-134, 1992. With permission from Elsevier Science.

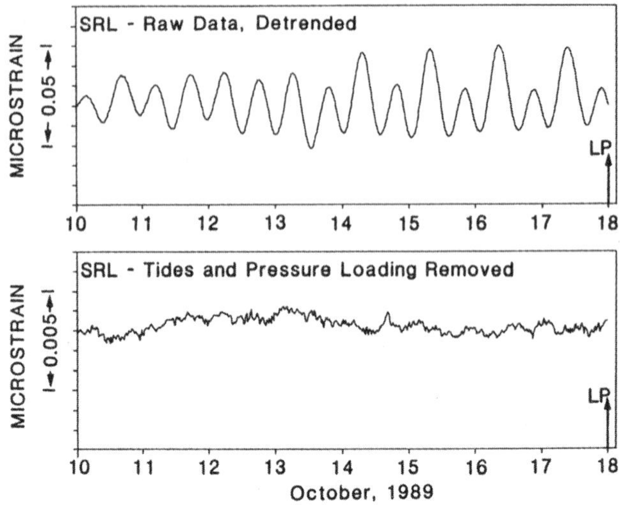

FIGURE 5.4 Preseismic strain observations of the 1989 Loma Prieta earthquake as recorded at San Juan Bautista. Upper panel shows a week of dilational strains with the oscillatory solid Earth tides. The lower plot presents the same record with Earth tides and atmospheric loading removed. The absence of a preseismic signal constrains the moment magnitude of an aseismic precursor to be no more than M 5.3, or less than 1 percent of the mainshock seismic moment. SOURCE: M.J.S. Johnston, A.T. Linde, and M.T. Gladwin, Near-field high-resolution strain measurements prior to the October 18, 1989, Loma Prieta Ms 7.1 earthquake, *Geophys. Res. Lett.*, **17**, 1777-1780, 1990. Copyright 1990 American Geophysical Union. Reproduced by permission of American Geophysical Union.

Several lines of evidence argue for a large nucleation zone (L_c greater than 100 meters). The low-frequency spectra for some earthquakes show a slow component that may precede the first detectable high-frequency waves by tens of seconds (*87*). For these events, there may be a gradual transition from aseismic nucleation to unstable rupture (*88*) (Figure 5.5). The character of the onset of microearthquakes suggests that very small events also begin with a slow onset that scales in duration with the overall source duration (*89*). The first arriving seismic waves of moderate to large events in the near field often show an initial phase of irregular growth (*90*). The duration of this phase shows a similar scaling with earthquake size as reported for the slow initial phase (Figure 5.6). If this phase represents the tail end of a process that is otherwise aseismic, then the dimensions of the nucleation zone are substantial.

Foreshocks provide the clearest evidence of a preparation process before at least some earthquakes. Approximately 40 percent of earthquakes

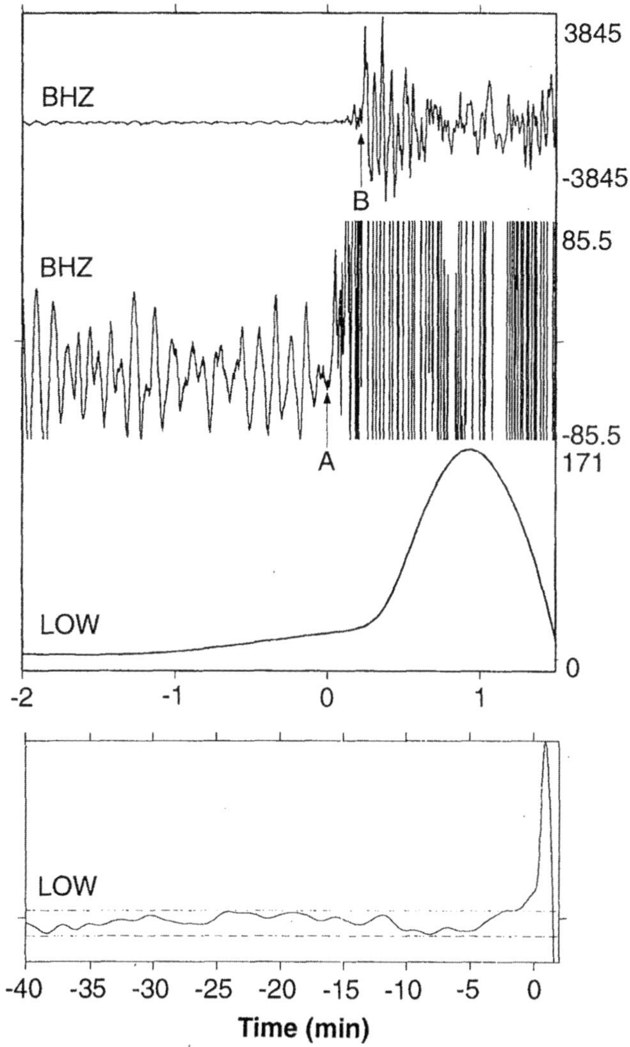

FIGURE 5.5 Vertical component *P*-wave seismograms from the 1994 Romanche Transform earthquake at low-noise GEOSCOPE station TAM. Top two panels show broadband trace at two magnifications and third panel shows a detided, low-pass filtered version of the same data revealing the precursory ramp beginning at least 1 minute before the high-frequency origin time. Lowermost panel shows filtered, detided data at a longer time scale indicating that the signal emerges from the background noise level (dashed lines). SOURCE: J. McGuire, P. Ihmle, and T. Jordan, Time-domain observations of a slow precursor to the 1994 Romanche Transform earthquake, *Science*, **274**, 82-85, 1996. Copyright 1996 American Association for the Advancement of Science.

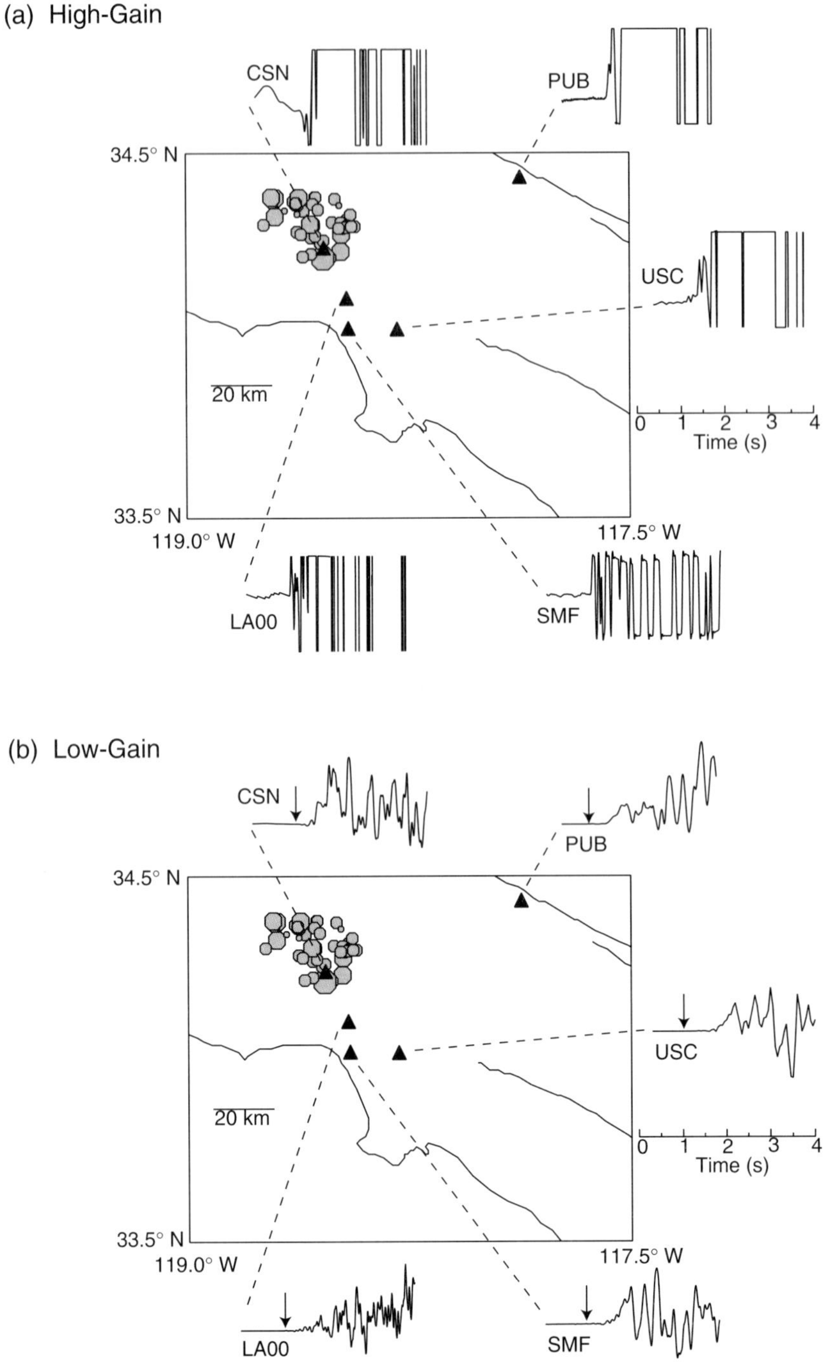
(a) High-Gain
CSN
PUB
34.5° N
USC
20 km
0 1 2 3 4
Time (s)
33.5° N
119.0° W
117.5° W
LA00
SMF
(b) Low-Gain
CSN
PUB
34.5° N
USC
20 km
0 1 2 3 4
Time (s)
33.5° N
119.0° W
117.5° W
LA00
SMF

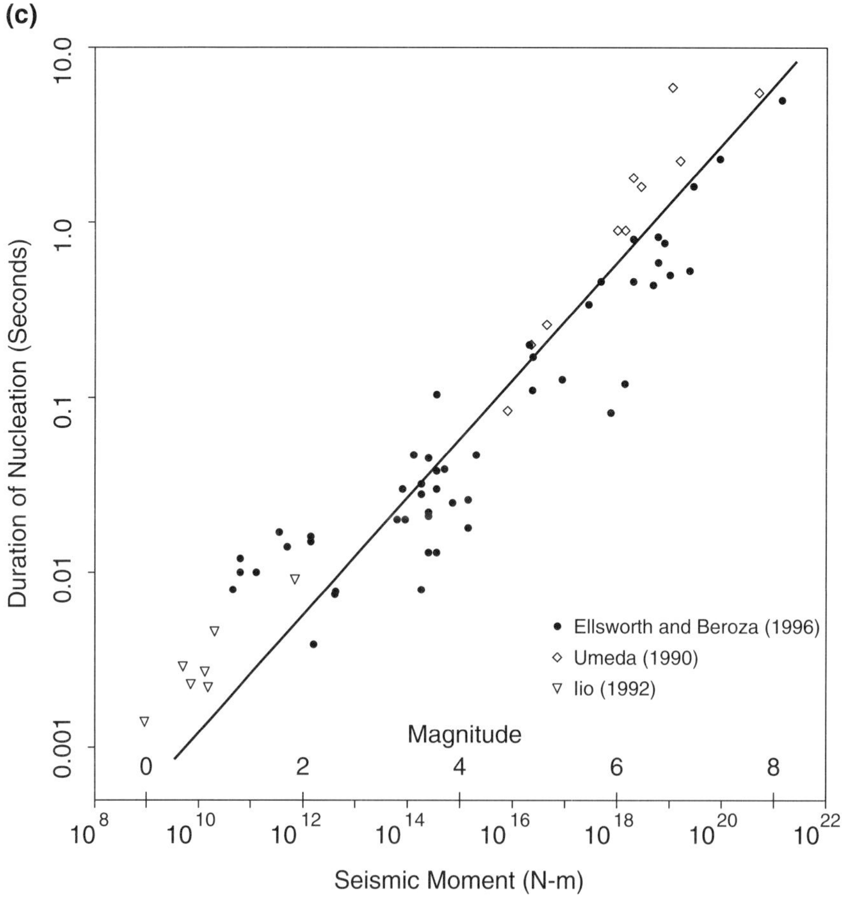

FIGURE 5.6 Left two panels show map view of the Northridge mainshock and aftershocks as filled octagons, with seismic stations as triangles. (a) Vertical component velocity seismograms of the initial *P* waves of the Northridge mainshock. SMF is a high-gain record that quickly clips, while the other traces are artificially clipped to simulate limited dynamic range recording and to show that the onset of the first *P* wave at this scale is abrupt. (b) Low-gain recordings of the initial *P* waves of the Northridge mainshock at the same sites. In each case the arrow indicates the first arriving waves shown in the upper panel. The onset, though abrupt, remains weak until about 0.5 second into the mainshock. Similar behavior is observed before other earthquakes. (c) The duration of the weak initial onset, or seismic nucleation phase, for earthquakes from several studies indicates that the seismic nucleation phase (vertical axis) scales with the seismic moment (horizontal axis) of the mainshock. SOURCE: G.C. Beroza and W.L. Ellsworth, Properties of the seismic nucleation phase, *Tectonophysics*, **261**, 209-227, 1996. With permission from Elsevier Science.

are preceded by at least one observable foreshock (*91*). Foreshock sequences are more common and are more protracted for earthquakes initiating at shallow depths, which is consistent with an expected decrease in frictional stability with decreasing normal stress (*92*). Foreshock frequency is observed to increase as t^{-1}, where t is the time before the mainshock (*93*). In at least some cases, foreshock sequences were unlikely to have triggered the mainshock (*94*). Instead, some other process, such as aseismic nucleation, may have driven both the foreshocks and the mainshocks to failure.

Earthquake nucleation may hold the key to whether or not earthquakes are predictable over the short term. If nucleation is so unstable that any small event could cascade into a large earthquake, then the prospects for deterministic earthquake prediction are grim because one would have to predict both the small initial earthquake and the fact that conditions would cause it to grow into a large earthquake. If, on the other hand, the nucleation process scales with earthquake size, the prospects for earthquake prediction are brighter. It is possible, even likely, that different faults will manifest different behaviors, with some (e.g., oceanic transforms) having different nucleation behavior than others. Understanding the nucleation process will require sensitive observations as close as possible to areas of likely earthquake initiation for a range of fault types and a number of large events. Current observational programs, with the exception of the Parkfield experiment (Section 2.6), are not designed to detect such phenomenon at the likely initiation points of significant events.

Rupture Propagation

Once nucleation occurs, rupture can propagate and expand in an earthquake. The mechanics of rupture propagation are complex and poorly understood for several previously discussed reasons. First, it is challenging to design laboratory measurements at the high sliding velocities and large displacements found in earthquakes. Second, physical phenomena that may be unimportant while the fault is locked or sliding slowly, such as shear heating of pore fluids or melting of fault-zone minerals, can become critically important at high slip speeds. Finally, in the near field, where the potential to make unobscured observations of the earthquake rupture process is highest, strong ground motion drives most seismic instrumentation off-scale. These factors have conspired to impede progress in understanding the mechanics of earthquake rupture; nevertheless, such an understanding is central to many of the most important goals of earthquake science, such as predicting the level and variability of strong ground motion, characterizing the nature of large earthquake recurrence, and understanding the extent to which earthquakes might be predictable.

The dynamics of earthquake rupture are usually described in the terminology of fracture mechanics (*95*). A common application of crack models to earthquake studies is to define relationships between seismological observations and dynamical parameters. The average offset on a fault $\overline{u}$ and its characteristic dimension L are related to the static stress drop $\Delta\sigma$ by the formula $\Delta\sigma = c_s G \overline{u} / L$, where c_s is a constant determined from crack theory, which depends on the fault type. In crack theory, the rupture velocity is a function of the fracture energy near the crack tip (*96*). The exact relationship depends on the crack geometry, but in general, rupture speed increases as the fracture energy decreases. The rupture velocity is generally much faster than the fault's particle velocity, the speed with which one side of the fault moves with respect to the other; typical values are 2 to 3 kilometers per second and 0.3 to 2.0 meters per second, respectively. The particle velocity can be related to the tectonic stress σ_0 driving the fault motion. Since fault motion is impeded by a frictional stress σf, the actual stress available for driving fault motion is the difference, $\sigma_e = \sigma_0 - \sigma f$, called the dynamic stress drop. The particle motion velocity V is given by $c_d \beta \sigma_e / G$, where β is the shear velocity, G is the rigidity, and c_d is a constant determined by the geometry of the fault. Particle velocities of about 1 meter per second imply that σ_e is of the order of 100 bars, or 10 megapascals (*97*).

Crack models are useful but must be applied with caution. The rupture velocity of large earthquakes is rarely constant and faults may rupture in a stop-and-go fashion. Cracks in ideally brittle materials have stress concentrations that are infinite at the sharp crack tip. In real materials, nonlinear deformations such as plastic flow eliminate this singularity by distributing the stress over a finite process zone. Various models have been advanced to describe this behavior (98), though their dynamical effects can usually be lumped into an effective value of K_c. This nonideal version of the critical stress intensity factor defines a material parameter called the fracture toughness.

Two main difficulties are encountered in the application of idealized crack mechanics to the earthquake problem. One lies in the assumption that the crack is cohesionless behind the crack tip, which implies that the stress drop during fracture is complete. On real faults, shear motion is impeded by friction, so that the stress drop is incomplete; in fact, the work against friction during fault slip turns out to be the dominant term in the energy balance. The second problem is the ad hoc treatment of what happens in the process zone at the edge of the crack, where an attempt must be made to stitch together two fundamentally different ways of describing material behavior, from the bulk rheology that governs the unfractured rock ahead of the crack tip to the surface friction that applies once the fracture has passed by. In these respects, the view of earthquakes as frictional instabilities is more appropriate.

An important research area is how ruptures in earthquakes compare with idealizations of rupture based on fracture mechanics. The notion that rupture in earthquakes propagates outward from the hypocenter was implicit in the recognition that earthquakes are caused by shear slip on faults, but it was not until the 1950s that the effects of rupture propagation on seismograms were first identified (*99*). Teleseismic and near-source estimates of average rupture velocity are consistently in the range of about 70 to 90 percent of the *S*-wave velocity (*100*). There is no evidence that rupture velocity varies with magnitude, from the very largest earthquakes to the very smallest earthquakes for which it can be determined (*101*). Rupture velocities that are a large fraction of the shear-wave velocity lead to pronounced directivity in strong ground motion, particularly for shear waves (*102*).

It is not clear why earthquakes should rupture at these velocities. The simplest models based on elastic-brittle fracture mechanics for a preexisting planar fault suggest that shear rupture ought to accelerate very quickly to a limiting velocity that depends on the mode of rupture: either the shear-wave velocity for antiplane rupture or the Rayleigh-wave velocity (about 92 percent of the shear-wave velocity) for in-plane rupture (*103*). The same models, predict that stresses for out-of-plane rupture will grow as the limiting velocity is approached, which should promote rupture bifurcation and a lower rupture velocity. For the most part, rupture is observed to propagate at velocities slightly below the limiting velocity for the elastic-brittle case. There are, however, important exceptions to this behavior.

Rupture velocity has locally exceeded the *S*-wave velocity for at least several earthquakes (*104*). Such supershear rupture velocities are expected in models that incorporate a process zone that fails under finite cohesive traction (*105*), and they have been observed recently in laboratory fracture experiments (*106*). If supershear rupture propagation should prove common, however, it would have important implications for strong ground motion. During an episode of supershear rupture propagation, an earthquake will form a Mach cone, the seismic equivalent of a sonic boom, but in the case of earthquakes, a high-amplitude wavefront will result (*107*), with the potential to contribute substantially to the level of damaging strong ground motion.

Slow earthquakes are seismic events for which the rupture and/or slip velocities are unusually low. They are identifiable by unusually strong seismic wave excitation at long periods (*108*). An important class of slow earthquakes is tsunami earthquakes, which generate tsunamis far larger than expected based on their magnitude (*109*). The devastation wrought by tsunami earthquakes can be extreme (*110*). Moreover, near the tsunamigenic source, there is little time for warning. Because of their tremendous destructive potential, it is extremely important to understand why such earthquakes occur. More generally, slow earthquakes are known to occur in many tectonic environments (*111*), but they are particularly common on oceanic transform faults. The fact that slow earthquakes are particularly common on transforms where sedimentary cover is negligible precludes rupture through, or slumping of, mechanically weak sediments as a uniform explanation for slow events. Their association with oceanic transforms may instead be related to properties of the relatively young, hot, and thin oceanic crust (*112*).

Silent earthquakes are slip episodes that occur so slowly that they do not generate short-period seismic waves and hence are not earthquakes in the usual sense of the word. The largest known silent earthquake was a precursor to the 1960 Chile earthquake. The slow component of this event at M 9.3 is larger than any other recorded earthquake except for the 1960 M 9.5 Chile mainshock that followed it. Because it did not radiate high-frequency seismic waves, the precursor was not even recognized until more than a decade later (*113*). There are now several spectacular examples of large silent earthquakes in Japan (*114*), as well as smaller silent earthquakes on the San Andreas fault system (*115*). A study of the Earth's longest-period free oscillations found excitations of the Earth's free oscillations that were not accounted for by known earthquake activity (*116*). More recently, it has been found that the Earth's free oscillations are continuously excited (*117*), although it is not yet clear what the source of this excitation is. If it is earthquake activity, then it requires a substantial revision of our view of faulting. Episodic slip would have to be common and more or less continuously occurring somewhere in the world. The source of continuous excitation could also be atmospheric, which if true offers new possibilities in seismology on other bodies of the solar system (*118*).

Fault creep, the steady motion of a fault without generation of seismic waves, can be episodic at the Earth's surface and occur in discrete events (*119*), but its behavior at depth is less well known. Some faults such as parts of the San Andreas, Hayward, and Calaveras faults in California seem to be creeping aseismically (*120*). Aseismic creep has also been called upon to explain postseismic deformation transients (see Section 4.2). It has long been known that seismicity on many of the Earth's major fault systems is insufficient to keep up with the rates of slip predicted from plate tectonics (*121*). To what extent this aseismic slip occurs continuously versus episodically remains an open question.

Another important aspect of earthquake rupture propagation is the rise time—the duration of slip at a point on the fault. The rise time has been determined for a few earthquakes for which adequate near-source strong-motion data are available (*122*), but for most large earthquakes the rise time is unresolved (*123*). If one supposes that the fault will not stop sliding until it receives information that allows it to heal from the farthest reaches of the fault plane (*124*), then the rise time should be proportional to the spatial extent of the fault. The rise time is much shorter than would be predicted given the length of the fault, and in some cases it is shorter than the width of the fault would predict as well (*125*).

Explanations for short rise times can be characterized as either dynamic or geometric. Dynamic explanations center on the notion that if the velocity dependence of friction is strong enough, it might lock the fault as the sliding velocity decreases, well before information propagates inward from the fault edges (*126*). Geometric explanations focus on smaller length scales in the faulting process, due to geometry or material properties of the fault that might cause the rise time to be short. In this case the rise time may be controlled by the dimension of the high-slip regions, rather than the overall fault dimensions. Quasi-dynamic models of earthquakes (*127*) support this point of view.

Whatever their cause, the combination of short rise time and high rupture velocity leads to strong shear-wave arrivals of short duration in the near field in which a broad range of frequencies arrive in phase. The strong pulse that results poses challenges for earthquake engineering (*128*), so that it is critical to determine what controls the behavior of these aspects of rupture propagation. Because our understanding of strong ground motion is based primarily on a limited number of moderate earthquakes (M < 7.0), an increased understanding of the factors that control strong ground motion, such as the rise time, is essential in efforts to extrapolate observations of strong ground motion in moderate earthquakes to larger earthquakes.

Slip on faults during earthquakes is known to be spatially variable. Early representations of earthquake sources as multiple point sources were motivated by observations that earthquakes are punctuated by a series of subevents that radiate energetically (*129*). A more general characterization of heterogeneity represents an earthquake by a continuous distribution of slip in space and time. This approach cannot be applied in a meaningful way to most earthquakes because of insufficient resolution (*130*); however, in the near field where high-frequency waves are not greatly attenuated and Green's functions vary strongly with position, detailed source imaging is possible. Extended-source models of rupture for several dozen earthquakes have been derived from strong-motion data (*131*) and show that both slip and rupture velocity in earthquakes are strongly heterogeneous in both space and time (*132*) (Figure 4.7).

Most extended-source models are kinematic in the sense that the slip distribution is specified without considering the stress on the fault and the fault-constitutive behavior it implies. Dynamic models explicitly account for the stress and attempt to characterize the behavior of the fault in terms of simple physical laws. One approach to reconcile kinematic and dynamic rupture modeling is termed quasi-dynamic modeling, in which dynamic rupture models are developed that reproduce kinematic models (*133*). Because modeling strong-motion and other data in quasi-dynamic models is indirect, there is no guarantee that the model will be consistent with the original data. A goal for the future is to estimate dynamic parameters directly from strong-motion data. Preliminary work in this area suggests that some dynamic parameters such as the slip weakening distance may be very difficult to resolve from surface measurements (*134*). Short of complete dynamic modeling, one can also recover aspects of fault rupture dynamics without developing a dynamic rupture model for the entire event (*135*).

It has long been recognized that earthquake rupture must be heterogeneous at small scale lengths to explain observed high-frequency ground motion (*136*). The acceleration spectra of earthquakes is observed to be constant above the corner frequency. Models of constant slip with smooth rupture propagation result in acceleration spectra that decay above the corner frequency (*137*) unless seismograms are dominated by the effects of rupture termination (*138*). A number of models have been developed to explain this observation (*139*). Seismologists have known that heterogeneous rupture should lead to enhanced radiation at frequencies of concern to earthquake engineering (*140*). There is now evidence to confirm this hypothesis. Areas of strong high-frequency generation are observed to correlate with areas of strong slip variations (*141*), and there are stochastic models of earthquake rupture that lead to realistic strong ground motions (*142*).

Rupture Arrest

Rupture will propagate along a fault in an earthquake until something stops it. For large earthquakes, the depth extent of seismic rupture is bounded from below by the depth of the transition from brittle to ductile behavior (*143*) and from above by the Earth's surface (*144*). What controls the horizontal extent of rupture in large earthquakes, or the spatial extent of smaller earthquakes that terminate before they reach the edges of the seismogenic zone, is less clear. Factors likely to influence the extent of rupture include fault geometry, variation of material properties, and stress heterogeneity.

The irregularities in geometry that occur at all scale lengths (*145*) have the potential to exert a strong control on earthquake rupture for earthquakes of all sizes (see Figure 3.2). At the surface, fault-zone irregularities can be mapped geologically. Such irregularities, particularly fault discontinuities, across which slip transfers from one surface to another, are thought to play an important role in controlling the maximum earthquake size on a particular fault system (*146*). This idea is supported by studies of fault segmentation as expressed both in surface faulting and in aftershock distributions (*147*). There are, however, clear observations of earthquakes that were not terminated by fault-zone discontinuities (*148*). The Landers earthquake provides a spectacular example (*149*). This earthquake started on the Johnson Valley fault and ruptured primarily to the north, then jumped across a discontinuity to the Homestead Valley fault and continued to rupture northward. It then jumped across yet another discontinuity and ruptured northward on the Emerson fault before stopping in the middle of a relatively straight fault segment. Given the potential utility of using fault segmentation to anticipate earthquake size, it is important to determine under what conditions a propagating rupture such as this will or will not jump from one fault segment to another. The ability to model the conditions under which an earthquake that ruptures toward a fault jog will terminate or breach the jog and continue to grow into a larger earthquake can help anticipate the size of future earthquakes. The possibility of multiple-segment ruptures has been included explicitly in assessments of earthquake probabilities in California (*150*).

One factor that determines how effectively a discontinuity will act to limit fault rupture is the distance between the offset fault segments. Empirical observations suggest that fault discontinuities with less than 1 kilometer of offset do not pose a strong impediment to rupture; whereas discontinuities with 1- to 5-kilometer offset terminated rupture some of the time, and discontinuities with offsets of 5 kilometers or more always terminated rupture (*151*). Two-dimensional numerical models of dynamic rupture interacting with a fault discontinuity are consistent with these observations (*152*).

Another factor is the sense of the discontinuity (i.e., whether a jog in a fault leads to extensional or compressional strains). Compressional jogs are more difficult to propagate across, because the normal stress will increase and because uplift to accommodate compressional strain within the jog will have to be done against gravity (*153*). Finite-difference modeling suggests that earthquakes are unlikely to propagate across compressional jogs with offsets greater than 3 kilometers or extensional jogs with offsets greater than 5 kilometers (*154*). Three-dimensional modeling of rupture across a fault discontinuity has refined the ability to model why some earthquakes terminate at segment boundaries while others cascade into multisegment ruptures and hence much larger earthquakes.

Variations in material properties may also exert a control on the extent of rupture in earthquakes. This is certainly true in the grossest sense at the brittle-ductile transition and at the Earth's surface, but variations in material properties may also be important either in the material adjacent to the fault or within the fault zone itself. There is strong evidence that material properties near the Earth's surface control rupture propagation through the shallowest layers (*155*). Further evidence that material in the vicinity of the fault zone may help control earthquake size comes from tomographic studies of velocity variations in the vicinity of recent large earthquakes (*156*). A possible interpretation is that these areas of the fault accumulate shear stress, while parts of the fault that are bordered by lower-velocity material may slip aseismically.

The observation that the creeping section of the San Andreas and Calaveras faults in California have areas of micro-earthquake activity interspersed with small zones that fail repeatedly in small earthquakes (*157*) suggests that material variations may cause some parts of the fault surface to fail in stick-slip while the rest of the fault creeps (*158*).

Accumulated stress is the fuel that provides the energy for earthquake faulting, and variations in stress may play an important role in controlling earthquake size. Rupture may stop when it propagates into a region that has very little pre-stress. Such a stress barrier is a means of terminating rupture (*159*) and is implicit in the stuck patch (*160*) or asperity model of earthquake behavior (*161*), in which highly stressed parts of the fault fail at high stress drop and the rupture stops within lower-stress areas on the surrounding fault. The termination of rupture on the Emerson fault in the Landers earthquake may provide an example of a rupture that stopped owing to low stress on the fault before the mainshock rupture (*162*). If the Emerson fault was far from failure before the Landers mainshock, then it may have terminated at shallow depth with rupture nucleating at shallow depth, but not propagating to greater depths or farther along the fault (Figure 5.7).

Understanding the factors that control the extent of earthquake rupture is extremely important. Fault geometry, material property variations, and stress variations, are all likely to play an important role. Moreover, these factors are interdependent, and it may be impossible to fully disentangle the effects of one from the others. The 1934 and 1966 Parkfield earthquakes illustrate this. The 1934 earthquake apparently did not rupture past the extensional fault jog in Cholame Valley, whereas the 1966 Parkfield earthquake did (*163*). The geometry did not change between 1934 and 1966; perhaps the fault jog was sufficient to terminate rupture in 1934, but not in 1966, because the fault to the south of the jog was closer to failure before the 1966 event than it was before the 1934 event.

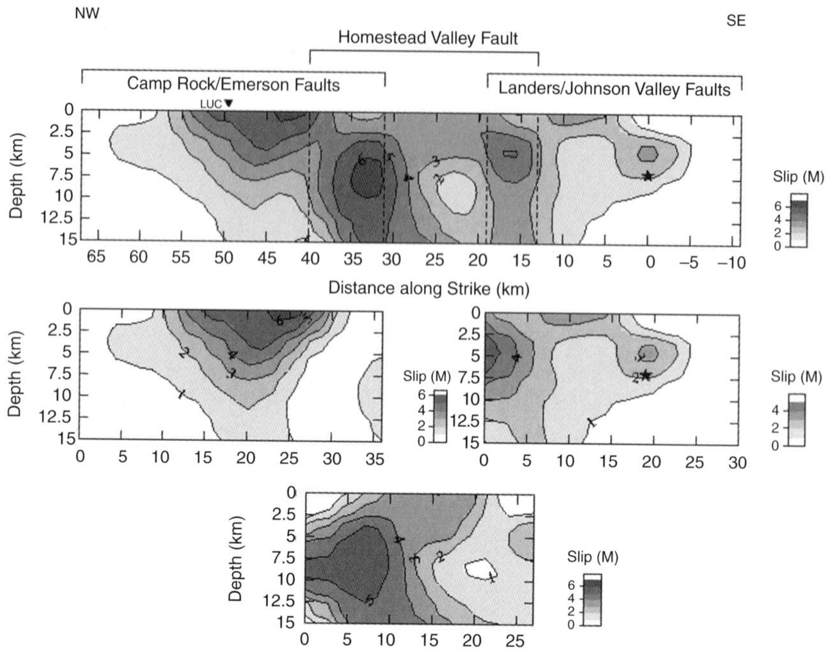

FIGURE 5.7 Slip distribution on three fault segments determined from strong-motion, teleseismic, and geodetic data for the 1992 Landers, California, earthquake. The combined, projected slip summed from the three fault segments is shown above the slip distributions for the individual segments. Note that there is ample slip near segment boundaries and that slip on the ends of the fault terminates within fault segments. SOURCE: D.J. Wald and T.H. Heaton, Spatial and temporal distribution of slip for the 1992 Landers, California earthquake, *Bull. Seis. Soc. Am.*, **84**, 668-691, 1994. Copyright Seismological Society of America.

Deep Earthquakes

Earthquakes below 70 kilometers present special research problems because fault ruptures at these depths cannot be explained by brittle fracture or friction (Figure 5.8; see Section 2.5). Although their depths limit the seismic hazard (*164*), these intermediate- and deep-focus events provide primary constraints on subduction-zone processes. According to plate-tectonic theory, slabs are colder and thus denser and stronger than the surrounding mantle; their sinking involves a balance between the gravitational forces that pull them down and the viscous resistance of the mantle to this penetration (*165*). The nonhydrostatic stresses engendered within cold slabs during the subduction process appear to be responsible

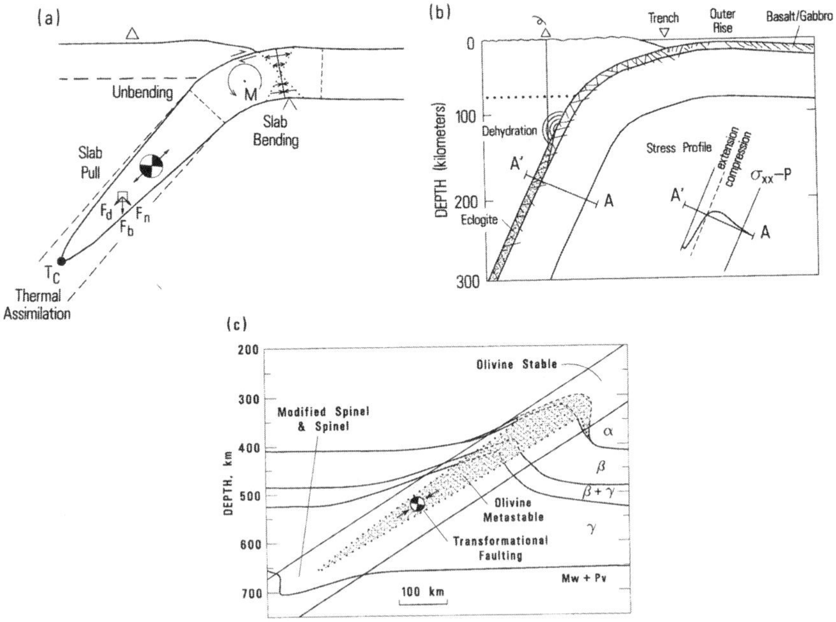

FIGURE 5.8 Conceptual models for earthquakes in descending lithospheric slabs. (a) Generic model for the stresses that cause intraplate earthquakes in slabs. The thermal buoyancy force F_b and its slab-normal and down-dip components, F_n and F_d, are shown for an element of volume. M is the near-trench bending moment. T_c is the critical temperature above which the slabs are too weak to support earthquakes. (b) Some aspects of slab structure that may contribute to earthquake occurrence in the intermediate-focus zone (70 to 300 kilometers), including zones of weakness from prior normal faulting of the oceanic lithosphere, bending stresses in the slab (inset diagram), dehydration-embrittlement, and the densification during the gabbro-to-eclogite phase transition. (c) Physical state of the deep-focus zone (300 to 700 kilometers) showing the equilibrium boundaries of the α, β, and γ phases of olivine and the phase transition of g-olivine to magnesiowüstite (Mw) and perovskite (Pv) (solid and dashed lines) and the region where α-olivine may be metastable (stippled region). According to one hypothesis, deep-focus earthquakes occur by transformational faulting inside the metastable wedge. SOURCE: Modified from S.H. Kirby, Intraslab earthquakes and phase-changes in subducting lithosphere, *Rev. Geophys. Suppl.*, **33**, 287-297, 1995. Copyright 1995 American Geophysical Union. Reproduced by permission of American Geophysical Union.

size? How short will the rise time be for much larger earthquakes in which the slip may exceed 10 meters? Is it limited by the geometry of the fault plane or the dynamics of friction at high slip velocities?

• What physical mechanisms explain the deep-focus earthquakes that occur in the descending lithosphere down to depths of nearly 700 kilometers? How do these mechanisms differ from shallow seismicity?

5.5 WAVE PROPAGATION

Earthquake damage is caused primarily by seismic waves. Seismic shaking is influenced heavily by the details of how seismic waves propagate through complex geological structures. In particular, strong ground motions can be amplified by trapping mechanisms in sedimentary basins and by wave multipathing along sharp geologic boundaries at basin edges, as well as by amplifications due to near-site properties. Although near-site effects such as liquefaction can be strongly nonlinear, most aspects of seismic-wave propagation are linear phenomena described by well-understood physics. Therefore, if the seismic source can be specified precisely and the wave velocities, density, and intrinsic attenuation are sufficiently well known, it is possible to predict strong motions by a forward calculation.

A conspicuous success of earthquake physics has been the development of computational techniques for describing the propagation of seismic waves. These techniques yield approximate solutions to the forward problem of seismic-wave propagation, which is to predict the wavefield as a function of position and time knowing the source and a model describing the Earth's elastic and anelastic constitutive properties (*180*). Such calculations can be used to predict the strong ground motions in the vicinity of an anticipated earthquake. Moreover, they provide the theoretical framework for solving the structural inverse problem (to estimate a set of constitutive parameters from recordings of the wavefield and knowledge of the source), as well as the source inverse problem (to estimate a set of source parameters from recordings of the wavefield and knowledge of the structure). The effects of source excitation and wave propagation are coupled in seismograms, which complicates their separation. Recent progress on solving these coupled inverse problems, outlined in the previous chapter, has enhanced the predictive capabilities of wavefield modeling. At present, numerical simulations using good propagation models can reproduce the recorded waveforms of low-frequency motions (less than 0.5 hertz) from events such as the 1994 Northridge earthquake and match the spectral amplitudes at higher frequencies with moderate success (*181*). However, matching the waveforms at higher frequencies will require much better seismological imaging of both the rupture process and the crustal structure.

For engineering applications, a high-priority goal is to determine the structure of high-risk, urbanized areas of the United States well enough to predict deterministically the surface motions from a specified seismic source at all frequencies up to at least 1 hertz and to formulate useful, consistent, stochastic representations of surface motions up to at least 10 hertz.

Theory and Numerical Methods

The Earth is almost spherical, and its internal layering is nearly concentric, at least on the gross scales of the mantle and core. Seismological research during the first 40 years of the twentieth century established the basic features in the radial distribution of seismic velocities and density, culminating in the Jeffreys-Bullen model. More recent work has refined these spherically symmetric global models, particularly with regard to the structure of the upper-mantle and midmantle transition zone (*182*), and has provided a number of regionalized estimates of the layering of the crust and upper mantle beneath both continents and oceans (*183*). For such one-dimensional Earth models, the partial differential equations of elastodynamics can be simplified to a set of ordinary differential equations, which can be solved numerically using various methods.

In the lowest frequency bands (0.0003 to 0.1 hertz), the most general and accurate techniques involve the representation of the displacement field in terms of the normal modes of the elastic structure (*184*). For compact seismic sources, theoretical seismograms synthesized from good one-dimensional Earth models by normal-mode summation can show remarkable agreement with observed seismograms. The normal-mode representation forms the basis for recovering earthquake source parameters from surface-wave and other low-frequency data. For three-dimensional Earth models where the deviations from spherical symmetry are relatively small, normal-mode perturbation theory provides general and efficient methods for computing theoretical seismograms, and it has been applied in many global tomographic studies to invert low-frequency seismic waveforms for three-dimensional Earth structure.

At high frequencies, the calculations become more difficult, and methods that approximate the waves as energy packets traveling along discrete ray paths are often employed (*185*). Ray theory is usually less accurate, but it provides an adequate representation of many seismic phases, especially the first arrivals at teleseismic distances, and it has been used extensively in algorithms for recovering source parameters from seismic waveforms. It is also the preferred representation in tomographic studies that image three-dimensional Earth structure from measurements of body-wave travel times.

The strongest ground motions during an earthquake are often generated by the trapping of waves within sedimentary basins and other three-dimensional structures (e.g., effects in Mexico City from the distant 1985 Michoacan earthquake), or by interference among elastic waves that have been diffracted along different paths at the edges of such structures (e.g., 1995 Hyogo-ken Nanbu earthquake). The semianalytical methods described above are often too inaccurate to describe the complexities observed in real seismograms in such cases, and seismologists have resorted to solving numerically the equations of motion for models discretized on two-dimensional and three-dimensional grids using finite-difference (*186*), finite-element (*187*), and pseudospectral techniques (*188*).

Crustal Waveguide Effects

As seismic waves propagate away from the fault, their intensity is reduced by geometrical spreading. For body waves in a uniform material, geometrical spreading reduces the amplitude in inverse proportion to distance (r^{-1}). For surface waves, the factor is $r^{-1/2}$. In a layered medium, the effects of spreading are complicated by scattering and internal reflections. When energy is reflected or scattered, it causes the energy to attenuate more than r^{-1} or $r^{-1/2}$ (*189*). However, internal reflections can also in result in enhanced ground motions at large distances from the hypocenter (*190*).

The advent of broadband seismometers in the late 1980s provided a vastly improved representation of actual ground motions against which to test wave propagation models. One such test, shown in Figure 5.9, demonstrates the ability of wave propagation models to reproduce the various body-wave and surface-wave phases recorded at a distance of about 160 kilometers from a small earthquake. The critical reflections from the Moho (*SmS* phases) that dominate the seismograms shown in Figure 5.9 have an important effect on the attenuation of strong ground motion from earthquakes. The arrival of these critical reflections, beginning at a distance of about 50 kilometers, causes a reduction in the rate of attenuation of ground motion out to distances of about 150 kilometers (Figure 5.10). Although the elevated ground-motion amplitudes in this distance range are usually not large enough by themselves to cause damage, they may produce damage if combined with the amplifying effects of soft soils. The destructive potential of these effects was demonstrated dramatically in the 1989 Loma Prieta earthquake (*191*) in which major damage was done to buildings and bridges in the San Francisco Bay area located 80 to 90 kilometers from the earthquake.

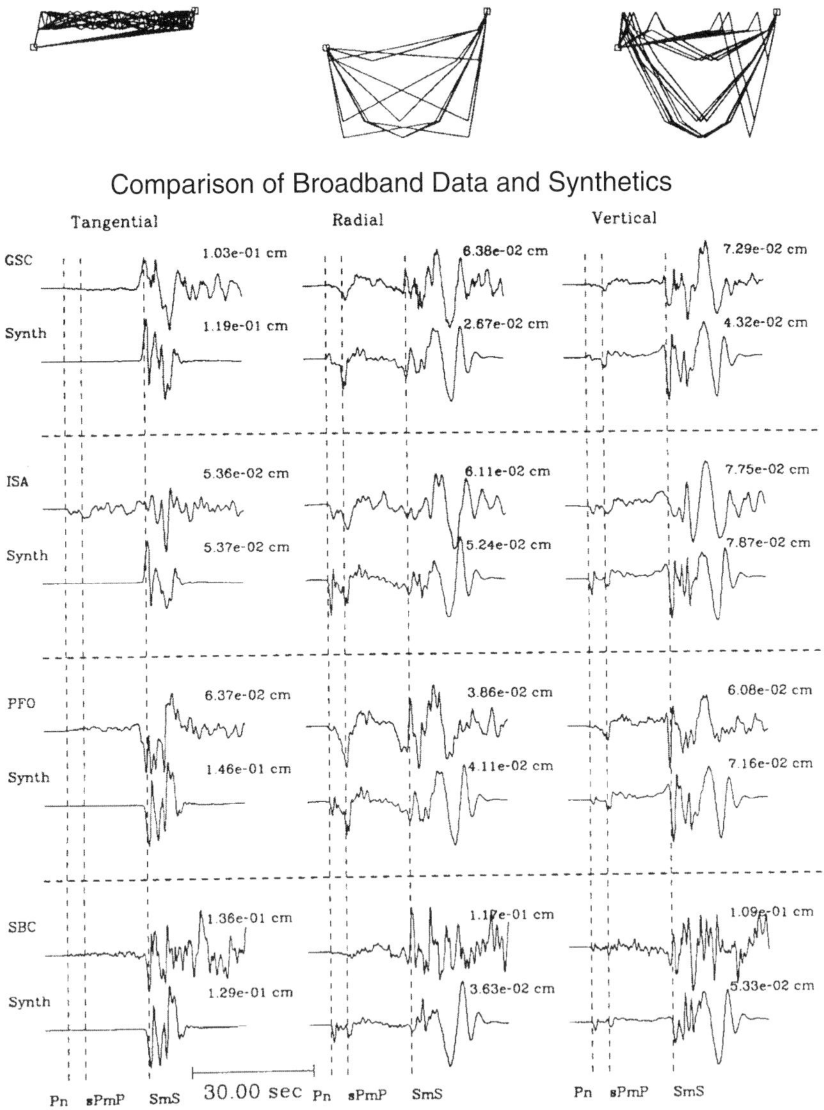

FIGURE 5.9 *Top*: Radial component paths of upgoing, downgoing, and surface-reflected downgoing rays. *Bottom*: Comparison of recorded and synthetic broadband displacement seismograms of the 1991 Sierra Madre earthquake. SOURCE: Modified from D. Helmberger, D. Dreger, R. Stead, and H. Kanamori, Impact of broadband seismology on the understanding of strong motions, *Bull. Seis. Soc. Am.*, **83**, 830-850, 1993. Copyright Seismological Society of America.

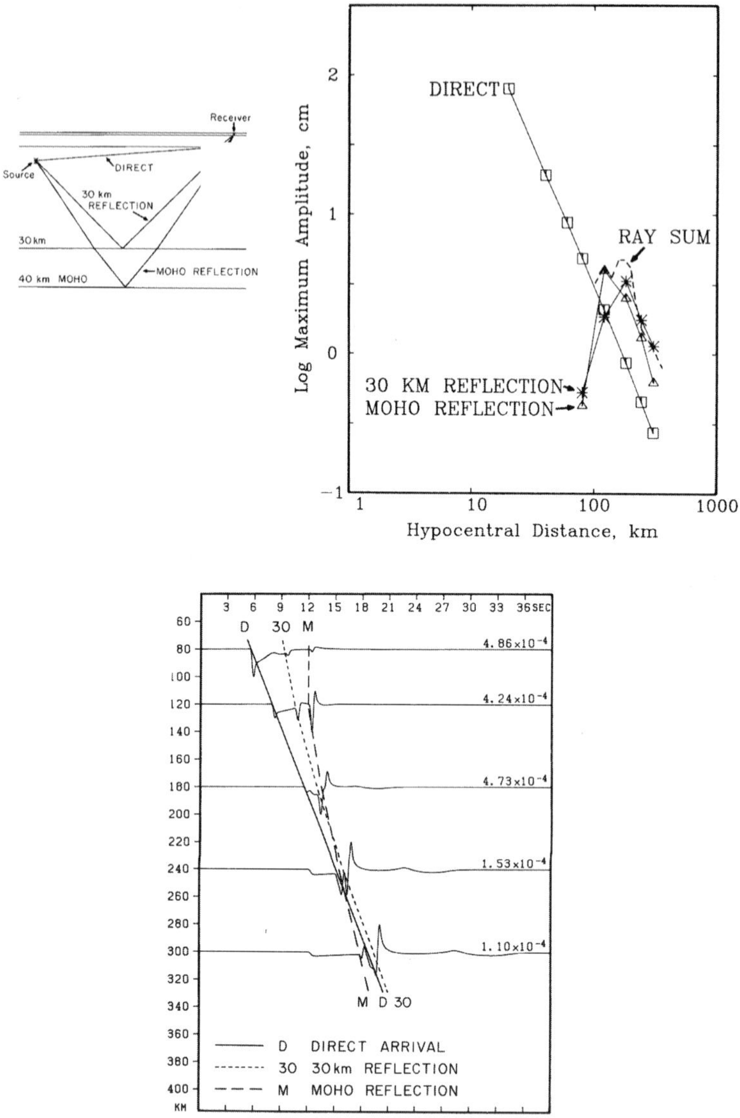

FIGURE 5.10 Effect of crustal structure on ground motion attenuation. Synthetic seismograms (*bottom*) calculated for a crustal waveguide (*top left*). The contributions of direct and reflected rays (*top right*) to the attenuation relation are shown in the top right panel. SOURCE: R.W. Burger, P.G. Somerville, J.S. Barker, R.B. Herrmann, and D.V. Helmberger, The effect of crustal structure on strong ground motion attenuation relations in eastern North America, *Bull. Seis. Soc. Am.*, **77**, 420-439, 1987. Copyright Seismological Society of America.

At larger distances (100 to 1000 kilometers), the effect of the crustal waveguide becomes increasing complex, and shear-wave arrivals are composed mainly of multiple reflections of *S* waves between the Moho and the surface (the *Lg* phase). This phenomenon is illustrated in the body-wave seismogram of the 1988 Saguenay earthquake recorded at a distance of 600 kilometers at Harvard, the first broadband recording of a moderate-magnitude earthquake in eastern North America (Figure 5.11). The synthetic seismograms calculated using a simple point source time function and a one-dimensional velocity model for the region provide a remarkably close fit to both the long-period and the short-period components of the *Pnl* and *Snl* body-wave phases generated by the crustal waveguide.

Effects of Sedimentary Basins

For many years, it has been known that ground motions on soil sites are typically stronger than those on rock sites due to the low shear moduli of the near-surface (upper 30 meters) sedimentary units (*192*). While this local effect has been recognized by engineers and incorporated into build-

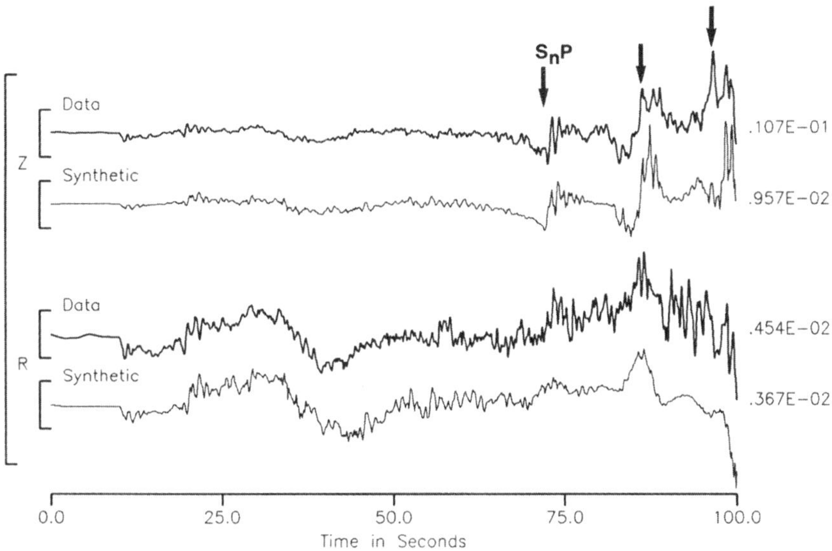

FIGURE 5.11 Comparison of recorded and synthetic broadband displacement seismograms of the 1988 Saguenay earthquake recorded at Harvard. SOURCE: C.K. Saikia, Modified frequency-wavenumber algorithm for regional seismograms using Filon's quadrature: Modeling of *Lg* waves in eastern North America, *Geophys. J. Int.*, **118**, 142-158, 1994. Reproduced by permission of Blackwell Publishing.

ing codes, recent observations and theoretical studies have demonstrated that a variety of complex wave propagation effects can also influence the ground motions on soil sites that are located in sedimentary basins. In many cases, the impact of the deeper basin structure is much greater than that due to the surficial site materials.

Seismic body waves entering a basin through its thickening edge can become trapped within the basin if postcritical incidence angles develop, generating surface waves whose amplitude and duration are significantly larger than those of the incoming body waves. This phenomenon is illustrated in Figure 5.12, which shows strong-motion velocity time histories of the 1994 Northridge earthquake recorded on a profile of stations, it begins in the San Fernando Valley, crosses the Santa Monica Mountains, and extends into the Los Angeles basin. The ground motions recorded on rock sites in the Santa Monica Mountains are brief and are dominated by the direct body waves. In contrast, the time histories recorded in the Los Angeles basin have long durations, and their peak velocities are associated not with the direct body waves but with surface waves generated at the northern edge of the Los Angeles basin. The ground motions were further amplified as they crossed the Santa Monica fault, which marks an abrupt deepening of the Los Angeles basin. This amplification is reflected dramatically in the damage distribution indicated by red-tagged buildings, which are concentrated immediately south of the fault scarp. The strong correlation of the damage pattern with the fault location indicates that the underlying basin-edge geology, not shallow soil conditions, is controlling the ground-motion response. The large amplification results from constructive interference of direct waves with the basin-edge generated surface waves. As described in Chapter 2, the 1995 Hyogo-ken Nanbu earthquake provided dramatic evidence for the destructive potential of basin-edge effects (*193*), manifested as severe damage in a narrow zone running parallel to the causative faults through Kobe and adjacent cities (Figure 2.22). These and other simulations of basin waves (*194*) demonstrate that it is now possible to perform simulations of strong ground motions in basin structures, and these demonstrations form the basis for the simulation of ground motions from scenarios of future earthquakes (*195*).

Wave propagation in fluid-saturated sediments can exhibit special complexities that cannot be modeled in terms of a single elastic continuum. Early work by Maurice Biot and recent research shows that a fluid-saturated porous solid can support two *P* waves as well as one *S* wave. The surface waves on such a medium have not been adequately studied. This is an important subject in which more research is necessary in terms of constitutive relations and boundary conditions (*196*).

Rupture Propagation Effects

During an earthquake, seismic waves are emitted from the slipping part of the fault behind the rupture front. Since the rupture velocity is usually close to the shear-wave velocity, the amplitude of the seismic waves ahead of the rupture front grows progressively as more energy is added by the propagating fault. This energy buildup results in a large pulse of motion at the arrival time of each kind of seismic wave in the seismogram that contains the cumulative effect of rupture on the fault (*197*). The radiation pattern of the shear dislocation causes the motions of the large pulse to be oriented perpendicular to the fault. Forward rupture directivity effects require two conditions: the rupture front propagates toward the site, and the slip direction is aligned with the site. These conditions are readily met at locations away from the epicenter in strike-slip faulting and are also met during dip-slip faulting in the region located up-dip of the hypocenter. The enormous destructive potential of near-fault ground motions was manifested in the 1994 Northridge and 1995 Hyogo-ken Nanbu earthquakes. In each of these earthquakes, peak ground velocities as high as 175 centimeters per second were recorded (Figure 5.13). The periods of the near-fault pulses recorded in both of these earthquakes were in the range of 1 to 2 seconds, comparable to the natural periods of structures such as bridges and midrise buildings, many of which were severely damaged. These near-fault recordings have led to revisions of building codes in the United States.

Anelastic Attenuation Effects

The effects of absorption are described by the quality factor Q, which is inversely proportional to the fractional loss of energy per wave cycle. In the Earth, the Q value depends on the frequency of the seismic wave and the properties of the rocks. Values of Q in the crust and lithosphere are much lower than those in the underlying mantle, and they vary significantly with the tectonic environment. In general, attenuation is lower in tectonically stable regions, so earthquakes cause damage at much greater distances in stable regions than in tectonically active regions. Also the frequency dependence of Q is greater in areas of active tectonics (*198*). Proposed absorption mechanisms in the crust include frictional sliding on cracks, thermoelastic effects, grain boundary deformation, and dissipation by fluid movement within cracks and pores. For depths less than 1 kilometer, there can be strong attenuation from open cracks in near-surface rocks and losses in unconsolidated soils, causing the intensity of ground motions to diminish with increasing frequency beyond about 5

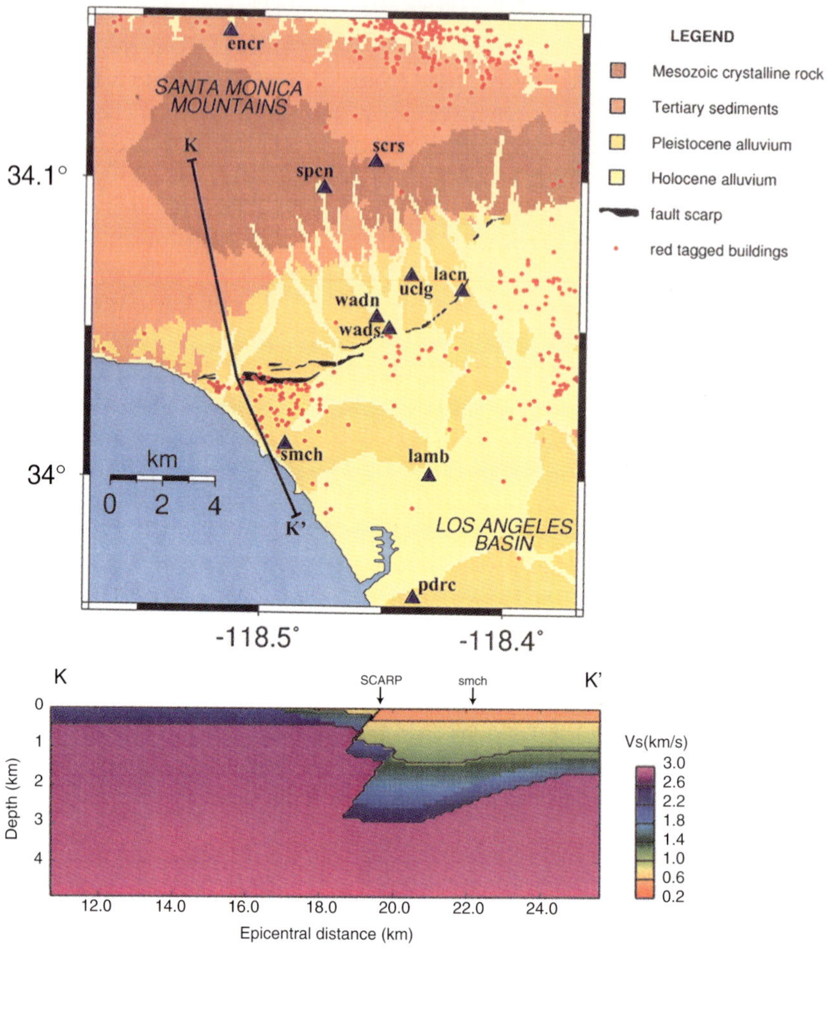

hertz. The distribution of Q for sedimentary rocks in basins influences the duration of shaking from strong earthquakes.

Nonlinear Site Effects

Soil response to strong shaking is a complex, nonlinear phenomenon that has long been investigated in laboratory experiments and in the field following large earthquakes (*199*). Laboratory tests clearly demonstrate nonlinear strain behavior in soils under dynamic loading. This nonlinearity is manifested by a reduction in shear modulus and an increase in

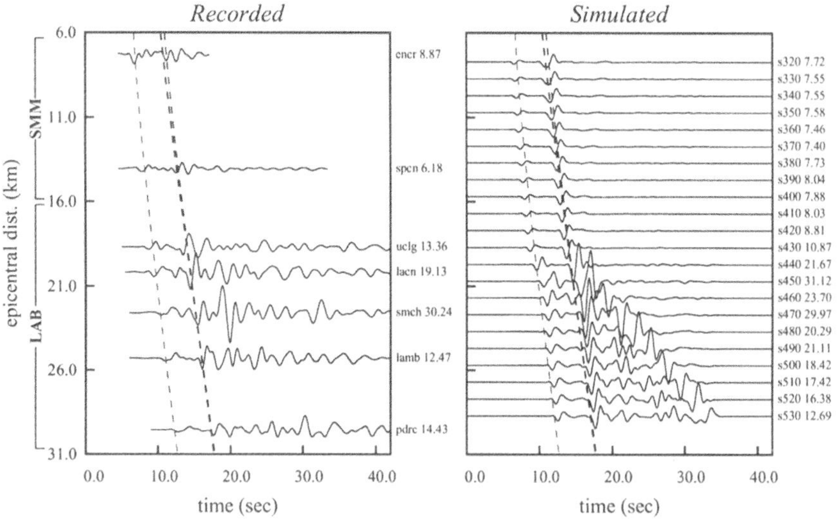

FIGURE 5.12 Basin effects in Santa Monica from the 1994 Northridge earthquake. Upper left panel is a geological map with the locations of seismic stations (triangles) and the scarp of the Santa Monica fault (arcuate black lines). Lower left panel is a sheer velocity model of the upper crust along map profile kk′. Upper panels compare the observed seismograms (left) with seismograms simulated from the model (right). The simulation indicates that the large pulse at station *smch* is caused by the focusing at the rock-sediment interface along the Santa Monica fault. SOURCE: Modified from R.W. Graves, A. Pitarka, and P.G. Somerville, Ground motion amplification in the Santa Monica area: Effects of shallow basin edge structure, *Bull. Seis. Soc. Am.*, **88**, 1224-1242, 1998. Copyright Seismological Society of America.

damping as shear strain levels increase beyond about 10^{-4} to 10^{-5} or as ground acceleration becomes greater than about 0.1*g*. This softening causes the fundamental period of the soil layer to lengthen. At higher frequencies, cyclic pore pressure increases may produce cyclic strain hardening, which is manifested by high-frequency spikes toward the end of the record, increasing the duration of the record and sometimes producing the largest accelerations. These effects have been demonstrated in the small number of data sets where surface and subsurface seismic data are available (*200*) to enable direct comparison between the motion in the rock or stiff soil (which is assumed to be linear) and the resulting motion

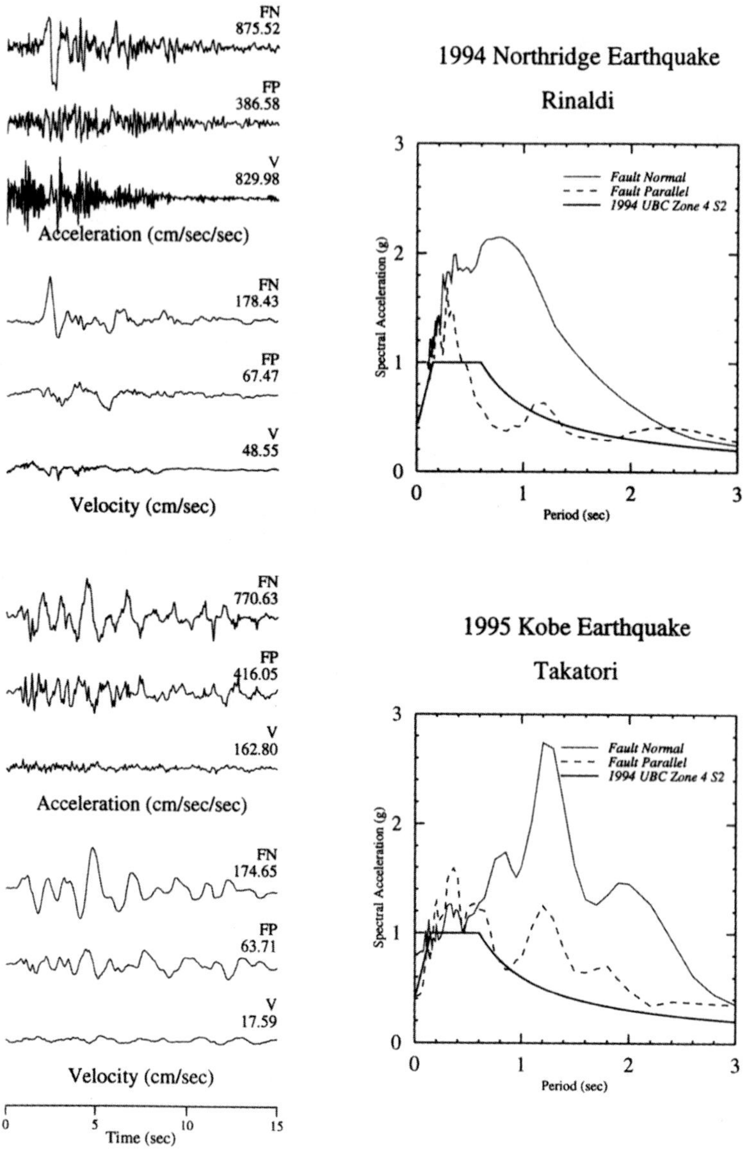

FIGURE 5.13 Recorded near-fault acceleration and velocity time histories and acceleration response spectra of the 1994 Northridge earthquake (*top*) and 1995 Hyogo-ken Nanbu earthquake (*bottom*), showing pulse-like motion on the strike-normal component of motion, which is much larger than the strike-parallel component at periods longer than about 0.4 second. SOURCE: P. Somerville, URS Corporation.

in the overlying softer soil. If nonlinear effects are important, then strong ground motions for large earthquakes can be difficult to predict from the measured accelerations during smaller events. Another complicating factor is that cohesionless soils are also subject to liquefaction and lateral spreading due to pore pressure effects (*201*). Nevertheless, numerical codes that account for soil nonlinearity are numerous, ranging from equivalent linear models to fully nonlinear models that also incorporate pore-pressure generation (*202*).

High-Frequency Ground Motions

Ground motions at frequencies above 1 hertz are the most damaging motions for small- and moderate-sized structures, and they also contain important information about the seismic source and details of stress on the fault plane. The character of high-frequency ground motions was documented from the analysis of the first strong-motion accelerograms (*203*). A key parameter is the corner frequency, which scales as the inverse of the rupture duration for events recorded in the far field or to the slip duration at a location on the fault for large events recorded by nearby seismometers. At low frequency, the displacement amplitude spectrum is constant with increasing frequency up to the corner frequency, where it changes slope and rolls off as the square of frequency (*204*). Correspondingly, the acceleration amplitude spectrum increases with frequency squared below the corner frequency and becomes flat above the corner frequency. Above 5 to 10 hertz, the acceleration spectrum declines rapidly with increasing frequency beyond a transition value denoted by f_{max} (*205*). Many theoretical studies have attempted to explain the flat portion of the acceleration spectrum (*206*). The spatial coherence of ground motions decreases rapidly with increasing frequency (*207*). Although the cause of this incoherence is not well understood, it may be due in part to focusing effects caused by irregular bedrock topography (*208*).

Key Questions

- How are the major variations in seismic-wave speeds related to geologic structures? How are these structures best parameterized for the purposes of wavefield modeling?
- What are the contrasts in shear-wave speed across major faults? Are the implied variations in shear modulus significant for dynamic rupture modeling? Do these contrasts extend into the lower crust and upper mantle?
- How are variations in the attenuation parameters related to wave speed heterogeneities? Is there a significant dependence of the attenua-

tion parameters on crustal composition or on frequency? How much of the apparent attenuation is due to scattering?

- What are the differences in near-fault ground motions from reverse, strike-slip, and normal faulting? In thrust faulting, how does energy trapped between the fault plane and free surface of the hanging-wall block amplify strong ground motions?
- How does the structure of sedimentary basins affect the amplitude and duration of ground shaking? How much of the amplification pattern in a basin is dependent on the location of the earthquake source? Can the structure of sedimentary basins be determined in sufficient detail to usefully predict the pattern of ground shaking for future large earthquakes?
- Are fault-parallel, low-velocity waveguides deep-seated features of faults? How continuous are they along strike and dip? Can studies of fault-zone trapped waves constrain the effective rheological parameters of the fault zone, such as effective fracture energy?
- Is the ability to model recorded seismograms limited mainly by heterogeneity in source excitation, focusing by geologic structure, or wavefield scattering?
- What role do small (sub-grid-scale) heterogeneities and irregular interfaces play in wave propagation at high frequencies? How do they depend on depth, geological formation, and tectonic structure? How important is multiple scattering in the low-velocity, uppermost layers? Can stochastic parameterizations be used to improve wavefield predictions?

5.6 SEISMIC HAZARD ANALYSIS

Although earthquakes cannot be predicted in the short term with any useful accuracy and generality and the feasibility of intermediate-term prediction is still an open question, the magnitudes and locations of larger events can be forecast over the long term, and their effects can be anticipated. Seismic hazard analysis (SHA), described in Chapter 3, involves the characterization of potential earthquake activity and associated ground motions in a form, either probabilistic or scenario based, that is useful for seismic design and emergency management. The scientific basis for probabilistic seismic hazard analysis (PSHA) is currently being improved by (1) the addition of denser and more precise geodetic and geologic data sets to seismic hazard characterization and (2) a better understanding of the geological controls on strong ground motions. The challenge is to recast the methodology of seismic hazard analysis in a way that more explicitly accounts for the dependence of earthquake phenomena on time.

Earthquake Forecasting

Historically, most seismic hazard analyses have assumed that earthquakes can be described as a Poisson process, for which the probability of at least one earthquake in a given time t is

$$P(t) = 1 - e^{-rt}, \tag{5.3}$$

where the rate parameter r can be estimated from the historical or prehistoric rate of earthquakes (e.g., a plot of magnitude versus frequency determined from seismic monitoring or paleoseismology). For a simple Poisson process, the probability per unit time is independent of absolute time and the time elapsed since the last event. These properties make the problem of earthquake forecasting a fairly straightforward exercise. For example, in the preparation of the recent USGS maps, probabilities were calculated with a Poisson model once the magnitude-frequency relation had been established for each of the seismic sources.

Driven by the development of plate tectonics, a growing catalog from seismic monitoring, and increasingly detailed measurements of historical seismic activity and fault slip, there is growing interest in time-dependent forecasting techniques for specific earthquakes. This work originated from the early models of earthquake recurrence that linked spatial and temporal aspects of seismicity with rates and pattern of fault slip. A simple time-dependent probabilistic model of the occurrence of large earthquakes has occasionally been implemented in PSHA in cases where the historical or paleoseismic record supports it. Compared to the Poisson formulation, the most important feature of this model is that the probability of occurrence of a similar event increases with time since the last one.

The challenge has been to identify data sets and develop physical models that might explain such time-dependent features of earthquake recurrence. Currently, these efforts are advancing on several fronts, some of which take very different tacks (e.g., seismic gaps versus earthquake clustering). Such differences may lead to discrepant forecasts. These points are illustrated below in discussions of characteristic earthquakes, seismic gaps, moment-rate budgeting, clustering, and stress-transfer effects.

Characteristic Earthquakes The characteristic earthquake hypothesis states that seismic moment release on an individual fault segment is dominated by a characteristic earthquake rupturing the entire length of the segment (i.e., the largest possible earthquake for that segment), and that moderate earthquakes within one magnitude range below the characteristic event may be rare or entirely absent. The model implicitly assumes that (1) rupture is limited to geometrically defined fault segments, (2) the displacement per event is constant at a point, (3) the slip rate along the fault is variable, and (4) slip deficits at the ends of fault are not "filled in" by slip from smaller earthquakes.

From the perspective of hazard assessments, this hypothesis offers tremendous simplification because only one earthquake scenario is considered (the characteristic earthquake) for each fault segment. Moreover, the size of this earthquake can be estimated from the length of the fault segment and moment-length scaling relations. In this way, the hypothesis reduces the dimension of the earthquake forecasting problem to one in which time is the only independent variable. Because of these simplifications, characteristic earthquakes have been incorporated into a large number of seismic hazards analyses (*209*).

Seismic Gap Hypothesis Building on the characteristic earthquake hypothesis, the seismic gap hypothesis addresses the distribution of these large events through time. To estimate earthquake likelihood for use in seismic hazard studies, the seismic gap model is implemented as follows. Mapped faults are divided into segments, and a characteristic magnitude is estimated for each segment. The slip rate on the fault is estimated from the displacement and age of features offset by the fault, and the characteristic slip is estimated from historical slip data or from regression relationships on magnitude and slip. The mean recurrence time is estimated from either the times of known earthquakes on the segment or the ratio of characteristic slip to fault slip rate. The probability distribution of recurrence times is estimated, and the conditional probability of an earthquake during some time interval is computed. The critical question to address is whether forecasted earthquakes in seismic gaps occur with greater probability than a simple random occurrence.

Moment-Rate Budgeting One approach to forecasting future earthquakes is to balance the long-term rates of fault slip and moment release as inferred from seismic monitoring. In practice, the application of moment-rate budgeting is difficult because the results are sensitive to the completeness of the historical catalog and the past distribution of earthquakes in space and time. Key, but controversial, assumptions of this method are that the long-term slip rates are representative of present rates and that the slip rates are completely seismogenic. The latter assumption may hold for crustal earthquakes, but it does not appear to be valid for many subduction zones, where significant aseismic deformation is occurring.

Clustering The term clustering is commonly used to describe concentrations of earthquakes in space, time, or both. Earthquakes are much more frequent in some places than in others, even along major faults or plate boundaries, as seen from inspection of maps of earthquake locations. The most abundant (and obvious) example of earthquake clustering in time and space is the occurrence of aftershocks (see Chapter 4). Indeed, many complete catalogs of earthquakes are dominated by aftershocks of moderate to large events (*210*). Thus, the occurrence of a single earthquake greatly increases the probability of another event in the same location.

Although the physical origin of clustering behavior is not clear, it has important implications for models of earthquake occurrence. Clustering suggests a causality between earthquakes that could change many of the assumptions that underlie seismic hazard assessments. In short to intermediate time scales, the most dangerous regions may be those that have recently experienced large earthquakes, rather than the locked portions of seismically active faults.

Stress Interactions Identifying the origins of clustering, or distinguishing among different models of earthquake recurrence, will require an explicit physical theory of seismic activity. Important components of such a theory will include an explicit model for stress evolution on major faults due to tectonic stress accumulation, previous earthquakes, and inelastic stress relaxation as well as the evolution of frictional strength on faults, the mechanical strength of materials off the fault, and the rupture of virgin rock required to accomplish finite displacements in a brittle medium. Such a complete theory will be difficult to develop and difficult to confirm experimentally because it requires a long span of accurate earthquake information including focal mechanisms. Furthermore, the stress model must include tectonic stress accumulation, for which there is no definitive model at present. However, significant progress has been made on parts of the theory.

An important first step was the development of expressions to calculate stresses everywhere in a homogeneous elastic half-space due to an arbitrary dislocation (*211*). These expressions allow calculations of the change in stress across any existing fault due to earthquakes causing known displacements. With this model, it has become routine to calculate stress changes for all earthquakes above M 5 in populated regions. Using these methods to calculate tectonic stress accumulations is more complex because it requires assumptions about strain partitioning throughout fault and plate boundary zones.

Application to Seismic Hazard Analysis There have been continuing efforts to utilize the understanding of earthquake forecasting to improve the capabilities of PSHA. To this end, the recent USGS ground-motion mapping study incorporated characteristic earthquakes for a limited number of fault segments. Other hazard analyses have also incorporated time-dependent probabilities (*212*). Time-dependent seismic hazard maps have been produced for California by the Southern California Earthquake Center and the California Division of Mines and Geology (CDMG) (*213*). The CDMG maps show substantial differences from the time-independent maps for certain faults.

Several efforts are also under way to produce urban seismic hazard maps that merge probabilistic seismic hazard assessment with site response and, in some cases, three-dimensional basin effects and rupture directivity. These maps, at scales of 1:24,000 to 1:50,000, could be used for engineering design purposes, loss estimation, and land-use planning.

Prediction of Strong Ground Motions

Seismic waves travel through a medium having a free surface, strong variations (usually increases) of seismic velocity with depth, large-scale lateral variations in seismic velocities related to mountains and sedimentary basins, small-scale lateral variations (scatterers), and dramatically different elastic properties at individual observation sites (local soil conditions). The wave trains generated by even very simple sources, such as explosions, can become highly complex due to propagation through such heterogeneous media. Source effects such as rupture directivity further add to the spatial variation of ground motions (see Section 5.5).

This large degree of variability in ground-motion characteristics presents a formidable challenge to earthquake engineers and engineering seismologists whose role is to characterize ground motions for the seismic design of structures. During the past two decades, careful studies of ground motions from well-recorded earthquakes, the application of rigorous representations of earthquake sources as shear dislocations, and the use of increasingly realistic methods of modeling seismic wave propagation through heterogeneous structures have resulted in a greatly enhanced ability to understand and predict the complex waveforms of strong ground-motion recordings. Common methods to estimate ground motions are summarized below.

Empirical Engineering Models of Strong Ground-Motion Attenuation A convenient collection of recent empirical ground-motion models was published in the 1997 January-February issue of *Seismological Research Letters* (*214*). These ground-motion models are for distinct tectonic categories of earthquakes: shallow crustal earthquakes in tectonically active regions, shallow crustal earthquakes in tectonically stable regions, and subduction-zone earthquakes. Subduction-zone earthquakes are further subdivided into those that occur on the shallow plate interface and those that occur at greater depths within the subducting plate. Significant differences exist in the ground-motion characteristics among these different earthquake categories, as illustrated in Figure 5.14.

The process of developing modern empirical ground-motion attenuation relations has become a routine endeavor. First, a comprehensive set of strong-motion data is compiled in which the following quantities are rigorously quantified or classified: earthquake category (e.g., crustal or subduction), seismic moment and moment magnitude, focal mechanism, geometry of the earthquake's rupture plane and distance of each recording station from this plane, and recording site conditions. Next, a complex functional form is usually selected and fit to the data. The equations that are developed relate ground-motion parameters (such as peak ground acceleration, response spectral acceleration, strong-motion duration) to the source parameters of magnitude and mechanism, the path parameters (usually source-to-site distance and sometimes focal depth), and local parameters (site geology and sometimes depth to basement rock).

Specification of Uncertainty in Ground-Motion Attenuation Models and SHA The complete description of a ground-motion parameter includes the central estimate of the parameter and its variability. The standard error in the predicted ground-motion level is relatively high; typically the median plus one standard deviation level of ground motion is about a factor of 1.5 to 2 greater than the median value (*215*).

Seismic hazard calculations for critical facilities include a comprehensive representation of uncertainty commonly separated into epistemic and aleatory components (*216*). Epistemic uncertainty is due to incomplete knowledge and data and, in principle, can be reduced by the collection of additional information. Aleatory uncertainty is due to the inherently unpredictable nature of future events and cannot be reduced. The total uncertainty is obtained from the combination of the epistemic and aleatory components. The epistemic uncertainty is usually represented by alternative branches on a logic tree, leading to alternative hazard curves. These alternative hazard curves can be used to define hazard curves at different confidence levels. Each hazard curve is produced from an integration over the aleatory component.

Characterization of Site Response Local geological conditions have a primary influence on the amplitude and frequency content of strong ground motions. In particular, the vertical gradient in shear-wave velocity (which generally increases rapidly with depth just below the surface) gives rise to motion amplification due to impedance contrast effects, which may be offset by the effects of viscoelastic damping and nonlinear response of the medium. The simplest way to account for effects of local

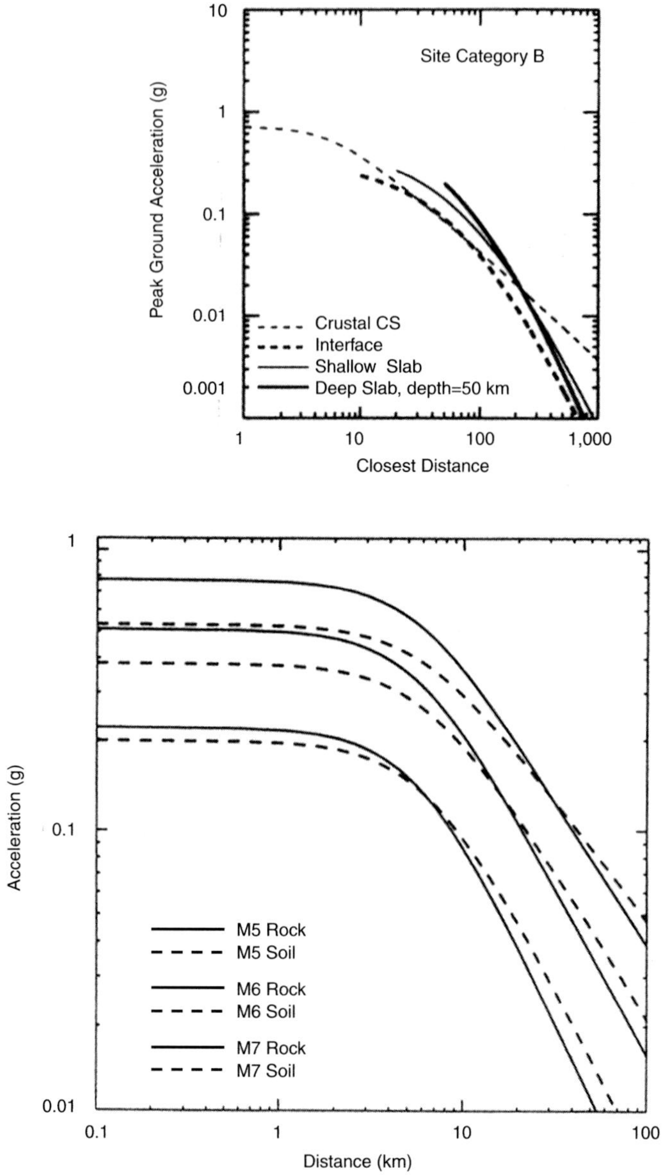

FIGURE 5.14 (*Top*) Peak acceleration attenuation relations for soil sites for M 6.5 earthquakes for a range of tectonic categories of earthquakes. SOURCE: P. Somerville, URS Corporation. (*Bottom*) Peak acceleration attenuation relations for crustal earthquakes showing the dependence on magnitude and site category. SOURCE: N.A. Abrahamson and K.M. Shedlock, Overview of ground-motion attenuation models, *Seis. Res. Lett.*, **68**, 9-23, 1997. Copyright Seismological Society of America.

geological conditions is to use empirical ground-motion attenuation relations for the site geology category (e.g., alluvium, rock) that are representative of the site. However, the response at a given site belonging to a broad category (e.g., "soil," "rock") is in general different from the average response of a large number of sites belonging to that category. Furthermore, the variability of the response between these many sites will, in general, be larger than the variability in response of a single site due to multiple earthquakes.

Another common procedure to estimate site response is with physically based models of vertical shear-wave propagation through a horizontally layered soil column whose properties, including shear-wave velocity, material damping, and density, have been determined from field and laboratory measurements. Some models include the nonlinear response of soils, which can have an important influence on the amplitude and frequency content of the ground motion. The important effect of nonlinear soil behavior on site response has been incorporated in the site response factors that are embodied in current building codes and provisions (*217*). In these codes and provisions, site response is represented by period- and amplitude-dependent factors derived from sets of recorded data and from analyses of site response based on nonlinear or equivalent-linear models of soil response.

Ground-Motion Prediction Using Seismological Models Based on developments in theoretical and computational seismology and on strong-motion recordings from a large number of major earthquakes that began with the 1979 Imperial Valley earthquake, much progress has been made in understanding the origin and composition of strong ground motion. In many instances, the causes of the large variations in strong ground-motion recordings are now understood. This understanding is being applied to the problem of constructing realistic earthquake scenarios (i.e., predicting ground motions from potential future earthquakes). The simplest seismologically based simulations treat strong motion as a time sequence of band-limited white noise. A Fourier spectral model of the ground motion is constructed, starting with a model of the source spectrum and modifying its shape by factors to represent wave propagation effects (*218*).

More complex methods have been developed that have a more rigorous basis in theoretical and computational seismology with fewer simplifications than the stochastic model. The earthquake source is represented as a shear dislocation on an extended fault plane, and the wave propagation is rigorously modeled by Green's functions computed for the seismic velocity structure, which contains the fault and the site, or by empirical Green's functions derived from strong-motion recordings of earthquakes smaller than the one being simulated. The ground-motion time history is calculated in the time domain using the elastodynamic representation theorem. This calculation involves integration over the fault surface of the convolution of the slip time function on the fault with the Green's function for the appropriate depth and distance. For structures having lateral variations in seismic velocities and densities, such as sedimentary basins, wave propagation is modeled numerically using finite difference methods (*219*).

To simulate broadband time histories using this Green's function-based approach, ground motions are computed separately in the short-period and long-period ranges and then combined into a single broadband time history. The use of different methods in these two vibrational period ranges is necessitated by the observation that ground motions are much more stochastic at short periods than at long periods. An example of broadband simulation of strong ground motions is shown in Figure 5.15, which compares the recorded and simulated ground motions at Arleta from the 1994 Northridge earthquake.

Because these seismologically based ground-motion models can include the specific source, path, and site conditions of interest, they can be used to generate ground-motion time histories, which augment the recorded data used to generate empirical models. Alternatively, they can be used as site-specific estimates that complement estimates based on empirical models. These seismological models have incorporated important characteristics such as rupture directivity, Moho reflections, and basin effects. Rupture directivity contributed greatly to the generation of peak ground velocities approaching 2 meters per second during the 1994 Northridge, California, and 1995 Kobe, Japan, earthquakes, and approaching 3 meters per second during the 1999 Taiwan event. As a result of these and previous earthquakes, rupture directivity effects have been incorporated in the specification of design ground motions in the 1997 Uniform Building Code and have produced large revisions in the code spectrum, as shown in Figure 5.16.

Key Questions

- What factors limit fault-rupture propagation? How valid are the characteristic earthquake models? What magnitude distributions are appropriate for different regions?
- Under what circumstances are large events Poissonian in time? What temporal models and distributions of recurrence intervals pertain to major plate boundary faults? Are these models and distributions different for stable continental regions?
- Can geodetic (Global Positioning System and interferometric synthetic aperture radar) measurements of deformation be employed to ac-

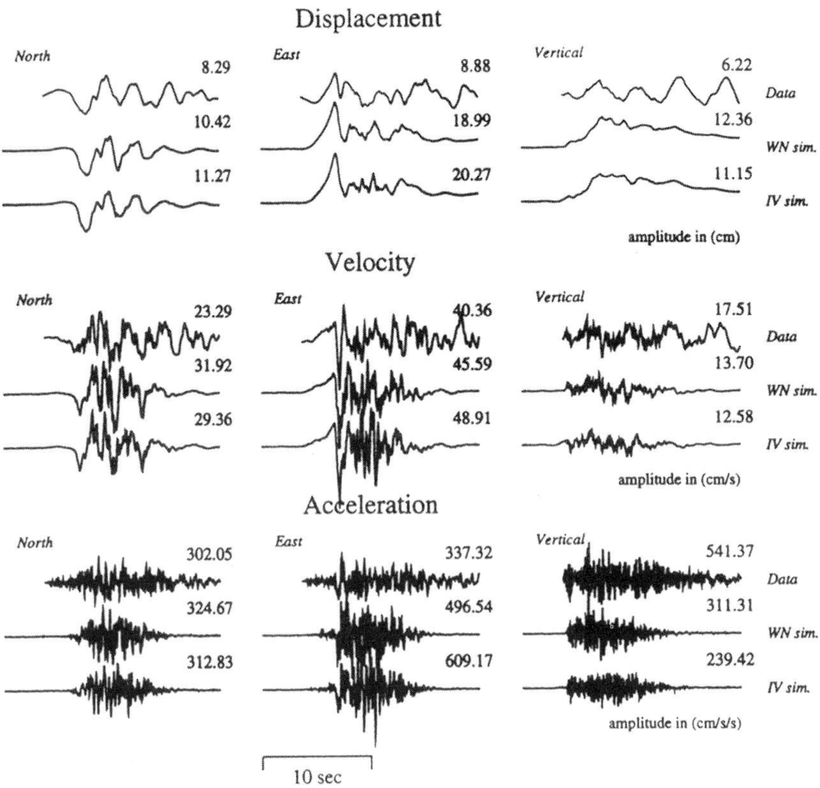

FIGURE 5.15 Comparison of recorded (*top row*) and simulated (*middle and bottom rows*) displacement, velocity, and accelerations at Arleta from the 1994 Northridge earthquake, plotted on a common scale with peak value given at the top left corner. SOURCE: P. Somerville, C.K. Saikia, D. Wald, and R. Graves, Implications of the Northridge earthquake for strong ground motions from thrust faults, *Bull. Seis. Soc. Am.*, **86**, S115-S125, 1996. Copyright Seismological Society of America.

curately constrain short- and long-term seismicity rates for use in seismic hazard assessment? How should geologic and paleoseismic data on faults best be used to determine earthquake recurrence rates?

• Can physics-based scenario simulations produce more accurate estimates of ground-motion parameters than standard attenuation relationships? Can these simulations be used to reduce the high residual variance in these relationships?

• What is the nature of near-fault ground motion? How do fault ruptures generate long-period directivity pulses? How do near-fault

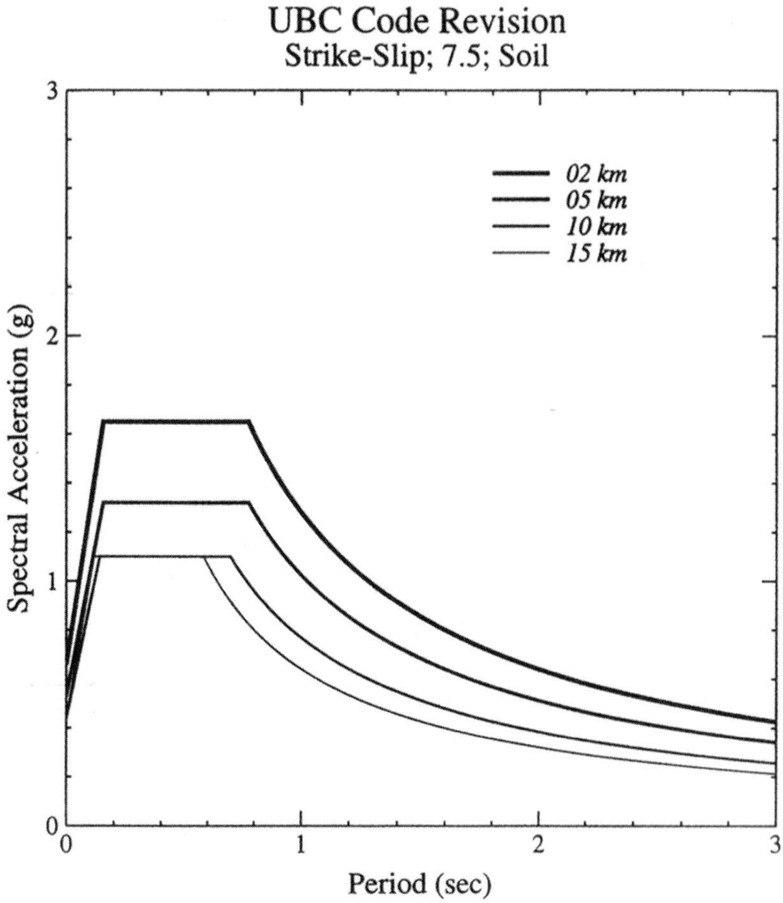

FIGURE 5.16 Design response spectra on soil sites at various distances from a Source Type A fault (maximum M 7.5 or larger) in the 1997 Uniform Building Code (UBC). The 1994 UBC was similar to the spectrum for 15 kilometers' distance. SOURCE: P.G. Somerville, URS Corporation.

ground motions differ between reverse and strike-slip faulting? Can these motions be predicted for scenario earthquakes?

- What are the earthquake source and strong ground-motion characteristics of large earthquakes (magnitudes greater than 7.5), for which there are few strong-motion recordings? Can the shaking from large earthquakes be predicted accurately from smaller events?
- How important is the nonlinear seismic response of stable soils in estimating strong ground motion?

NOTES

1. For a precise mathematical definition of classical dynamical systems, see V.I. Arnold and A. Avez, *Ergodic Problems of Classical Mechanics*, W.A. Benjamin, New York, 286 pp., 1984. A comprehensive treatment is provided by A. Katok and B. Hasselblatt, *Introduction to the Modern Theory of Dynamical Systems*, Cambridge University Press, Cambridge, U.K., 802 pp., 1995.

2. Earth Systems Science Committee, *Earth System Science: A Program for Global Change*, NASA Advisory Council, Washington, D.C., 208 pp., 1988.

3. The stress tensor, measured in units of pascals (1 pascal = 1 newton per square meter), specifies the vector forces acting at a point across an arbitrarily oriented plane in the solid. For fluids of the atmosphere and ocean, the stress tensor reduces to a single number, the pressure.

4. Models of self-organization in earthquakes have also been used to explain similar processes in other areas such as brain dynamics in neurobiology (J.J. Hopfield, Neurons, dynamics, and computation, *Physics Today*, 40-46, February 1994) and charge density waves in solids (S. Brown and G. Gruner, Carge and spin density waves, *Sci. Am.*, **270**, 50-57, 1994).

5. B.B. Mandelbrot, *The Fractal Geometry of Nature*, W.H. Freeman, New York, 468 pp., 1982.

6. Evidence that fault networks have a self-similar structure has been presented by many authors, including D.L. Turcotte, A fractal model for crustal deformation, *Tectonophysics*, **132**, 261-269, 1986; C.A. Aviles, C.H. Scholz, and J. Boatwright, Fractal analysis applied to characteristic segments of the San Andreas fault, *J. Geophys. Res.*, **92**, 331-344, 1987; P.G. Okubo and K. Aki, Fractal geometry in the San Andreas fault system, *J. Geophys. Res.*, **92**, 345-355, 1987; G.C.P. King, R.S. Stein, and J.B. Rundle, The growth of geological structures by repeated earthquakes: 1. Conceptual framework, *J. Geophys. Res.*, **93**, 13,307-13,318, 1988; W.L. Power, T.E. Tullis, and J.D. Weeks, Roughness and wear during brittle faulting, *J. Geophys. Res.*, **93**, 15,268-15,278, 1988; T. Hirata, Fractal dimension of fault systems in Japan; Fractal structure in rock fracture geometry at various scales, *Pure Appl. Geophys.*, **131**, 157-170, 1989; M.C. Robertson, C.G. Sammis, M. Sahimi, and A.J. Martin, Fractal analysis of three-dimensional spatial distributions of earthquakes with a percolation interpretation, *J. Geophys. Res.*, **100**, 609-620, 1995; W.L. Power and T.E. Tullis, A review of the fractal character of natural fault surfaces with implications for friction and the evolution of fault zones, in *Fractals in the Earth Sciences*, P. Lapointe and C. Barton, eds., Plenum, New York, pp. 89-105, 1995; and G. Ouilion, C. Castaing, and D. Sornette, Hierarchical geometry of faulting, *J. Geophys. Res.*, **101**, 5477-5487, 1996.

7. The principle of self-organized criticality was first articulated by P. Bak, C. Tang, and K. Wiesenfeld (Self-organized criticality: An explanation, *Phys. Rev. Lett.*, **59**, 381-384, 1987) as an explanation for $1/f$ noise in dynamical systems; for a popular account of the development of the theory, see P. Bak, *How Nature Works: The Science of Self-Organized Criticality*, Springer-Verlag, New York, 226 pp., 1996.

8. P. Bak and C. Tang, Earthquakes as a self-organized critical phenomenon, *J. Geophys. Res.*, **94**, 15,635-15,637, 1989; A. Sornette and D. Sornette, Self-organized criticality and earthquakes, *Europhys. Lett.*, **9**, 197-202, 1989; Z. Olami, H.J. Feder, and K. Christensen, Self-organized criticality in a continuous nonconservative cellular automation modelling earthquakes, *Phys. Rev. Lett.*, **68**, 1244-1247, 1992.

9. A cellular automaton is an array of identically programmed automata, or "cells," which interact with one another (e.g., T.J. Bossomaier and D.G. Green, *Patterns in the Sand—Computers, Complexity and Life*, Allen and Unwin, Sydney, 224 pp., 1998).

10. R. Burridge and L. Knopoff, Model and theoretical seismicity, *Bull. Seis. Soc. Am.*, **57**, 341-371, 1967.

11. G. Narkounskaia and D.L. Turcotte, A cellular-automata, slider-block model for earthquakes; I, Demonstration of chaotic behavior for a low order system, *Geophys. J.*, **11**, 250-258, 1992; D. Turcotte, *Fractals and Chaos is Geology and Geophysics*, Cambridge University Press, Cambridge, U.K., pp. 125-136, 1992.

12. J.R. Rice, Spatio-temporal complexity of slip on a fault, *J. Geophys. Res.*, **98**, 9885-9907, 1993.

13. J.M. Carlson, J.S. Langer, B.E. Shaw, and C. Tang, Intrinsic properties of a Burridge-Knopoff model of an earthquake fault, *Phys. Rev. A*, **44**, 884-897, 1991; J.S. Langer, J.M. Carlson, C.R. Myers, and B.E. Shaw, Slip complexity in dynamic models of earthquake faults, *Proc. Natl. Acad. Sci.*, **93**, 3825-3829, 1996.

14. B.E. Shaw and J.R. Rice, Existence of continuum complexity in the elastodynamics of repeated fault ruptures, *J. Geophys. Res.* **105**, 791-810, 2000; G. Zheng and J.R. Rice, Conditions under which velocity-weakening friction allows a self-healing versus a cracklike mode of rupture, *Bull. Seis. Soc. Am.*, **88**, 1466-1483, 1998.

15. J.S. Langer, Instabilities and pattern formation in crystal growth, *Rev. Mod. Phys.*, **52**, 1-28, 1980; W. Kurz and D.J. Fisher, *Fundamentals of Solidification*, Trans Tech Publications, Aedermansdorf, Switzerland, 319 pp., 1998.

16. D.C. Samuels and D. Kivotides, A damping length scale for superfluid turbulence, *Phys. Rev. Lett.*, **83**, 5306-5309, 1999.

17. N.I. Muskhelishvili, *Singular Integral Equations: Boundary Problems of Function Theory and Their Application to Mathematical Physics* (trans. by J.R.M. Radok), P. Noordhoff, Groningen, Netherlands, 447 pp., 1953; R. Hill, A landmark in the theory of elasticity, *Nature*, **174**, 713-714, 1954.

18. The subject is reviewed by Y. Ben-Zion and C.G. Sammis, Characterization of fault zones, *Pure Appl. Geophys.*, in press, 2002.

19. P.J. Tackley, Self-consistent generation of tectonic plates in three-dimensional mantle convection, *Earth Planet. Sci. Lett.*, **157**, 9-22, 1998; R. Trompert and U. Hansen, Mantle convection simulations with rheologies that generate plate-like behavior, *Nature*, **395**, 686-689, 1998.

20. S. Stein and E.A. Okal, Seismicity and tectonics of the Ninety East Ridge area: Evidence for internal deformation of the Indian plate, *J. Geophys. Res.*, **83**, 2233-2245, 1978; D.A. Wiens, C. DeMets, R.G. Gordon, S.A. Stein, D. Argus, J.F. Engeln, P. Lundgren, D. Quible, C.S. Stein, S. Weinstein, and D. Woods, A diffuse plate boundary model for Indian Ocean tectonics, *Geophys. Res. Lett.*, **12**, 429-432, 1985.

21. D.A. Lockner and J.D. Byerlee, An earthquake instability model based on faults containing high fluid-pressure compartments, *Pure Appl. Geophys.*, **145**, 717-745, 1995.

22. G. Hongqing, The clustering feature of strong earthquakes in China and its implications, *J. Earthquake Prediction Res.*, **3**, 215-225, 1994.

23. T.K. Rockwell, S. Lindvall, M. Herzberg, D. Murbach, T. Dawson, G. Berger, and D.J. Huntley, Paleoseismology of the Johnson Valley, Kickapoo and Homestead Valley faults of the Eastern California Shear Zone, *Bull. Seis. Soc. Am.*, **90**, 1200-1236, 2000.

24. W.L. Ellsworth, Characteristic earthquakes and long-term earthquake forecasts: Implications of central California seismicity, in *Urban Disaster Mitigation: The Role of Science and Technology*, F.Y. Cheng and M.S. Sheu, eds., Elsevier, Oxford, U.K., pp. 1-14, 1995.

25. D.P. Schaff, G.C. Beroza, and B.E. Shaw, Postseismic response of repeating aftershocks, *Geophys. Res. Lett.*, **25**, 4549-4552, 1998.

26. For example, see D.P. Schwartz and K.J. Coppersmith, Fault behavior and characteristic earthquakes: Examples from the Wasatch and San Andreas fault zones, *J. Geophys. Res.*, **89**, 5681-5698, 1984; S.C. Lindvall, T.K. Rockwell, and K.W. Hudnut, Evidence for prehistoric earthquakes on the Superstition Hills fault from offset geomorphic features, *Bull. Seis. Soc. Am.*, **79**, 342-361, 1989; D.P. Schwartz, Paleoseismicity, persistence of segments, and

temporal clustering of large earthquakes—Examples from the San Andreas, Wasatch, and Lost River fault zones, in *Proceedings of Conference XLV; A Workshop on Fault Segmentation and Controls of Rupture Initiation and Termination*, D.P. Schwartz and R.H. Sibson, eds., U.S. Geological Survey Open File Report 89-315, Reston, Va., pp. 361-375, 1989.

27. A.A. Barka, Slip distribution along the North Anatolian fault associated with large earthquakes of the period 1939 to 1967, *Bull. Seis. Soc. Am.*, **86**, 1238-1254, 1996; K. Sieh, The repetition of large-earthquake ruptures, *Proc. Natl. Acad. Sci.*, **93**, 3764-3771, 1996; A.P. Thomas and T.K. Rockwell, A 300-550 year history of slip on the Imperial fault near the U.S.-Mexico border: Missing slip at the Imperial fault bottleneck, *J. Geophys. Res.*, **101**, 5987-5997, 1996; R.S. Stein, A.A. Barka, and J.H. Dieterich, Progressive failure on the North Anatolian fault since 1939 by earthquake stress triggering, *Geophys. J. Int.*, **128**, 594-604.

28. S.G. Wesnousky, The Gutenberg-Richter or characteristic earthquake distribution: Which is it? *Bull. Seis. Soc. Am.*, **84**, 1940-1959, 1994; Y. Kagan and S. Wesnousky, The Gutenberg-Richter or characteristic earthquake distribution, Which is it? Discussion and reply, *Bull. Seis. Soc. Am.*, **86**, 274-291, 1996.

29. J.F. Pacheco, C.H. Scholz, and L.R. Sykes, Changes in frequency-size relationship from small to large earthquakes, *Nature*, **355**, 71-73, 1992.

30. The issue here is how slip scales with fault dimension. Elastic dislocation theory implies that slip is proportional to the minimum diameter of the earthquake and thus saturates when the earthquake is large enough to rupture through the thickness of the elastic lithosphere (B. Romanowicz and J.B. Rundle, On scaling relations for large earthquakes, *Bull. Seis. Soc. Am.*, **83**, 1294-1297, 1993). However, some authors maintain that data from field observations of surface slip support the position that slip for large earthquakes continues to grow with earthquake length (C.H. Scholz, A reappraisal of large earthquake scaling, *Bull. Seis. Soc. Am.*, **84**, 215-218, 1994). P.M. Mai and G.C. Beroza (Source scaling properties from finite-fault rupture models, *Bull. Seis. Soc. Am.*, **90**, 605-614, 2000) analyzed a smaller, but better resolved, earthquake data set consisting of heterogeneous slip models and confirmed that slip in large strike-slip earthquakes continues to grow with length well after rupture width saturates. B.E. Shaw and C.H. Scholz (Slip-length scaling in large earthquakes: Observations and theory and implications for earthquake physics, *Geophys. Res. Lett.*, **28**, 2633-2636, 2001) use dynamic rupture modeling to show that these observations are consistent with their model, which predicts that slip does not saturate until the rupture length approaches 10 times the seismogenic thickness.

31. The Coulomb stress increment ΔS caused by a fault dislocation on another fault surface with specified orientation is defined by $\Delta S = \Delta|\tau| - \mu\,\Delta\sigma$, where $\Delta|\tau|$ is the change in the magnitude of the shear stress acting on the fault surface due to the dislocation, $\Delta\sigma$ is the change in the fault normal stress, and μ is the coefficient of friction across the fault.

32. S.C. Jaumé and L.R. Sykes, Change in the state of stress on the southern San Andreas fault resulting from the California earthquake sequence of April to June 1992, *Science*, **258**, 1325-1328, 1992; R.S. Stein, G.C.P. King, and J. Lin, Change in failure stress on the southern San Andreas fault system caused by the magnitude = 7.4 Landers earthquake, *Science*, **258**, 1328-1332, 1992; R.W. Simpson and P.A. Reasenberg, Earthquake-induced static stress changes on central California faults, in *Loma Prieta, California Earthquake of October 17, 1989—Tectonic Processes and Models*, R.W. Simpson, ed., U.S. Geological Survey Professional Paper 1550-F, Denver, Colo., F55-F89, 1994; G.C.P. King, R.S. Stein, and J. Lin, Static stress changes and the triggering of earthquakes, *Bull. Seis. Soc. Am.*, **84**, 935-953, 1994; R.A. Harris and R.W. Simpson, In the shadow of 1857—Effect of the great Ft. Tejon earthquake on the subsequent earthquakes in southern California, *Geophys. Res. Lett.*, **23**, 229-232, 1996; L.M. Jones and E. Hauksson, The seismic cycle in southern California: Precursor or response? *Geophys. Res. Lett.*, **24**, 469-472, 1997.

33. T. Parsons, S. Toda, R. Stein, A. Barka, and J.H. Dieterich, Heightened odds of large

earthquakes near Istanbul; An interaction-based probability calculation, *Science*, **288**, 661-665, 2000.

34. Pioneering Japanese work on stress changes and seismicity includes K. Yamashina, Induced earthquakes in the Izu Peninsula by the Izu-Hanto-Oki earthquake of 1974, Japan, *Tectonophysics*, **51**, 139-154, 1979; T. Kato, K. Rybicki, and K. Kasahara, Mechanical interaction between neighboring active faults—An application to the Altera fault, central Japan, *Tectonophysics*, **144**, 181-188, 1987; Y. Okada and K. Kasahara, Earthquake of 1987, off Chiba, central Japan and possible triggering of eastern Tokyo earthquake of 1988, *Tectonophysics*, **172**, 351-364, 1990. For more recent studies in Japan and other countries, see summary by R.A. Harris (Introduction to special section: Stress triggers, stress shadows, and implications for seismic hazard, *J. Geophys. Res.*, **103**, 24,347-24,358, 1998).

35. J.H. Dieterich, Earthquake nucleation on faults with rate- and state-dependent strength, *Tectonophysics*, **211**, 115-134, 1992.

36. D.P. Hill, P.A. Reasenberg, A. Michael, W. Arabaz, G.C. Beroza, J.N. Brune, D. Brumbaugh, S. Davis, D. DePolo, W.L. Ellsworth, J. Gomberg, S. Harmsen, L. House, S.M. Jackson, M. Johnston, L. Jones, R. Keller, S. Malone, S. Nava, J.C. Pechmann, A. Sanford, R.W. Simpson, R.S. Smith, M. Stark, M. Stickney, S. Walter, and J. Zollweg, Seismicity remotely triggered by the magnitude 7.3 Landers, California, earthquake, *Science*, **260**, 1617-1623, 1993.

37. P.A. Reasenberg, and L.M. Jones, Earthquake hazard after a mainshock in California, *Science*, **243**, 1173-1176, 1989.

38. J.H. Dieterich and B. Kilgore, Implications of fault constitutive properties for earthquake prediction, *Proc. Natl. Acad. Sci.*, **93**, 3787-3794, 1996.

39. J.H. Dieterich, Probability of earthquake recurrence with non-uniform stress rates and time-dependent failure, *Pure Appl. Geophys.*, **126**, 589-617, 1988.

40. R.S. Stein, A.A. Barka, and J.H. Dieterich, Progressive failure on the North Anatolian fault since 1939 by earthquake stress triggering, *Geophys. J. Int.*, **128**, 594-604, 1997; T. Parsons, S. Toda, R. Stein, A. Barka, and J.H. Dieterich, Heightened odds of large earthquakes near Istanbul; An interaction-based probability calculation, *Science*, **288**, 661-665, 2000.

41. S. Toda, R.S. Stein, P.A. Reasenberg, J.H. Dieterich, and A. Yoshida, Stress transferred by the 1995 $M_W = 6.9$ Kobe, Japan, shock: Effect on aftershocks and future earthquake probabilities, *J. Geophys. Res.*, **103**, 24,543-24,565, 1998.

42. L.R. Sykes and S. Jaumé (Seismic activity on neighboring faults as a long-term precursor to large earthquakes in the San Francisco Bay Area, *Nature*, **348**, 595-599, 1990) found that large earthquakes are preceded by a cluster of intermediate-sized events (within two magnitude units of the main shock) in a large surrounding region. L. Knopoff, T. Levshina, V.I. Keilis-Borok, and C. Mattoni (Increased long-range intermediate-magnitude earthquake activity prior to strong earthquakes in California, *J. Geophys. Res.*, **101**, 5779-5796, 1996) showed that all 11 earthquakes in California since 1941 with magnitudes greater than 6.8 were preceded by an increase in the rate of occurrence of earthquakes with magnitudes greater than 5.1 in the appropriate tectonic domain. For a recent summary and earlier references, see S.C. Jaumé and L.R. Sykes, Evolving towards a critical point; A review of accelerating seismic moment/energy release prior to large and great earthquakes, *Pure Appl. Geophys.*, **155**, 279-306, 1999.

43. C.G. Bufe and D.J. Varnes, Predictive modeling of the seismic cycle of the greater San Francisco Bay region, *J. Geophys. Res.*, **98**, 9871-9883, 1993; C.G. Bufe, S.P. Nishenko, and D.J. Varnes, Seismicity trends and potential for large earthquakes in the Alaska-Aleutian region, *Pure Appl. Geophys.*, **142**, 83-99, 1994.

44. D. Sornette and C.G. Sammis, Complex critical exponents from renormalization group theory of earthquakes: Implications for earthquake predictions, *J. Phys. I.*, **5**, 607-619, 1995; C.G. Sammis, D. Sornette, and H. Saleur, Complexity and earthquake forecasting, *Reduction and Predictability of Natural Disasters, SFI Studies in the Sciences of Complexity*, vol.

XXV, J.B. Rundle, W. Klein, and D.L. Turcotte, eds., Addison-Wesley, Reading, Mass., pp. 143-156, 1996.

45. R.J. Geller, D.D. Jackson, Y.Y. Kagan, and F. Mulargia, Earthquakes cannot be predicted, *Science*, **275**, 1616-1617, 1997.

46. C. Sammis and S. Smith, Seismic cycles and the evolution of stress correlation in cellular automation models of finite fault networks, *Pure Appl. Geophys.*, **155**, 307-334, 1999; H. Saleur, C.G. Sammis, and D. Sornette, Renormalization group theory of earthquakes, *Nonlinear Processes Geophys.*, **3**, 102-109, 1996; G. Zoller, S. Hainzl, and J. Kurths, Observation of growing correlation length as an indicator for critical point behavior prior to large earthquakes, *J. Geophys. Res.*, **106**, 2167-2175, 2001.

47. E.G. Triep and L.R. Sykes, Frequency of occurrence of moderate to great earthquakes in intracontinental regions: Implications for change in stress, earthquake prediction, and hazard assessments, *J. Geophys. Res.*, **102**, 9923-9948, 1997; D.D. Bowman, G. Ouillon, C.G. Sammis, A. Sornette, and D. Sornette, An observational test of the critical earthquake concept, *J. Geophys. Res.*, **103**, 24,359-24,372, 1998; R. Robinson, A test of the precursory accelerating moment release model on some recent New Zealand earthquakes, *Geophys. J. Int.*, **140**, 568-576, 2000; D.D. Bowman and G.C.P. King, Accelerating seismicity and stress accumulation before large earthquakes, *Geophys. Res. Lett.*, **28**, 4039-4042, 2001.

48. Y. Ben-Zion, K. Dahmen, V. Lyakhovsky, D. Ertas, and A. Agnon, Self-driven mode switching of earthquake activity on a fault system, *Earth Planet. Sci. Lett.*, **172**, 11-21, 1999; J.B. Rundle, K.F. Tiampo, W. Klein, and J.S. Sá Martins, Self-organization in leaky threshold systems: The influence of near-mean field dynamics and its implications for earthquakes, neurobiology, and forecasting, *Proc. Nat. Acad. Sci.*, **99**, 2524-2521, 2002.

49. R.A. Harris, Introduction to special section: Stress triggers, stress shadows, and implications for seismic hazard, *J. Geophys. Res.*, **103**, 24,347-24,358, 1998.

50. N. Lapusta, J.R. Rice, Y. Ben-Zion, and G. Zheng, Elastodynamic analysis for slow tectonic loading with spontaneous rupture episodes on faults with rate- and state-dependent friction, *J. Geophys. Res.*, **105**, 23,765-23,789, 2000.

51. Y.-G. Li, K. Aki, D. Adams, A. Hasemi, and W.H.K. Lee, Seismic guided waves trapped in the fault zone of the Landers, California, earthquake of 1992, *J. Geophys. Res.*, **99**, 11,705-11,722, 1994; Y.-G. Li, K. Aki, J.E. Vidale, and F. Xu, Shallow structure of the Landers fault zone from explosion-generated trapped waves, *J. Geophys. Res.*, **104**, 20,257-20,275, 1999.

52. M.J. Unsworth, P.E. Malin, G.D. Egbert, and J.R. Booker, Internal structure of the San Andreas fault at Parkfield, California, *Geology*, **25**, 359-362, 1997; M.J. Unsworth, G. Egbert, and J. Booker, High-resolution electromagnetic imaging of the San Andreas fault in central California, *J. Geophys. Res.*, **104**, 1131-1150, 1999.

53. F.M. Chester and J.S. Logan, Composite planar fabric of gouge from the Punchbowl fault, California, *J. Struct. Geol.*, **9**, 631-634, 1987; W.L. Power and T.E. Tullis, Euclidian and fractal models for the description of rock surface roughness, *J. Geophys. Res.*, **96**, 415-424, 1991; F.M. Chester, J.P. Evans, and R.L. Biegel, Internal structure and weakening mechanisms of the San Andreas fault, *J. Geophys. Res.*, **98**, 771-786, 1993; J.P. Evans and F.M. Chester, Fluid-rock interactions in faults of the San Andreas system: Inferences from San Gabriel fault rock geochemistry and microstructures, *J. Geophys. Res.*, **100**, 13,007-13,020, 1995; F.M. Chester and J.S. Chester, Ultracataclasite structure and friction processes of the Punchbowl fault, San Andreas system, California, *Tectonophysics*, **295**, 199-221, 1998; J.P. Evans, Z.K. Shipton, M.A. Pachell, S.J. Lim, and K. Robeson, The structure and composition of exhumed faults, and their implication for seismic processes, in *Proceedings of the 3rd Conference on Tectonic Problems of the San Andreas System*, G. Bokelmann and R. Kovach, eds., Stanford University Publications in Geological Science 21, Stanford, pp. 67-81, 2000.

54. S. Wesnousky, Seismicity as a function of cumulative geologic offset: Some observa-

tions from southern California, *Bull. Seis. Soc. Am.*, **80**, 1374-1381, 1990; S.G. Wesnousky, The Gutenberg-Richter or characteristic earthquake distribution: Which is it? *Bull. Seis. Soc. Am.*, **84**, 1940-1959, 1994; M.W. Stirling, S.G. Wesnousky, and K. Shimazaki, Fault trace complexity, cumulative slip, and the shape of the magnitude-frequency distribution for strike-slip faults; A global survey, *Geophys. J. Int.*, **124**, 833-868, 1996.

55. F.M. Chester, J.P. Evans, and R.L. Biegel, Internal structure and weakening mechanisms of the San Andreas fault, *J. Geophys. Res.*, **98**, 771-786, 1993; F.M. Chester and J.S. Chester, Ultracataclasite structure and friction processes of the Punchbowl fault, San Andreas system, California, *Tectonophysics*, **295**, 199-221, 1998; J.P. Evans, Z.K. Shipton, M.A. Pachell, S.J. Lim, and K. Robeson, The structure and composition of exhumed faults, and their implication for seismic processes, in *Proceedings of the 3rd Conference on Tectonic Problems of the San Andreas System*, G. Bokelmann and R. Kovach, eds., Stanford University Publications in Geological Science 21, Stanford, pp. 67-81, 2000.

56. R. Harris and S.M. Day, Effects of a low velocity zone on a dynamic rupture, *Bull. Seis. Soc. Am.*, **87**, 1267-1280, 1997.

57. Crack propagation is limited by the Rayleigh speed for mode II slip and the shear speed for mode III slip; see, for example, L.B. Freund, *Dynamic Fracture Mechanics*, Cambridge University Press, Cambridge, U.K., 563 pp., 1990.

58. A.N.B. Poliakov, R. Dmowska, and J.R. Rice, Dynamic shear rupture interactions with fault bends and off-axis secondary faulting, *J. Geophys. Res.*, in press, 2002.

59. Such interpretation had been mentioned earlier by R.M. Stesky (Mechanisms of high temperature frictional sliding in Westerly granite, *Canadian J. Earth Sci.*, **15**, 361-375, 1977) and was implicit in F.M. Chester (Effects of temperature on friction: Constitutive equations and experiments with fault gouge, *J. Geophys. Res.*, **99**, 7247-7261, 1994). This interpretation was suggested explicitly, in the rate and state framework, in the following papers: F. Heslot, T. Baumberger, B. Perrin, B. Caroli, and C. Caroli, Creep, stick-slip, and dry friction dynamics: Experiments and a heuristic model, *Phys. Rev. E*, **49**, 4973-4988, 1994; Y. Bréchet and Y. Estrin, The effect of strain rate sensitivity on dynamic friction of metals, *Scripta Met. Mat.*, **30**, 1449-1454, 1994; T. Baumberger, Contact dynamics and friction at a solid-solid interface: Material versus statistical aspects, *Solid State Comm.*, **102**, 175-185, 1997; N.H. Sleep, Application of a unified rate and state friction theory to the mechanics of fault zones with strain localization, *J. Geophys. Res.*, **102**, 2875-2895, 1997; B.N.J. Persson, *Sliding Friction, Physical Principles and Applications*, Springer-Verlag, Berlin, 462 pp., 1998; T. Baumberger, P. Berthoud, and C. Caroli, Physical analysis of the state- and rate-dependent friction law. II. Dynamic friction, *Phys. Rev. B*, **60**, 3928-3939, 1999; N. Lapusta, J.R. Rice, Y. Ben-Zion, and G. Zheng, Elastodynamic analysis for slow tectonic loading with spontaneous rupture episodes on faults with rate- and state-dependent friction, *J. Geophys. Res.*, **105**, 23,765-23,789, 2000; J.R. Rice, N. Lapusta, and K. Ranjith, Rate and state dependent friction and the stability of sliding between elastically deformable solids, *J. Mech. Phys. Solids*, **49**, 1865-1898, 2001; M. Nakatani, Conceptual and physical clarification of rate and state friction: Frictional sliding as a thermally activated rheology, *J. Geophys. Res.*, **106**, 13,347-13,380, 2001.

60. N. Lapusta, J.R. Rice, Y. Ben-Zion, and G. Zheng, Elastodynamic analysis for slow tectonic loading with spontaneous rupture episodes on faults with rate- and state-dependent friction, *J. Geophys. Res.*, **105**, 23,765-23,789, 2000.

61. J.H. Dieterich and B.D. Kilgore, Direct observation of frictional contacts: New insights for state-dependent properties, *Pure Appl. Geophys.*, **143**, 283-302, 1994; J.H. Dieterich and B.D. Kilgore, Imaging surface contacts: Power law contact distributions and contact stresses in quartz, calcite, glass and acrylic plastic, *Tectonophysics*, **256**, 219-239, 1996.

62. Y. Bréchet and Y. Estrin, The effect of strain rate sensitivity on the dynamic friction of metals, *Scripta Metallurgica et Materialia*, **30**, 1449-1454, 1994.

63. L.A. Reinen, J.D. Weeks, and T.E. Tullis, The frictional behavior of serpentine: Impli-

cations for aseismic creep on shallow crustal faults, *Geophys. Res. Lett.*, **18**, 1921-1924, 1991; L.A. Reinen, T.E. Tullis, and J.D. Weeks, Two-mechanism model for frictional sliding of serpentine, *Geophys. Res. Lett.*, **19**, 1535-1538, 1992. When constitutive parameters of the two mechanisms are fit to experiments on serpentine, Reinen and others infer stable velocity strengthening at slip rates up to about 0.1 millimeter per second (3000 millimeters per year), but unstable velocity weakening at faster rates. This provides a good candidate for a region of fault zone that normally creeps aseismically in response to tectonic forcing but, when subjected to high stress by rupture of a neighboring part of the fault, can join in slipping as part of a large earthquake (Y. Ben-Zion and J.R. Rice, Dynamic simulations of slip on a smooth fault in an elastic solid, *J. Geophys. Res.*, **102**, 17,771-17,784, 1997).

64. M.L. Blanpied, D.A. Lockner, and J.D. Byerlee, Fault stability inferred from granite sliding experiments at hydrothermal conditions, *Geophys. Res. Lett.*, **18**, 609-612, 1991; M.L. Blanpied, D.A. Lockner, and J.D. Byerlee, Frictional slip of granite at hydrothermal conditions, *J. Geophys. Res.*, **100**, 13,045-13,064, 1995; F.M. Chester and N. Higgs, Multimechanism friction constitutive model for ultrafine quartz gouge at hypocentral conditions, *J. Geophys. Res.*, **97**, 1859-1870, 1992; F.M. Chester, Effects of temperature on friction: Constitutive equations and experiments with fault gouge, *J. Geophys. Res.*, **99**, 7247-7261, 1994.

65. F.M. Chester, A rheologic model for wet crust applied to strike-slip faults, *J. Geophys. Res.*, **100**, 13,033-13,044, 1995.

66. A. Tsutsumi and T. Shimamoto, High-velocity frictional properties of gabbro, *Geophys. Res. Lett.*, **24**, 699-702, 1997; J.G. Spray, Viscosity determinations of some frictionally generated silicate melts; Implications for fault zone rheology at high strain rates, *J. Geophys. Res.*, **98**, 8053-8068, 1993; J.G. Spray, Pseudotachylyte controversy; Fact or fiction? *Geology*, **23**, 1119-1122, 1995.

67. S.C. Lim and M.F. Ashby, Wear mechanism maps, *Acta Metallurgica*, **35**, 1-24, 1987; S.C. Lim, M.F. Ashby, and J.F. Brunton, The effect of sliding conditions on the dry friction of metals, *Acta Metallurgica*, **37**, 767-772, 1989; J.R. Rice, Flash heating at asperity contacts and rate-dependent friction, *Eos, Trans. Am. Geophys. Union*, **80**, F681, 1999.

68. F.M. Chester and J.S. Chester, Ultracataclasite structure and friction processes of the Punchbowl fault, San Andreas system, California, *Tectonophysics*, **295**, 199-221, 1998.

69. R.H. Sibson, Generation of pseudotachylyte by ancient seismic faulting, *Geophys. J. R. Astron. Soc.*, **43**, 775-794, 1975; J.G. Spray, Pseudotachylyte controversy; Fact or fiction? *Geology*, **23**, 1119-1122, 1995; S.K. Ray, Transformation of cataclastically deformed rocks to pseudotachylyte by pervasion of frictional melt; Inferences from clast-size analysis, *Tectonophysics*, **301**, 283-304, 1999.

70. G. Zheng and J.R. Rice, Conditions under which velocity-weakening friction allows a self-healing versus a cracklike mode of rupture, *Bull. Seis. Soc. Am.*, **88**, 1466-1483, 1998; S.B. Nielsen and J.M. Carlson, Rupture pulse characterization; Self-healing, self-similar, expanding solutions in a continuum model of fault dynamics, *Bull. Seis. Soc. Am.*, **90**, 1480-1497, 2000.

71. A. Cochard and R. Madariaga, Dynamic faulting under rate-dependent friction, *Pure Appl. Geophys.*, **142**, 419-445, 1994; A. Cochard and R. Madariaga, Complexity of seismicity due to highly rate dependent friction, *J. Geophys. Res.*, **101**, 25,321-25,336, 1996; G. Perrin, J.R. Rice, and G. Zheng, Self-healing slip pulse on a frictional surface, *J. Mech. Phys. Solids*, **43**, 1461-1495, 1995; N.M. Beeler and T.E. Tullis, Self-healing slip pulse in dynamic rupture models due to velocity-dependent strength, *Bull. Seis. Soc. Am.*, **86**, 1130-1148, 1996; R. Madariaga and A. Cochard, Dynamic friction and the origin of the complexity of earthquake sources, *Proc. Natl. Acad. Sci.*, **93**, 3819-3824, 1996.

72. T.H. Heaton, Evidence for and implications of self-healing pulses of slip in earthquake rupture, *Phys. Earth Planet. Int.*, **64**, 1-20, 1990.

73. S.M. Day, Three-dimensional finite difference simulation of fault dynamics: Rectan-

gular faults with fixed rupture velocity, *Bull. Seis. Soc. Am.*, **72**, 705-727, 1982; E. Johnson, On the initiation of unidirectional slip, *Geophys. J. Int.*, **101**, 125-132, 1990; G.C. Beroza and T. Mikumo, Short slip duration in dynamic rupture in the presence of heterogeneous fault properties, *J. Geophys. Res.*, **101**, 22,449-22,460, 1996.

74. J. Weertman, Unstable slippage across a fault that separates elastic media of different elastic constants, *J. Geophys. Res.*, **85**, 1455-1461, 1980; D.J. Andrews and Y. Ben-Zion, Wrinkle-like slip pulse on a fault between different materials, *J. Geophys. Res.*, **102**, 553-571, 1997; Y. Ben-Zion and D.J. Andrews, Properties and implications of dynamic rupture along a material interface, *Bull. Seis. Soc. Am.*, **88**, 1085-1094, 1998.

75. M.D. Zoback, M.L. Zoback, V.S. Mount, J. Suppe, J.P. Eaton, J.H. Healy, D. Oppenheimer, P. Reasenberg, L. Jones, C.B. Raleigh, I.G. Wong, O. Scotti, and C. Wentworth, New evidence of the state of stress on the San Andreas fault system, *Science*, **238**, 1105-1111, 1987.

76. C.H. Scholz, Evidence for a strong San Andreas fault, *Geology*, **28**, 163-166, 2000.

77. C. Morrow, B. Radney, and J. Byerlee, Frictional strength and the effective pressure law of montmorillonite and illite clays, in *Fault Mechanics and Transport Properties of Rocks*, B. Evans and T.-F. Wong, eds., Academic Press, London, pp. 69-88, 1992.

78. S.H. Hickman, R.H. Sibson, and R.L. Bruhn, Introduction to special session: Mechanical involvement of fluids in faulting, *J. Geophys. Res.*, **100**, 12,831-12,840, 1995; M. Brudy, M.D. Zoback, K. Fuchs, F. Rummel, and J. Baumgärtner, Estimation of the complete stress tensor to 8 km depth in the KTB scientific drill holes: Implications for crustal strength, *J. Geophys. Res.*, **102**, 18,453-18,475, 1997.

79. P. Mora and D. Place, Numerical simulation of earthquake faults with gouge: Toward a comprehensive explanation for the heat flow paradox, *J. Geophys. Res.*, **103**, 21,067-21,089, 1998.

80. G.G. Adams, Steady sliding of two elastic half-spaces with friction reduction due to interface stick-slip, *J. Appl. Mech.*, **65**, 470-475, 1998. A. Cochard and J.R. Rice (Fault rupture between dissimilar materials: Ill-posedness, regularization and slip-pulse response, *J. Geophys. Res.*, **105**, 25,891-25,907, 2000) found that the simplest model of the process, incorporating continuum elastodynamics in the solids and Coulomb friction at the interface, was mathematically ill-posed (i.e., had no solution). They were able to show that a finite-time response of friction to rapid change of normal stress, revealed in oblique shock impact experiments (V. Prakash and R.J. Clifton, Pressure-shear plate impact measurement of dynamic friction for high speed machining applications, in *Proceedings of VII International Congress on Experimental Mechanics*, Society of Experimental Mechanics, Bethel, Conn., 556-564, 1992; V. Prakash, Frictional response of sliding interfaces subjected to time varying normal pressures, *Tribology, Trans. ASME*, **120**, 97-102, 1998), regularized the problem.

81. J.N. Brune, S. Brown, and P.A. Johnson, Rupture mechanism and interface separation in foam rubber models of earthquakes: A possible solution to the heat flow paradox and the paradox of large overthrusts, *Tectonophysics*, **218**, 59-67, 1993; A. Anooshehpoor and J.N. Brune, Frictional heat generation and seismic radiation in a foam rubber model of earthquakes: Faulting, friction, and earthquake mechanics. Part 1, *Pure Appl. Geophys.*, **142**, 735-747, 1994.

82. J.H. Dieterich, Modeling of rock friction, 2. Simulation of preseismic slip, *J. Geophys. Res.*, **84**, 2169-2175, 1979; M. Ohnaka, Y. Kuwahara, K. Yamamoto, and T. Hirasawa, Dynamic breakdown processes and the generating mechanism for high-frequency elastic radiation during stick-slip instabilities, in *Earthquake Source Mechanics*, S. Das, J. Boatwright, and C.H. Scholz, eds., American Geophysical Union, Maurice Ewing Series 6, Geophysical Monograph 37, Washington, D.C., pp. 13-24, 1986.

83. D.J. Andrews, Rupture velocity of plane strain shear cracks, *J. Geophys. Res.*, **81**, 5679-5687, 1976; J.H. Dieterich, A model for the nucleation of earthquake slip, in *Earthquake Source Mechanics*, S. Das, J. Boatwright, and C.H. Scholz, eds., American Geophysical Union,

Maurice Ewing Series 6, Geophysical Monograph 37, Washington, D.C., pp. 37-47, 1986; S. Das, and C.H. Scholz, Theory of time-dependent rupture in the earth, *J. Geophys. Res.*, **86**, 6039-6051, 1981.

84. An upper bound of approximately 1 percent was found by M.J.S. Johnston, A.T. Linde, M.T. Gladwin, and R.D. Borcherdt (Fault failure with moderate earthquakes, *Tectonophysics*, **144**, 189-206, 1987) in their study of seven moderate earthquakes in California. In the case of the 1989 Loma Prieta earthquake, the constraint from borehole strainmeter observations is that any precursory slip could not have accounted for more than about 0.3 percent of the co-seismic moment. This corresponds to a bound of about $M \leq 5.3$ for an aseismic precursor found by M.J.S. Johnston, A.T. Linde, and M.T. Gladwin (Near-field high resolution strain measurements prior to the October 18, 1989, Loma Prieta M_S 7.1 earthquake, *Geophys. Res. Lett.*, **17**, 1777-1780, 1990).

85. R.E. Abercrombie and P.C. Leary (Source parameters of small earthquakes recorded at 2.5 km depth, Cajon Pass, southern California: Implications for earthquake scaling, *Geophys. Res. Lett.*, **20**, 1511-1514, 1993) inferred these small source dimensions from the corner frequencies of the radiated waves, which they were able to measure because the waves recorded by the borehole instrument did not suffer propagation through the highly attenuative material near the Earth's surface.

86. E. Richardson and T.H. Jordan, Seismicity in deep gold mines of South Africa: Implications for tectonic earthquakes, *Bull. Seis. Soc. Am.*, **92**, 1766-1782, 2002.

87. A formalism for detecting slow precursors from the amplitude and phase-shift spectra of low-frequency seismic waves was developed by T.H. Jordan (Far-field detection of slow precursors to fast seismic ruptures, *Geophys. Res. Lett.*, **18**, 2019-2022, 1991); J.J. McGuire, P.F. Ihmlé, and T.H. Jordan (Time-domain observations of a slow precursor to the 1994 Romanch transform earthquake, *Science*, **274**, 82-85, 1996); and J.J. McGuire and T.H. Jordan (Further evidence for the compound nature of slow earthquakes; The Prince Edward Island earthquake of April 28, 1997, *J. Geophys. Res.*, **105**, 7819-7827, 2000).

88. P.F. Ihmlé, P. Harabaglia, and T.H. Jordan (Teleseismic detection of a slow precursor to the great 1989 Macquarie Ridge earthquake, *Science*, **261**, 177-183, 1993) report observations of a large slow precursor with a moment magnitude of 7.6 and a duration of more than 100 seconds for the 1989 Macquarie Ridge earthquake. This observation is controversial, however, because the slow precursor was detected as an anomaly by indirect measurements of the phase delay in the frequency domain. S. Kedar, S. Watada, and T. Tanimoto (The 1989 Macquarie Ridge earthquake; Seismic moment estimation from long-period free oscillations, *J. Geophys. Res.*, **99**, 17,893-17,908, 1994) studied isolated normal-mode peaks for the Macquarie Ridge earthquake that were selected to be free of strong coupling effects. They found large amplitude anomalies, suggesting that the earthquake was about 50 percent larger than indicated by surface-wave analysis; however, they argued that the phase and possibly the amplitude anomalies might be attributable to finite-source effects, rather than to a slow precursor. This seems unlikely, however; an earlier study of the radial free oscillation singlet ${}_0S_0$ by J. Park (Radial mode observations from the 5/23/89 Macquarie Ridge earthquake, *Geophys. Res. Lett.*, **17**, 1005-1009, 1990) also showed a phase advance that could be attributed to a slow precursor, but not to either mode-coupling or finite-source effects. P. Ihmlé and T.H. Jordan (Teleseismic search for slow precursors to large earthquakes, *Science*, **266**, 1547-1551, 1994) subsequently analyzed the low-frequency spectra of 107 large, shallow-focus earthquakes and found that of the 20 events with detectable slow precursors, 19 were slow earthquakes associated with the oceanic ridge-transform system. The oceanic ridge-transform system of plate boundaries appears to be rich in these types of earthquake anomalies, perhaps owing to the presence of elevated temperatures and hydrated upper-mantle rocks (serpentinized peridotites) at shallow depths.

89. Observations of a slow initial phase that scales with earthquake size for micro-earthquakes in Japan (2.7 < M < –0.7) have been made in Japan by Y. Iio (Slow initial phase of the *P*-wave velocity pulse generated by microearthquakes, *Geophys. Res. Lett.*, **19**, 477-480, 1992; Observations of the slow initial phase generated by microearthquakes: Implications for earthquake nucleation and propagation, *J. Geophys. Res.*, **100**, 15,333-15,349, 1995). Although the waveforms were likely to have been distorted by attenuation, it was argued that the slow initial phase seen in these earthquakes was a source effect and could be explained in terms of a slip-weakening model. A study of the Ridgecrest, California, sequence (J. Mori and H. Kanamori, Initial rupture of earthquakes in the 1995 Ridgecrest, California sequence, *Geophys. Res. Lett.*, **23**, 2437-2440, 1996) found no evidence for a slow initial phase at somewhat larger magnitudes.

90. W.L. Ellsworth and G.C. Beroza (Seismic evidence for an earthquake nucleation phase, *Science*, **268**, 851-855, 1995) found this behavior for the set of about 50 earthquakes occurring in a number of tectonic environments. They put forward two possible models to explain their observations: the pre-slip model and the cascade model. In the pre-slip model, the seismic nucleation phase represents the very last stages of an otherwise aseismic nucleation process described above. In the cascade model (J.N. Brune, Implications of earthquake triggering and rupture propagation for earthquake prediction based on premonitory phenomena, *J. Geophys. Res.*, **84**, 2195-2198, 1979; Y. Fukao and M. Furumoto, Hierarchy in earthquake size distribution, *Phys. Earth Planet. Int.*, **37**, 149-168, 1985), the seismic nucleation phase can be viewed as a succession of small events leading from the first small subevent to the largest subevent in the earthquake. It is difficult to test whether this scaling applies to the aseismic part of the nucleation process because the preceding signals would be undetectable using seismic waves. The seismic moment of the seismic nucleation phase was found to average about 0.5 percent of the mainshock seismic moment. Thus, the aseismic part of the process, which is likely much smaller, would be undetectable using the borehole strainmeter observations currently available. In contrast to the findings for smaller earthquakes and large oceanic earthquakes, the seismic nucleation phase does not appear deficient in high-frequency radiation; rather, it is weak or erratic in that the growth of the earthquake is initially hesitant.

91. Foreshocks are small earthquakes occurring before but spatially near the ensuing large earthquake's hypocenter. Foreshocks are otherwise similar to other earthquakes, having no obvious properties that would allow them to be identified as foreshocks before the subsequent large earthquake occurs. Using various definitions of what constitutes a foreshock, studies have found that earthquakes with foreshocks include 42 percent of large earthquakes worldwide (L.M. Jones and P. Molnar, Some characteristics of foreshocks and their possible relation to earthquake prediction and premonitory slip on faults, *J. Geophys. Res.*, **84**, 3596-3608, 1979); 35 percent for M ≥ 5.0 earthquakes in the San Andreas fault system (L.M. Jones, Foreshocks (1966-1980) in the San Andreas system, California, *Bull. Seis. Soc. Am.*, **74**, 1361-1380, 1984); and 44 percent of M ≥ 5.0 earthquakes in the western United States (R. Abercrombie and J. Mori, Occurrence patterns of foreshocks to large earthquakes in the western United States, *Nature*, **381**, 303-307, 1996). In many cases the mainshocks were preceded by only a single foreshock. If one defines a foreshock sequence as five or more events preceding a mainshock in the space and time window used by Abercrombie and Mori (op. cit., 1996), then only 19 percent of the mainshocks in their population were preceded by foreshock sequences.

92. R. Abercrombie and J. Mori, Occurrence patterns of foreshocks to large earthquakes in the western United States, *Nature*, **381**, 303-307, 1996.

93. By studying foreshock sequences from all earthquakes in their study, L.M. Jones and P. Molnar (Some characteristics of foreshocks and their possible relation to earthquake prediction and premonitory slip on faults, *J. Geophys. Res.*, **84**, 3596-3608, 1979) found that

the temporal growth of foreshock rates increased inversely with the approach time of the mainshock. This temporal behavior is consistent with a model of foreshocks as an accelerating sequential failure of strongly interacting asperities under static fatigue.

94. D.A. Dodge, G.C. Beroza, and W.L. Ellsworth (Detailed observations of California foreshock sequences: Implications for the earthquake initiation process, *J. Geophys. Res.*, **101**, 22,371-22,392, 1996) studied the mechanics of foreshock-mainshock interaction for six foreshock sequences by obtaining very accurate relative event locations through a combination of waveform cross-correlation to estimate improved relative arrival times and joint-hypocentral determination with station corrections to obtain highly precise earthquake locations. From these they found that the foreshocks were unlikely to have triggered the mainshock through shear, normal, or mean stress changes. The same study also found that the dimension of the foreshock zones followed the same scaling, inferred independently for the seismic nucleation phase, suggesting that the two phenomena may be related.

95. In its engineering context, fracture usually means the creation of discontinuities in a continuous medium due to stress—the formation of cracks. Under ordinary conditions, all real materials have preexisting cracks. Stresses tend to concentrate around the tip of these cracks, and favorably oriented cracks become unstable and grow. Actual fracturing occurs in a process zone at the edge of a crack as it propagates into the medium. Ahead of the fracture front, the medium is continuous and its response to stress is described by a bulk rheology, which can be elastic, viscoelastic, or plastic. Behind the front, the traction across the crack face is described by a friction law. In an ideally brittle material, the bulk rheology is elastic, the cracks are cohesionless, and transition from the continuous to the fractured state occurs instantaneously (i.e., the process zone has zero width). For a planar crack in an ideally brittle material, the stresses are proportional to $K/\sqrt{r}$, where r is the distance from the crack tip and K is a stress intensity factor that depends on the mode of cracking and increases with the square root of the crack length. A crack will grow if the stress intensity factor exceeds some critical value K_c, whose square scales with the specific energy needed to create new fracture area. Engineers distinguish among three modes of crack propagation: I. tensile opening, in which crack wall displacement is normal to the crack; II. in-plane shearing, in which the displacement is in the plane of the crack and perpendicular to the crack edge; and III. antiplane shearing, in which the displacement is in the plane of the crack and parallel to the crack edge. Mode II corresponds to dip-slip faulting, and Mode III to strike-slip faulting. A comprehensive discussion of fracture mechanics is given by B.R. Lawn, *Fracture of Brittle Solids*, 2nd ed., Cambridge University Press, Cambridge, U.K., 378 pp., 1993.

96. B.V. Kostrov and S. Das, *Principles of Earthquake Source Mechanics*, Cambridge University Press, Cambridge, U.K., 286 pp., 1988.

97. J.N. Brune, Tectonic stress and the spectra of seismic shear waves from earthquakes, *J. Geophys. Res.*, **75**, 4997-5009, 1970.

98. D.S. Dugdale, Yielding of steel sheets containing slits, *J. Mech. Phys. Solids*, **8**, 100-115, 1960; G.I. Barenblatt, The mathematical theory of equilibrium cracks in brittle fracture, *Adv. Appl. Mech.*, **7**, 55-80, 1962.

99. B. Gutenberg (Magnitude determination for larger Kern County shocks, 1952; Effects of station azimuths and calculation methods, *Calif. Div. Mines and Geol. Bull.*, **171**, 171-175, 1955) determined the direction of rupture propagation in the 1952 Kern County earthquake from the azimuthal asymmetry of the surface-wave radiation. Rupture velocity and fault length were first determined for the 1960 Chile earthquake using phase shifts of normal modes (H. Benioff, F. Press, and S. Smith, Excitation of free oscillations of the Earth by earthquakes, *J. Geophys. Res.*, **66**, 605-619, 1961). Subsequent work determined the same properties using spectral ratios of surface waves (A. Ben-Menahem and M.N. Toksoz, Source mechanism from the spectra of long period seismic surface waves, 1. The Mongolian earthquake of December 4, 1957, *J. Geophys. Res.*, **67**, 1943-1955, 1962), the timing and location of

subevents within a mainshock rupture (M. Wyss and J.N. Brune, The Alaska earthquake of 28 March 1964: A complex multiple rupture, *Bull. Seis. Soc. Am.*, **57**, 1017-1023, 1967), and the duration of the far-field *P*-wave pulse (G. Bollinger, Determination of earthquake fault parameters from long period *P* waves, *J. Geophys. Res.*, **73**, 785-807, 1968).

100. For example, average rupture velocities in this range were reported for: the 1966 Parkfield earthquake (K. Aki, Seismic displacements near a fault, *J. Geophys. Res.*, **73**, 5959-5976, 1968), the 1971 San Fernando earthquake (M. Bouchon, A dynamic source model for the San Fernando earthquake, *Bull. Seis. Soc. Am.*, **68**, 1555-1576, 1978), the 1979 Imperial Valley earthquake (A.H. Olson, and R.J. Apsel, Finite faults and inverse theory with applications to the 1979 Imperial Valley earthquake, *Bull. Seis. Soc. Am.*, **72**, 1969-2001, 1982), the 1984 Morgan Hill earthquake (S.H. Hartzell and T.H. Heaton, Rupture history of the 1984 Morgan Hill, California, earthquake from the inversion of strong motion records, *Bull. Seis. Soc. Am.*, **76**, 649-674, 1986), the 1989 Loma Prieta earthquake (G. Beroza, Near-source modeling of the Loma Prieta earthquake: Evidence for heterogeneous slip and implications for earthquake hazard, *Bull. Seis. Soc. Am.*, **81**, 1603-1621, 1991), and the 1994 Northridge earthquake (D.J. Wald, T.H. Heaton, and K.W. Hudnut, The slip history of the 1994 Northridge, California, earthquake determined from strong-motion, teleseismic, GPS, and leveling data, *Bull. Seis. Soc. Am.*, **86**, S49-S70, 1996).

101. A. Frankel, J.B. Fletcher, F.L. Vernon, L.C. Haar, J. Berger, T.C. Hanks, and J.N. Brune, Rupture characteristics and tomographic source imaging of M_L approximately 3 earthquakes near Anza, Southern California, *J. Geophys. Res.*, **91**, 12,633-12,650, 1986.

102. P.G. Somerville, N.F. Smith, R.W. Graves, and N.A. Abrahamson, Modification of empirical strong ground motion attenuation relations to include the amplitude and duration effects of rupture directivity, *Seis. Res. Lett.*, **68**, 199-222, 1997. This effect is particularly important since *S* waves have much larger amplitudes than *P* waves for shear dislocation sources and because rupture velocities are a much larger fraction of the *S*-wave velocity than of the *P*-wave velocity. Aside from angle-dependent terms (i.e., what is called the radiation pattern), the amplitude of the *S* wave relative to the *P* wave in the elastodynamic Green's function differs by the inverse cube of their respective velocities (K. Aki and P.G. Richards, *Quantitative Seismology: Theory and Methods*, vol 1., W.H. Freeman, San Francisco, p. 81, 1981). Since the Earth can be approximated as a Poisson solid at seismogenic depths, the ratio of the *P*-wave velocity a to the *S*-wave velocity b is $\sqrt{3}$, so that the ratio of the *S*-wave to the *P*-wave amplitude should be $3^{3/2}$ or approximately 5.1.

103. The stress concentration ahead of the crack tip for a dynamically propagating crack will be modified, relative to the stress concentration of a nonpropagating crack tip, by a factor that decreases monotonically to zero as a limiting velocity for that mode of fracture propagation is approached (L.B. Freund, *Dynamic Fracture Mechanics*, Cambridge University Press, Cambridge, U.K., 563 pp., 1990). In the case of antiplane failure, in which the sense of shear slip is perpendicular to the direction of rupture propagation, the limiting velocity is the shear-wave velocity. In the case of in-plane failure, in which the direction of shear slip is parallel to the direction of rupture propagation, the limiting velocity is the Rayleigh-wave velocity.

104. R.J. Archuleta (A faulting model for the 1979 Imperial Valley earthquake, *J. Geophys. Res.*, **89**, 4559-4585, 1984) found supershear rupture during the 1979 Imperial Valley, California, earthquake. G.C. Beroza and P. Spudich (Linearized inversion for fault rupture behavior: Application to the 1984 Morgan Hill, California, earthquake, *J. Geophys. Res.*, **93**, 6275-6296, 1988) found supershear rupture on a part of the fault that also suffered high slip during the 1984 Morgan Hill, California, earthquake. M. Bouchon, N. Toksoz, H. Karabulut, M.-P. Bouin, M. Dietrich, M. Aktar, and M. Edie (Seismic imaging of the Izmit rupture inferred from near-fault recordings, *Geophys. Res. Lett.*, **27**, 3013-3016, 2000) found supershear rupture over a substantial part of the fault for the 1999 Izmit, Turkey, earthquake.

105. Numerical simulations of a growing in-plane fracture under a slip-weakening fracture criterion (D.J. Andrews, Rupture velocity of plane strain shear cracks, *J. Geophys. Res.*, **81**, 5679-5687, 1976) revealed that rupture would first accelerate to the Rayleigh-wave velocity and then jump to supershear velocities. It was subsequently found by R. Burridge, G. Conn, and L.B. Freund (The stability of a rapid mode II shear crack with finite cohesive traction, *J. Geophys. Res.*, **84**, 2210-2222, 1979) that in-plane fracture could propagate stably at less than the Rayleigh-wave velocity, or between the *S*- and the *P*-wave velocities.

106. A.J. Rosakis, O. Samudrala, and D. Coker, Cracks faster than the shear wave speed, *Science*, **284**, 1337-1340, 1999.

107. A. Ben-Menahem and S.J. Singh, *Seismic Waves and Sources*, Springer Verlag, New York, p. 236, 1981.

108. H. Benioff and F. Press (Progress report on long period seismographs, *Geophys. J. R. Astron. Soc.*, **1**, 208-215, 1958) first recognized this signature for events recorded on recently improved long-period instrumentation. Examples of slow earthquakes include the 1946 Aleutian Islands earthquake (H. Kanamori, Mechanism of tsunami earthquakes, *Phys. Earth Planet. Int.*, **6**, 346-359, 1972); an event on June 6, 1960, on the Chilean transform (H. Kanamori and G.S. Stewart, A slow earthquake, *Phys. Earth Planet. Int.*, **18**, 167-179, 1979), and other transform fault earthquakes (H. Kanamori and G.S. Stewart, Mode of the strain release along the Gibbs fracture zone, Mid-Atlantic Ridge, *Phys. Earth Planet Int.*, **11**, 312-332, 1976; E.A. Okal and L.M. Stewart, Slow earthquakes along oceanic fracture zones: Evidence for asthenospheric flow away from hotspots? *Earth Planet. Sci. Lett.*, **57**, 75-87, 1982). Because the body-wave magnitude m_b and the surface-wave magnitude M_S are calculated at periods of approximately 1 and 20 seconds, respectively, slow earthquakes have substantially larger seismic moment than other earthquakes of comparable magnitude.

109. H. Kanamori (Mechanism of tsunami earthquakes, *Phys. Earth Planet. Int.*, **6**, 346-359, 1972) found that the 1946 Aleutian Islands and the 1896 Sanriku, Japan, earthquakes generated anomalously large tsunamis relative to their magnitude. See also the systematic work on tsunami versus earthquake size by K. Abe (Tsunami and mechanism of great earthquakes (with comment), *Phys. Earth Planet. Int.*, **7**, 143-153, 1973). Tsunami earthquakes have been observed most often in the accretionary prism of subduction zones. In addition to the 1896 Sanriku and 1946 Aleutian Islands events, tsunami earthquakes have been found in the Kurile Islands and Peru by A.M. Pelayo and D.A. Wiens (The November 20, 1960 Peru tsunami earthquake; Source mechanism of a slow event, *Geophys. Res. Lett.*, **17**, 661-664, 1990; Tsunami earthquakes: Slow thrust-faulting events in the accretionary wedge, *J. Geophys. Res.*, **97**, 15,321-15,337, 1992). Their slow nature has been attributed to relatively low velocity rupture propagation along the basal decollement of the accretionary prism (A.M. Pelayo and D.A. Wiens, Tsunami earthquakes: Slow thrust-faulting events in the accretionary wedge, *J. Geophys. Res.*, **97**, 15,321-15,337, 1992) or earthquake-triggered slumping. For example, a small tsunami occurred in Monterey Bay in the aftermath of the 1989 Loma Prieta, California, earthquake. This tsunami was more easily explained as having been generated by the slumping of sediments than by elastic deformation of the seafloor by the earthquake. (K.-F. Ma, K. Satake, and H. Kanamori, The origin of the tsunami excited by the 1989 Loma Prieta earthquake—Faulting or slumping, *Geophys. Res. Lett.*, **18**, 637-640, 1991). There are, however, examples of tsunami earthquakes in subduction zones that are inconsistent with both of these models. The 1992 Nicaragua earthquake, which caused a large and destructive tsunami with a local amplitude of 10 meters, occurred in an area with no accretionary prism (H. Kanamori and M. Kikuchi, The 1992 Nicaragua earthquake; A slow tsunami earthquake associated with subducted sediments, *Nature*, **361**, 714-716, 1993) and presumably was caused by slow deformation of the seafloor.

110. The 1896 Sanriku, Japan, tsunami earthquake was particularly damaging (A. Ima-

mura, *Theoretical and Applied Seismology*, Maruzen Company, Tokyo, 358 pp., 1937) as quoted by D. Myles (*The Great Waves*, McGraw Hill, New York, 206 pp., 1985). The tsunami reached its greatest height of 94.5 feet at Yoshihama. The official death toll was 27,122.

111. For example, a slow earthquake was identified in the Santa Maria Basin, California, by H. Kanamori and E. Hauksson (A slow earthquake in the Santa Maria Basin, California, *Bull. Seis. Soc. Am.*, **82**, 2087-2096, 1992).

112. P.F. Ihmlé and T.H. Jordan, Teleseismic search for slow precursors to large earthquakes, *Science*, **266**, 1547-1551, 1994.

113. The slow precursor to the 1960 Chile earthquake was documented by H. Kanamori and J.J. Cipar (Focal process of the great Chilean earthquake May 22, 1960, *Phys. Earth Planet Int.*, **9**, 128-136, 1974) from the Benioff strainmeter record. Further evidence to support their interpretation of this event was presented by I.L. Cifuentes and P.G. Silver (Low-frequency source characteristics of the great 1960 Chilean earthquake, *J. Geophys. Res.*, **94**, 643-663, 1989) who found the signature of a large, slow precursor in the spectra of the Earth's free oscillations.

114. K. Heki, S.I. Miyazake, and H. Tsuji (Silent fault slip following an interplate thrust earthquake at the Japan trench, *Nature*, **386**, 595-598, 1997) found that the M 7.6 Sanriku earthquake of the coast of Japan was followed by silent postseismic slip that released seismic moment comparable to that of the mainshock. H. Hirose, K. Hirahara, F. Kimata, N. Fujii, and S.-I. Miyazake (A slow thrust slip event following the two 1996 Hyuganada earthquakes beneath the Bungo Channel, southwest Japan, *Geophys. Res. Lett.*, **26**, 3237-3240, 1999) discovered a silent earthquake of about M 6.8 with a duration of approximately one year.

115. A.T. Linde, M. Gladwin, M.J.S. Johnston, R.L. Gwyther, and R.G. Bilham (A slow earthquake sequence on the San Andreas fault, *Nature*, **282**, 65-68, 1996) discovered an extremely slow earthquake sequence of approximately one-week duration in strainmeter and creepmeter records of the San Andreas fault near San Juan Bautista, California. The sequence was punctuated by several episodes of accelerated slip and a few microearthquakes. The moment magnitude of the aseismic component of the sequence was approximately 4.9.

116. G.C. Beroza and T.H. Jordan, Searching for slow and silent earthquakes using free oscillations, *J. Geophys. Res.*, **95**, 2485-2510, 1990. P.M. Shearer (Global seismic event detection using a matched filter on long-period seismograms, *J. Geophys. Res.*, **99**, 13,713-13,725, 1994) found that some slow undetected earthquakes occurred on oceanic transforms.

117. N. Suda, K. Nawa, and Y. Fukao, Earth's background free oscillations, *Science*, **279**, 2089-2091, 1998.

118. If the source of normal mode excitation is atmospheric, it may occur on other terrestrial planets and provide a convenient way to study their internal structure (T. Tanimoto, J. Um, K. Nishida, and N. Kobayashi, Earth's continuous oscillations observed on seismically quiet days, *Geophys. Res. Lett.*, **25**, 1553-1556, 1998).

119. R.L. Wesson, Dynamics of fault creep, *J. Geophys. Res.*, **93**, 8929-8951, 1988.

120. See, for example the review by W. Thatcher, Present-day crustal movements and the mechanics of cyclic deformation, in *San Andreas Fault System, California*, R.E. Wallace, ed., U S. Geological Survey Professional Paper 1515, Washington, D.C., pp. 189-205, 1990.

121. J.N. Brune, Seismic moment, seismicity and rate of slip along major fault zones, *J. Geophys. Res.*, **73**, 777-784, 1968; G.F. Davies and J.N. Brune, Regional and global fault slip rates from seismicity, *Nature*, **229**, 101-107, 1971.

122. T.H. Heaton, Evidence for and implications of self-healing pulses of slip in earthquake rupture, *Phys. Earth Planet. Int.*, **64**, 1-20, 1990.

123. K. Aki (Earthquake mechanism, *Tectonophysics*, **13**, 423-446, 1972) showed that teleseismic observations are not sensitive to the rise time.

124. Fault healing was first modeled by R. Madariaga (Dynamics of an expanding circular fault, *Bull. Seis. Soc. Am.*, **66**, 639-666, 1976) using a finite-difference approach. Healing of an expanding circular rupture was initiated simultaneously at the edges, and subsequently propagated inward toward the center of the fault at a large fraction of the shear-wave velocity. The fault did not stop slipping at interior points until information that the rupture had stopped propagated inward from the edges.

125. S.M. Day, Three-dimensional finite difference simulation of fault dynamics: Rectangular faults with fixed rupture velocity, *Bull. Seis. Soc. Am.*, **72**, 705-727, 1982.

126. T.H. Heaton, Evidence for and implications of self-healing pulses of slip in earthquake rupture, *Phys. Earth Planet. Int.*, **64**, 1-20, 1990.

127. Quasi-dynamic models are dynamic models constructed to reproduce the principal features of kinematic models, which were previously obtained by modeling seismic data. Examples of quasi-dynamic models with short rise times include the following: H. Quin, Dynamic stress drop and rupture dynamics of the October 15, 1979 Imperial Valley, California, earthquake, *Tectonophysics*, **175**, 93-117, 1990; T. Miyatake, Dynamic rupture processes of inland earthquakes in Japan; Weak and strong asperities, *Geophys. Res. Lett.*, **19**, 1041-1044, 1992; E. Fukuyama and T. Mikumo, Dynamic rupture analysis; Inversion for the source process of the 1990 Izu-Oshima, Japan, earthquake (M = 6.5), *J. Geophys. Res.*, **98**, 6529-6542, 1993; G.C. Beroza and T. Mikumo, Short slip duration in dynamic rupture in the presence of heterogeneous fault properties, *J. Geophys. Res.*, **101**, 22,449-22,460, 1996.

128. T.H. Heaton, J.F. Hall, D.J. Wald, and M.W. Halling, Response of high-rise and base-isolated buildings to a hypothetical Mw 7.0 blind thrust earthquake, *Science*, **267**, 206-211, 1995.

129. Early references suggesting such behavior include E. Vesanen (On the character interpretation of seismograms, *Ann. Acad. Sci. Fenn.*, **AIII**, 5, 1942); T. Usami, *Quart. J. Seis.*, **21**, 1-13, 1956); S.S. Miyamura, R. Omote, R. Teisseyre, and E. Vesanen, Multiple shocks and earthquake series pattern, *Bull. Int. Inst. Seis. Earthquake Eng.*, **2**, 71-92, 1965); and M. Båth (*Seis. Bull.*, Uppsala, February 4, 1965). M. Wyss and J.N. Brune (The Alaska earthquake of 28 March 1964: A complex multiple rupture, *Bull. Seis. Soc. Am.*, **57**, 1017-1023, 1967) found that the Good Friday earthquake of 1964 consisted of six subevents of higher than average slip and that, based on the timing of these subevents, rupture propagated primarily to the southwest from the epicenter. Studies of moderate to large earthquakes in the far field using *P*- and *S*- wave pulse shapes often depict earthquake rupture with multiple point sources (H. Kanamori and G.S. Stewart, Seismological aspects of the Guatemala earthquake of February 4, 1976, *J. Geophys. Res.*, **83**, 3427-3434, 1978), which have been termed asperities (T. Lay and H. Kanamori, An asperity model of great earthquake sequences, in *Earthquake Prediction—An International Review*, D.W. Simpson and P.G. Richards, eds., American Geophysical Union, Maurice Ewing Series, 4, Washington, D.C., pp. 579-592, 1980). For an isolated point source or even for multiple point sources, it is often sufficient to allow for the finite duration through the moment rate function, without considering the spatial extent of each source.

130. For a thorough review of source tomography in the far field, see the review by L. Ruff, Tomographic imaging of seismic sources, in *Seismic Tomography*, G. Nolet, ed., D. Reidel, Boston, pp. 339-366, 1987. The pulse shape of far-field body waves can be used to infer the spatial and temporal distribution of slip on the fault; however, only in the case of particularly large earthquakes is it possible to resolve spatial and temporal variation of slip on the fault plane at teleseismic distances. See, for example the study of the 1985 Michoacan, Mexico, earthquakes by C. Mendoza (Coseismic slip of two large Mexican earthquakes from teleseismic body waveforms: Implications for asperity interaction in the Michoacan plate boundary segment, *J. Geophys. Res.*, **98**, 8197-8210, 1993). Moreover, it is theoretically impossible to reconstruct the slip distribution from the far-field data alone. In the Fraun-

hoffer approximation (i.e., to the extent that the only difference in the Green's function from different parts of the fault is a difference in phase) (K. Aki and P.G. Richards, *Quantitative Seismology: Theory and Methods*, vol 2., W.H. Freeman, San Francisco, pp. 803-805, 1981), there are a range of wavenumbers that are theoretically impossible to recover. Hence, a complete recovery of the source history is impossible.

131. The first such study by K. Aki (Seismic displacements near a fault, *J. Geophys. Res.*, **73**, 5359-5376, 1968) modeled a single near-source record of the 1966 Parkfield earthquake, specifying the depth extent and length of the fault, the rupture velocity, and the average slip. Near-field ground motion data can be supplemented by teleseismically recorded data (S.H. Hartzell and T.H. Heaton, Inversion of strong ground motion and teleseismic waveform data for the fault rupture history of the 1979 Imperial Valley, California earthquake, *Bull. Seis. Soc. Am.*, **73**, 1553-1583, 1983) and geodetic data (D.J. Wald, and T.H. Heaton, Spatial and temporal distribution of slip for the 1992 Landers, California earthquake, *Bull. Seis. Soc. Am.*, **84**, 668-691, 1994).

132. For example, the 1979 Imperial Valley earthquake (A.H. Olson, and R.J. Apsel, Finite faults and inverse theory with applications to the 1979 Imperial Valley earthquake, *Bull. Seis. Soc. Am.*, **72**, 1969-2001, 1982; S.H. Hartzell and T.H. Heaton, Inversion of strong ground motion and teleseismic waveform data for the fault rupture history of the 1979 Imperial Valley, California earthquake, *Bull. Seis. Soc. Am.*, **73**, 1553-1583, 1983), the 1989 Loma Prieta earthquake (see the October 1991 special issue of the *Bulletin of the Seismological Society of America*), the 1994 Northridge earthquake (see the February 1996 special issue of the *Bulletin of the Seismological Society of America*), the 1995 Hyogo-ken Nanbu earthquake (D.J. Wald, A preliminary dislocation model for the 1995 Kobe (Hyogo-ken Nanbu), Japan, earthquake determined from strong motion and teleseismic waveforms, *Seis. Res. Lett.*, **66**, 22-28, 1995); and the 1992 Landers earthquake (D.J. Wald and T.H. Heaton, Spatial and temporal distribution of slip for the 1992 Landers, California earthquake, *Bull. Seis. Soc. Am.*, **84**, 668-691, 1994; B.P. Cohee and G.C. Beroza, Slip distribution of the 1992 Landers earthquake and its implications for earthquake source mechanics, *Bull. Seis. Soc. Am.*, **84**, 692-712, 1994; F. Cotton and M. Campillo, Frequency domain inversion of strong motions; Application to the 1992 Landers earthquake, *J. Geophys. Res.*, **100**, 3961-3975, 1995).

133. Examples of quasi-dynamic rupture models include a study of the 1979 Imperial Valley earthquake (H. Quin, Dynamic stress drop and rupture dynamics of the October 15, 1979 Imperial Valley, California, earthquake, *Tectonophysics*, **175**, 93-117, 1990) and the 1984 Morgan Hill earthquake (G.C. Beroza and T. Mikumo, Short slip duration in dynamic rupture in the presence of heterogeneous fault properties, *J. Geophys. Res.*, **101**, 22,449-22,460, 1996). M. Bouchon (The state of stress on some faults of the San Andreas system as inferred from near-field strong motion data, *J. Geophys. Res.*, **102**, 11,731-11,744, 1997) studied the stress drop and strength excess for several earthquakes. A quasi-dynamic rupture model was derived for the 1995 Hyogo-ken Nanbu earthquake by S. Ide and M. Takeo (Determination of constitutive relations of fault slip based on seismic wave analysis, *J. Geophys. Res.*, **102**, 27,379-27,391, 1997) and for the 1992 Landers earthquake by K.B. Olsen, R. Madariaga, R.J. Archuleta (Three-dimensional dynamic simulation of the 1992 Landers earthquakes, *Science*, **278**, 834-838, 1997) using a slip-weakening constitutive model. Quasi-dynamic models are parameterized in terms such as the strength excess (how close the fault is to failure before the earthquake begins), the dynamic stress drop (difference between the initial stress and the residual sliding friction), and the slip-weakening displacement (amount of slip required to reach the sliding frictional stress).

134. M. Guatteri and P. Spudich, What can strong motion data tell us about slip-weakening friction laws?, *Bull. Seis. Soc. Am.*, **90**, 98-116, 2000.

135. As earthquake rupture evolves, it exerts a time-varying stress field on the points around it, including portions of the fault that are slipping. As shown by P.K.P. Spudich (On

the inference of absolute stress levels from seismic radiation, *Tectonophysics*, **211**, p. 99-106, 1992) under certain assumptions the response of the fault to this load can be used to estimate the absolute stress level if the slip direction at a point on the fault varies as a function of time during rupture. Application of this technique to the 1995 Hyogo-ken Nanbu earthquake has shown promising results that match borehole stress data where available and suggest a low level of absolute stress (P. Spudich, M. Guatteri, I. Otsuki, and J. Minagawa, Use of fault striations and dislocation models to infer tectonic shear stress during the 1995 Hyogo-ken Nanbu (Kobe) earthquake, *Bull. Seis. Soc. Am.*, **88**, 413-427, 1998; M. Guatteri and P. Spudich, Coseismic temporal changes of slip direction; The effect of absolute stress on dynamic rupture, *Bull. Seis. Soc. Am.*, **88**, 777-789, 1998).

136. Early studies of strong ground motion characterized it as emanating from the source in short impulses (G.W. Housner, Characteristics of strong-motion earthquakes, *Bull. Seis. Soc. Am.*, **37**, 19-31, 1947; G.W. Housner, Properties of strong ground motion earthquakes, *Bull. Seis. Soc. Am.*, **45**, 197-218, 1955; W.T. Thompson, Spectral aspect of earthquakes, *Bull. Seis. Soc. Am.*, **49**, 91-98, 1959). A stochastic model for earthquake slip was formulated by N.A. Haskell (Total energy and energy spectral density of elastic wave radiation from propagating faults, 2, A statistical source model, *Bull. Seis. Soc. Am.*, **56**, 125-140, 1966) in a study of the radiated seismic energy. Haskell characterized earthquake slip as a spatial random field using both a correlation length and a correlation time to specify the autocorrelation function.

137. R. Madariaga (High-frequency radiation from crack (stress drop) models of earthquake faulting, *Geophys. J. R. Astron. Soc.*, **51**, 625-651, 1977) demonstrated that smooth rupture propagation produces relatively little high-frequency ground motion.

138. As, for example, a model of circular rupture and arrest studied by R. Madariaga (Dynamics of an expanding circular fault, *Bull. Seis. Soc. Am.*, **66**, 639-666, 1976) in which rupture was stopped artificially and simultaneously around the entire circumference of the fault.

139. The spectral behavior of ground motion above the corner frequency is often cast in terms of the displacement spectra (K. Aki, Scaling law of seismic spectrum, *J. Geophys. Res.*, **72**, 1217-1231, 1967). A spectrum that is flat in acceleration will decay as f^{-2} in displacement (J.N. Brune, Tectonic stress and the spectra of seismic shear waves from earthquakes, *J. Geophys. Res.*, **75**, 4997-5009, 1970).

140. R. Madariaga, High frequency radiation from dynamic earthquake fault models, *Ann. Geophys.*, **1**, 17-23, 1983; A.S. Papageorgiou, and K. Aki, A specific barrier model for the quantitative description of inhomogeneous faulting and the prediction of strong ground motion; Part I, Description of the model, *Bull. Seis. Soc. Am.*, **73**, 953-978, 1983.

141. Y. Zeng, K. Aki, and T.-L. Teng, Mapping of the high-frequency source radiation for the Loma Prieta earthquake, California, *J. Geophys. Res.*, **98**, 11,981-11,993, 1993.

142. See, for example, C.K. Saikia and P.G. Somerville (Simulated hard-rock motions in Saint Louis, Missouri, from large New Madrid earthquake (Mw ≥ 6.5), *Bull. Seis. Soc. Am.*, **87**, 123-139, 1997) or Y. Zeng and J.G. Anderson (A composite source modeling of the 1994 Northridge earthquake using Genetic Algorithm, *Bull. Seis. Soc. Am.*, **86**, 71-83, 1996).

143. C.H. Scholz, Microfracturing and the inelastic deformation of rock in compression, *J. Geophys. Res.*, **73**, 1417-1432, 1968.

144. Shear rupture is naturally limited by the Earth's surface. There is also evidence that rupture through the shallowest sedimentary layers of the Earth's crust may be delayed and occur as aseismic afterslip. The decay of afterslip with time is consistent with steady-state velocity strengthening friction on this part of the fault (C. Marone and C. Scholz, The depth of seismic faulting and the upper transition from stable to unstable slip regimes, *Geophys. Res. Lett.*, **15**, 621-624, 1988).

145. J.S. Tchalenko, Similarities between shear zones of different magnitudes, *Geol. Soc.*

Am. Bull., **81**, 1625-1640, 1970; W.L. Power, T.E. Tullis, S. Brown, G.N. Boitnott, and C.H. Scholz, Roughness of natural fault surfaces, *Geophys. Res. Lett.*, **14**, 29-32, 1987.

146. For example, fault segmentation was used to estimate the size of future earthquakes in the report by the Working Group on California Earthquake Probabilities (*Probabilities of Large Earthquakes Occurring in California on the San Andreas Fault*, U.S. Geological Survey Open-File Report 88-398, Reston, Virginia, 62 pp., 1988).

147. For examples of earthquakes that may have been terminated by a fault discontinuity, see the following studies: W.H. Bakun, R.M. Stewart, C.G. Bufe, and S.M. Marks, Implication of seismicity for failure of a section of the San Andreas fault, *Bull. Seis. Soc. Am.*, **70**, 185-201, 1980; A.G. Lindh and D.M. Boore, Control of rupture by fault geometry during the 1966 Parkfield earthquake, *Bull. Seis. Soc. Am.*, **71**, 95-116, 1981; P. Reasenberg and W.L. Ellsworth, Aftershocks of the Coyote Lake, California, earthquake of August 6, 1979: A detailed study, *J. Geophys. Res.*, **87**, 10,637-10,655, 1982; A. Barka and K. Kadinsky-Cade, Strike-slip fault geometry in Turkey and its influence on earthquake activity, *Tectonics*, **7**, 663-684, 1988.

148. Examples of earthquakes that ruptured through fault discontinuities include the Borrego Mountain, California (M.M. Clark, Surface rupture along the Coyote Creek Fault, in *The Borrego Mountain Earthquake*, U.S. Geological Survey Professional Paper 787, Reston, Va., pp. 55-86, 1972), the Erzincan, Turkey (A. Barka and K. Kadinsky-Cade (Strike-slip fault geometry in Turkey and its influence on earthquake activity, *Tectonics*, **7**, 663-684, 1988), and the 1966 Parkfield, California, earthquakes (P. Segall and Y. Du, How similar were the 1934 and 1966 Parkfield earthquakes? *J. Geophys. Res.*, **98**, 4527-4538, 1993).

149. K.E. Sieh, and 19 others, Near-field investigations of the Landers earthquake sequence, April to July 1992, *Science*, **260**, 171-176, 1993.

150. Working Group on California Earthquake Probabilities, Seismic hazards in southern California: Probable earthquakes, 1994-2024, *Bull. Seis. Soc. Am.*, **85**, 379-439, 1995; Working Group on California Earthquake Probabilities, *Earthquake Probabilities in the San Francisco Bay Region: 2000 to 2030—A Summary of Findings*, U.S. Geological Survey Open File Report 99-517, Reston, Va., 46 pp., 1999.

151. Based on a study of Ms ≥ 6.8 earthquakes on the North Anatolian fault in Turkey. See A. Barka and K. Kadinsky-Cade (Strike-slip fault geometry in Turkey and its influence on earthquake activity, *Tectonics*, **7**, 663-684, 1988).

152. R.A. Harris and S.M Day, Dynamics of fault interaction; Parallel strike-slip faults, *J. Geophys. Res.*, **98**, 4461-4472, 1993.

153. A possible complicating factor is the effect of pore fluids. If pore fluids are present, they will tend to drop when an extensional jog is stressed. The drop in fluid pressure will tend to lock the fault by counteracting the drop in normal stress across it (R. Sibson, Stopping of earthquake ruptures at dilational fault jogs, *Nature*, **316**, 248-251, 1985).

154. R.A. Harris and S.M. Day, Dynamic three-dimensional simulations of earthquakes on en echelon faults, *Geophys. Res. Lett.*, **26**, 2089-2092, 1999.

155. C. Marone and C. Scholz, The depth of seismic faulting and the upper transition from stable to unstable slip regimes, *Geophys. Res. Lett.*, **15**, 621-624, 1988.

156. A.J. Michael and D. Eberhart-Phillips (Relations among fault behavior, subsurface geology, and three-dimensional velocity models, *Science*, **253**, 651-654, 1991) studied five large earthquakes in California and found variations in the three-dimensional velocity structure at seismogenic depths that tended to correlate with regions of high mainshock slip. Other studies (A. Michelini and T.V. McEvilly, Seismological studies at Parkfield, I, Simultaneous inversion for velocity structure and hypocenters using cubic B-splines parameterization, *Bull. Seis. Soc. Am.*, **81**, 524-552, 1991; C. Nicholson and J.M. Lees, Travel-time tomography in the northern Coachella Valley using aftershocks of the 1986 ML 5.9 North Palm Springs earthquake, *Geophys. Res. Lett.*, **19**, 1-4, 1992; W. Foxall, A. Michelini, and T.V.

McEvilly, Earthquake travel time tomography of the southern Santa Cruz Mountains; Control of fault rupture by lithological heterogeneity of the San Andreas fault zone, *J. Geophys. Res.*, **98**, 17,691-17,710, 1993) have noted a similar correlation.

157. D.H. Oppenheimer, W.H. Bakun, and A.G. Lindh (Slip partitioning of the Calaveras Fault, California, and prospects for future earthquakes, *J. Geophys. Res.*, **93**, 9007-9026, 1988) found areas on the southern Calaveras fault that failed repeatedly during small to moderate earthquakes. These same areas were devoid of microearthquakes in the interseismic period and hence presumably stuck. Much of the slip on the southern Calaveras fault appears to take place aseismically as indicated by the postseismic transient to the 1984 Morgan Hill, California, earthquake, which was larger than that of the mainshock itself (W.H. Prescott, N.E. King, and G. Guohua, Preseismic, coseismic, and postseismic deformation associated with the 1984 Morgan Hill, California, earthquake, in *The 1984 Morgan Hill, California Earthquake*, J.H. Bennet and R.W. Sherburne, eds., California Division of Mines and Geology Special Publication 68, Sacramento, pp. 137-148, 1986).

158. D.P. Schaff, B. Shaw, and G.C. Beroza, Postseismic response of repeating aftershocks, *Geophys. Res. Lett.*, **25**, 4549-4552, 1998; T.E. Tullis, Perspective—Deep slip rates on the San Andreas fault, *Science*, **285**, 671-672, 1999.

159. M.I. Husseini, D.B. Jovanovich, M.J. Randall, and L.B. Freund, The fracture energy of earthquakes, *Geophys. J. R. Astron. Soc.*, **43**, 367-385, 1975.

160. R.L. Wesson and W.L. Ellsworth, Seismicity preceding moderate earthquakes in California, *J. Geophys. Res.*, **78**, 8527-8546, 1973.

161. T. Lay and H. Kanamori, An asperity model of great earthquake sequences, in *Earthquake Prediction—An International Review*, D.W. Simpson and P.G. Richards, eds., American Geophysical Union, Maurice Ewing Series, 4, Washington, D.C., pp. 579-592, 1980.

162. Low prestress levels on the Emerson fault inferred from dynamic rupture modeling (M. Bouchon, M. Campillo, and F. Cotton, Stress field associated with the rupture of the 1992 Landers, California, earthquake and its implications concerning the fault strength at the onset of the earthquake, *J. Geophys. Res.*, **103**, 21,091-21,097, 1998) and the lack of post-mainshock shear stress across the Emerson fault based on a stress inversion of the aftershock focal mechanisms (E. Hauksson, State of stress from focal mechanisms before and after the 1992 Landers earthquake sequence, *Bull. Seis. Soc. Am.*, **84**, 917-934, 1994) support this hypothesis. Three-dimensional simulations of dynamic rupture across a fault discontinuity indicate that the rupture on the second fault segment will nucleate at shallow depth because the normal stress is lower and because the free surface amplifies the dynamic stress field. Analyses of strong-motion (B.P. Cohee and G.C. Beroza, Slip distribution of the 1992 Landers earthquake and its implications for earthquake source mechanics, *Bull. Seis. Soc. Am.*, **84**, 692-712. 1994; D.J. Wald and T.H. Heaton, Spatial and temporal distribution of slip for the 1992 Landers, California earthquake, *Bull. Seis. Soc. Am.*, **84**, 668-691, 1994) and geodetic (J. Freymueller, N.E. King, and P. Segall, Co-seismic slip distribution of the 1992 Landers earthquake, *Bull. Seis. Soc. Am.*, **84**, 646-659, 1994) data indicate that mainshock slip on the Emerson fault was shallow compared with slip on the other segments. This signature is reflected in the aftershock depths as well, which are found to be shallow over the transition from the Homestead Valley to the Emerson faults (K.R. Felzer and G.C. Beroza, Deep structure of a fault discontinuity, *Geophys. Res. Lett.*, **26**, 2121-2124, 1999). If the Emerson fault was far from failure before the Landers mainshock, then it may have terminated at shallow depth with rupture nucleating at shallow depth but not propagating to greater depths or further along the fault.

163. A point made by R.A. Harris and S.M. Day (Dynamic three-dimensional simulations of earthquakes on en echelon faults, *Geophys. Res. Lett.*, **26**, 2089-2092, 1999) based on the comparison of the 1934 and 1966 earthquakes reported by P. Segall and Y. Du (How similar were the 1934 and 1966 Parkfield earthquakes?, *J. Geophys. Res.*, **98**, 4527-4538, 1993).

164. Earthquakes below crustal depths in descending slabs can shake the surface hard enough to cause deadly secondary effects. For example, the 1970 intermediate-focus Peru earthquake (M 8.0), which occurred at a focal depth of 64 kilometers, initiated the huge Huascarán slide (Figure 3.3) that killed 60,000 people. Seismic waves propagate efficiently below the asthenosphere (i.e., below about 300-kilometer depth) and within the deeper parts of the thickened cratonic lithosphere; the giant M 8.2 deep-focus earthquake 660 kilometers beneath Bolivia was felt by people as far away as Canada.

165. M. Vassilou and B. Hager, Subduction zone earthquakes and stress in slabs, *Pure Appl. Geophys.*, **128**, 547-624, 1988.

166. B. Isacks and P. Molnar, Distribution of stresses in the descending lithosphere from a global survey of focal-mechanism solutions of mantle earthquakes, *Rev. Geophys. Space Phys.*, **9**, 103-174, 1971.

167. C. Frohlich, The nature of deep focus earthquakes, *Ann. Rev. Earth Planet. Sci.*, **17**, 227-254, 1989; H.W. Green, II and H. Houston, The mechanics of deep earthquakes, *Ann. Rev. Earth Planet. Sci.*, **23**, 169-213, 1995.

168. P.B. Stark and C. Frohlich, Depths of the deepest earthquakes, *J. Geophys. Res.*, **90**, 1859-1869, 1985.

169. The problem of initiating shear instabilities at very high pressures was broached by D. Griggs and J. Handin (Observations on fracture and a hypothesis of earthquakes, in *Rock Deformation*, D. Griggs and J. Handin, eds., Geological Society of America Memoir 79, Boulder, Colo., pp. 347-364, 1960). Early experimental and theoretical work focused on the volumetric instabilities in polymorphic phase transitions (e.g., P.W. Bridgman, Polymorphic transitions and geologic phenomena, *Am. J. Sci.*, **243A**, 90-97, 1945; F.F. Evison, On the occurrence of volume change at the earthquake source, *Bull. Seis. Soc. Am.*, **57**, 9-25, 1967, L. Liu, Phase transformations, earthquakes, and the descending lithosphere, *Phys. Earth Planet. Int.*, **32**, 226-240, 1983). The volumetric-instability hypothesis was encouraged by seismological evidence that the great 1970 Columbia deep-focus earthquake radiated energy with a significant isotropic component (A. Dziewonski and J.F. Gilbert, Temporal variation of the seismic moment tensor and the evidence of precursive compression for two deep earthquakes, *Nature*, **257**, 185-188, 1974); however, it is now believed that the isotropic component of deep-focus source mechanisms is small compared to the shear component (D. Russakoff, G. Ekstrom, and J. Tromp, A new analysis of the great 1970 Colombia earthquake and its isotropic component, *J. Geophys. Res.*, **102**, 20,423-20,434, 1997).

170. D. Griggs, The sinking lithosphere and the focal mechanism of deep earthquakes, in *The Nature of the Solid Earth,* E.C. Robertson, ed., McGraw-Hill Inc., New York, pp. 361-384, 1972; M. Ogawa, Shear instability in a viscoelastic material as the cause of deep focus earthquakes, *J. Geophys. Res.*, **92**, 13,801-13,810, 1987; B.E. Hobbs and A. Ord, Plastic instabilities: Implications for the origin of intermediate and deep focus earthquakes, *J. Geophys. Res.*, **93**, 10,521-10,540, 1988.

171. C.B. Raleigh, Tectonic implications of serpentinite weakening, *Geophys. J. R. Astr. Soc.*, **14**, 113-118, 1967; M.S. Paterson, *Experimental Rock Deformation—The Brittle Field,* Springer, Berlin, 254 pp., 1978.

172. S.H. Kirby, Localized polymorphic phase transitions in high-pressure faults and applications to the physical mechanisms of deep earthquakes, *J. Geophys. Res.*, **93**, 13,789-13,800, 1987; C. Meade and R. Jeanloz, Acoustic emissions and shear instabilities during phase transformations in Si and Ge at ultra high pressures, *Nature*, **339**, 616-618, 1989; H.W.I. Green and P.B. Burnley, A new self-organizing mechanism for deep-focus earthquakes, *Nature*, **341**, 733-737, 1989.

173. H.W.I. Green and P.B. Burnley, A new self-organizing mechanism for deep-focus earthquakes, *Nature*, **341**, 733-737, 1989; C. Meade and R. Jeanloz, Deep-focus earthquakes

and recycling of water into the Earth's mantle, *Science*, **252**, 68-72, 1991; C. Frohlich, A break in the deep, *Nature*, **368**, 100-101, 1994.

174. S.H. Kirby, Intraslab earthquakes and phase-changes in subducting lithosphere, *Rev. Geophys. Suppl.*, **33**, 287-297, 1995.

175. H.W. Green, T.E. Young, D. Wlaker, and C. Scholz, Anti-crack assisted faulting at very high pressure in natural olivine, *Nature*, **348**, 720-722, 1990; H.W. Green, Solving the paradox of deep earthquakes, *Sci. Am.*, **271**, 64-71, 1994.

176. H.W. Green II and H. Houston, The mechanics of deep earthquakes, *Ann. Rev. Earth Planet. Sci.*, **23**, 169-213, 1995.

177. The March 9, 1994, Tonga earthquake (M_W 7.6, 564 kilometers deep) was recorded by an array of nine instruments deployed directly over the hypocenter (D.A. Wiens, J.J. McGuire, P.J. Shore, M.G. Bevis, K. Draunidao, G. Prasad, and S. Helu, A deep earthquake aftershock sequence and implications for the rupture mechanism of deep earthquakes, *Nature*, **372**, 540-543, 1995). On June 9, 1994, the M_W 8.3 636-kilometer depth Bolivian earthquake was recorded by 26 portable instruments deployed close to the epicenter (P.G. Silver, S.L. Beck, T.C. Wallace, C. Meade, S.C. Myers, S.E. James, and R. Kuehnel, Rupture characteristics of the deep Bolivian earthquake of 9 June 1994 and the mechanism of deep-focus earthquakes, *Science*, **268**, 69-73, 1995). The Bolivian earthquake was the largest deep earthquake ever recorded.

178. In the case of the Bolivian earthquake, the rupture extended about 60 kilometers in the horizontal direction through the thermal boundary layer of the subducting lithosphere (P.G. Silver, S.L. Beck, T.C. Wallace, C. Meade, S.C. Myers, S.E. James, and R. Kuehnel, Rupture characteristics of the deep Bolivian earthquake of 9 June 1994 and the mechanism of deep-focus earthquakes, *Science*, **268**, 69-73, 1995).

179. H. Kanamori, D.L. Anderson, and T.H. Heaton, Frictional melting during the rupture of the 1994 Bolivian earthquake, *Science*, **279**, 839-842, 1998.

180. In an isotropic Earth, where properties do not depend on direction, three independent elastic parameters are needed to describe seismic-wave propagation. These are usually taken to be compressional-wave speed v_p, shear-wave speed v_s, and mass density ρ; alternatives to the first two are the compressional modulus $K = \rho\,(v_p^2 - 4v_s^2/3)$ and shear modulus $G = \rho v_s^2$ Two anelastic parameters describe isotropic attenuation, such as the inverses of the nondimensional compressional-wave and shear-wave quality factors, Q_p^{-1} and Q_s^{-1}. In most regions of the Earth, the attenuation of pure compression appears to be negligible, which implies that $Q_p^{-1} \approx (4/9)Q_s^{-1}$. An anelastic, isotropic Earth model is thus described by the quadruple function of position (ρ, v_p, v_s, Q_s^{-1}).

181. K.B. Olsen, Site amplification in the Los Angeles Basin from 3D modeling of ground motions, *Bull. Seis. Soc. Am.*, **90**, S77-S94, 2000.

182. A modern standard is the spherically symmetric Preliminary Reference Earth Model (PREM) of A.M. Dziewonski and D.L. Anderson (Preliminary reference Earth model, *Phys. Earth Planet. Int.*, **25**, 297-356, 1981).

183. L.C. Pakiser and W.D. Mooney, eds., *Geophysical Framework of the Continental United States*, Geological Survey of America Memoir 172, Boulder, Colo., 826 pp. 1989.

184. The normal modes can either be standing waves (free oscillations) or traveling waves (surface-wave overtones). An exhaustive discussion of normal mode methods applied to seismology is given by F.A. Dahlen and J. Tromp, *Theoretical Global Seismology*, Princeton University Press, Princeton, N.J., 1025 pp., 1998.

185. A comprehensive treatment of ray-theoretic methods can be found in K. Aki and P.G. Richards, *Quantitative Seismology: Theory and Methods*, vols. I & II, W.H. Freeman, San Francisco, 932 pp., 1980.

186. Codes based on the fourth-order staggered velocity-stress formulation, for example, have proven efficient and flexible (e.g., R.W. Graves, Simulating seismic wave propa-

gation in 3d elastic media using staggered-grid finite differences, *Bull. Seis. Soc. Am.*, **86**, 1091-1106, 1996).

187. Features include parallelism (K.B. Olsen, R. Madariaga, and R.J. Archuleta, Three-dimensional dynamic simulation of the 1992 Landers earthquakes, *Science*, **278**, 834-838, 1997; H. Bao, J. Bielak, O. Ghattas, D.R. O'Hallaron, L.F. Kallivokas, J. R. Shewchuk, and J. Xu, Large-scale simulation of elastic wave propagation in heterogeneous media on parallel computers, *Comput. Meth. Appl. Mech. Eng.*, **152,** 85-102, 1998), realistic anelastic losses (S.M. Day, Efficient simulation of constant Q using coarse-grained memory variables, *Bull. Seis. Soc. Am.*, **88**, 1051-1062, 1998; S.M. Day and C.R. Bradley, Memory-efficient simulation of anelastic wave propagation, *Bull. Seis. Soc. Am.*, **91**, 520-531, 2001), discontinuous gridding (multiple structured grids coupled at simple interfaces; S. Aoi and H. Fujiwara, 3D finite-difference method using discontinuous grids, *Bull. Seis. Soc. Am.*, **89**, 918-930, 1999), unstructured meshing (Bao et al., op. cit.), memory optimization (Graves, op. cit.), surface topography, propagating kinematic earthquake sources, an interface to optional rupture dynamics modules, and links to unified structural representations, with options for user-defined model modifications.

188. D. Komatitsch and J. Tromp, Introduction to the spectral element method for three-dimensional seismic wave propagation, *Geophys. J. Int.*, **139**, 806-822, 1999.

189. This steep decay of amplitude with distance from deeper sources was observed for sites within about 40 kilometers of the Loma Prieta earthquake, a relatively deep crustal source.

190. Waves travelling downward from the hypocenter can be critically reflected by layer interfaces below the hypocenter, causing relatively large amplitudes at the surface for certain distance ranges corresponding to critical reflection at these interfaces, such that the reflected waves are larger than the direct waves. One example of this is the SmS phase, which is an *S* wave critically reflected from the crust-mantle interface (Moho). This SmS phase was one factor responsible for increased ground motions and damage about 90 kilometers away from the Loma Prieta earthquake. The ground motion at these distances was actually greater than at some closer sites with similar geology. Analysis of ground-motion data from aftershocks of the Northridge earthquake demonstrates that reflections from midcrustal interfaces can increase the amplitude of seismic waves from shallow sources at certain distances less than 100 kilometers.

191. Moho reflections were partially responsible for the strong ground motions that damaged San Francisco after the 1989 Loma Prieta earthquake (P.G. Somerville and J. Yoshimura, The influence of critical Moho reflections on strong ground motions recorded in San Francisco and Oakland during the 1989 Loma Prieta earthquake, *Geophys. Res. Lett.* **17**, 1203-1206, 1990).

192. Generally, the shear-wave velocity and density increase with depth in surficial geological materials. As seismic waves propagate to the surface, the decreasing impedance (given by the product of shear-wave velocity and density) gives rise to a corresponding increase in amplitude of seismic waves, so that within the linear range, the less stiff the surface materials, the higher is the ground motion. Current building codes (1997 NEHRP Provisions and 1997 UBC) include ground-motion amplification factors in two period ranges that are related to the shear-wave velocity averaged over the top 30 meters (the typical depth of geotechnical borings). These amplification factors also account for nonlinear effects, which generally reduce the elastic ground motions at short periods.

193. H. Kawase, The cause of the damage belt in Kobe: "The basin-edge effect," constructive interference of the direct S-wave with the basin-induced diffracted/Rayleigh waves, *Seis. Res. Lett.*, **67**, 25-34, 1996; A. Pitarka, K. Irikura, T. Iwata, and H. Sekiguchi, Three-dimensional simulation of the near-fault ground motion for the 1995 Hyogo-ken Nanbu (Kobe), Japan, earthquake, *Bull. Seis. Soc. Am.*, **88**, 428-440, 1998.

194. D.J. Wald and R.W. Graves, The seismic response of the Los Angeles basin, California, *Bull. Seis. Soc. Am.*, **88**, 337-356, 1998.

195. A. Frankel, Three-dimensional simulations of ground motions in the San Bernardino Valley, California for hypothetical earthquakes on the San Andreas fault, *Bull. Seis. Soc. Am.*, **83**, 1020-1041, 1993; K.B. Olsen, R.J. Archuleta, and J.R. Matarese, Magnitude 7.75 earthquake on the San Andreas fault; Three-dimensional ground motion in Los Angeles, *Science*, **270**, 1628-1632, 1995; K.B. Olsen and R.J. Archuleta, Three-dimensional simulation of earthquakes on the Los Angeles fault system, *Bull. Seis. Soc. Am.*, **86**, 575-596, 1996; R.W. Graves, Three-dimensional finite-difference modeling of the San Andreas fault: Source parameterization and ground motion levels, *Bull. Seis. Soc. Am.*, **88**, 881-897, 1998; T. Sato, R.W. Graves, and P.G. Somerville, 3-D finite difference simulations of long period strong motions in the Tokyo metropolitan area during the 1990 Odawara earthquake (Mj 5.1) and the great Kanto earthquake (Ms 8.2) in Japan, *Bull. Seis. Soc. Am.*, **89**, 579-607, 1999.

196. P.N. Sahay, T.J.T. Spanos, and V. de la Cruz, Seismic wave propagation in inhomogeneous and anisotropic porous media, *Geophys. J. Int.*, **145**, 209-223, 2001.

197. P.G. Somerville, N.F. Smith, R.W. Graves, and N.A. Abrahamson, Modification of empirical strong ground motion attenuation relations to include the amplitude and duration effects of rupture directivity, *Seis. Res. Lett.*, **68**, 199-222, 1997.

198. Present research addresses how the Q values and their frequency dependence relate to the physical state of the lithosphere and to the physical mechanisms that cause attenuation. It has been proposed that the regional differences in crustal Q are caused by differences in crustal temperature, crack properties, continuity of geologic structures, and/or scatterer properties. See H.M. Benz, A. Frankel, and D.M. Boore, Regional Lg attenuation for the continental United States, *Bull. Seis. Soc. Am.*, **87**, 606-619, 1997.

199. H.B. Seed and I.M. Idriss, Analyses of ground motions at Union Bay, Seattle during earthquakes and distant nuclear blasts, *Bull. Seis. Soc. Am.*, **60**, 125-136, 1970; M. Zeghal and A.-W. Elgamal, Analysis of site liquefaction using earthquake records, *J. Geotech. Engr.*, **120**, 996-1017, 1994; E.H. Field, P.A. Johnson, I.A. Beresnev, and Y.H. Zeng, Nonlinear ground-motion amplification by sediments during the 1994 Northridge earthquake, *Nature*, **390**, 599-602, 1997; J. Aguirre and K. Irikura, Nonlinearity, liquefaction and velocity variation, of soft soil layers in Port Island, Kobe, during the Hyogo-ken Nanbu earthquake, *Bull. Seis. Soc. Am.*, **87**, 1244-1258, 1997.

200. K.-L. Wen, Nonlinear soil response in ground motions, *Earthquake Eng. Struct. Dynamics*, **23**, 599-608, 1994; I.A. Beresnev, K.-L. Wen, and Y.T. Yeh, Nonlinear soil amplification—Its corroboration in Taiwan, *Bull. Seis. Soc. Am.*, **85**, 496-515, 1995; J. Aguirre and K. Irikura, Nonlinearity, liquefaction and velocity variation of soft soil layers in Port Island, Kobe, during the Hyogo-ken Nanbu earthquake, *Bull. Seis. Soc. Am.*, **87**, 1244-1258, 1997.

201. In general, stiff soils have higher thresholds and can exhibit quasi-linear response; soft, cohesionless soils have lower nonlinear thresholds. The frequency of vibration is also a critical parameter. At low frequencies, response spectral ratios are unaffected by strain amplitudes. At intermediate frequencies, including those usually most prominent in large-earthquake spectra, damping dominates and strong motion-soil amplification is less than it is for weak motions. At high frequencies, reduction in the shear modulus dominates and soil amplification increases with increased dynamic strain levels. The transition frequencies will vary depending on soil type and thickness and the input seismic spectrum.

202. K. Arulanandan and R.F. Scott, *Verification of Numerical Procedures for the Analysis of Soil Liquefaction Problems*, 2 vols, Balkema, Rotterdam, 1801 pp., 1994.

203. Housner reported in the 1940s that strong-motion records could be described as white noise. He attributed the complex character of the records to inhomogeneity on the fault. In many respects, these basic concepts have withstood the test of time and the accumulation of more data. However, as described above, deterministic features of strong-

ground-motion waveforms, including some produced by source processes such as rupture directivity and others produced by seismic-wave propagation phenomena, such as the trapping of waves in basins, have since been recognized and quantified.

204. The f^{-2} falloff in the displacement amplitude spectrum was recognized in spectra from regional and teleseismic events by K. Aki (Scaling law of seismic spectrum, *J. Geophys. Res.*, **72**, 1217-1231, 1967). The low-frequency spectral level is proportional to the seismic moment. Greater stress drops produce a larger high-frequency spectral level for a given seismic moment. It is generally observed that stress drops are approximately uniform over a wide range of earthquake sizes (see Section 2.5).

205. It has been shown that the falloff can be explained by attenuation in the near-surface material beneath a site. f_{max} is observed to correlate with the site geology, with soil sites having lower f_{max} values than rock sites. The falloff above f_{max} can be described using a frequency-independent Q. Typically, this is parameterized by k, which is related to the slope of the spectral falloff on a log-linear plot. Although it was initially proposed that f_{max} is produced by a characteristic length scale of the rupture process, f_{max} is observed to increase for seismometers located in boreholes, indicating that it is at least in part an artifact of near-surface attenuation. For small earthquakes (M lower than 4), the source corner frequency is often obscured by the effects of near-surface attenuation.

206. J.N. Brune (Tectonic stress and the spectra of seismic shear waves from earthquakes, *J. Geophys. Res.*, **75**, 4997-5009, 1970) derived this behavior from a earthquake model in which a stress pulse propagated along the fault. He was the first to relate the corner frequency to the radius of rupture and derived the relation between stress drop, seismic moment, and corner frequency. Madariaga (Dynamics of an expanding circular fault, *Bull. Seis. Soc. Am.*, **66**, 639-666, 1976) showed that a simple dynamic model of crack nucleation could not produce enough high frequency for an f^{-2} falloff and suggested that the f^{-2} falloff was caused by the stopping phase of an earthquake rupture. More recently, the hypothesis that the high-frequency level of the acceleration spectrum is a manifestation of the complexity of the rupture process has been explored. In this view, smaller-scale variations of stress along the fault plane produce higher frequencies of radiated ground motion. Recently, several investigators have proposed fractal models of fault stress heterogeneity. In these models the stress on the fault is a random, self-similar variable with a fluctuation spectrum whose spectral amplitude is proportional to the wavelength raised to some power. Asperities on the fault that produce subevents are described with a power-law distribution of sizes. In separate studies, T. Hanks (*b* values and $\omega^{-\gamma}$ seismic source models; Implications for tectonic stress variations along active crustal fault zones and the estimation of high-frequency strong ground motion, *J. Geophys. Res.*, **84**, 2235-2242, 1979), D.J. Andrews (A stochastic fault model, 2. Time-independent case, *J. Geophys. Res.*, **86**, 10,821-10,834, 1981), and A. Frankel (High frequency spectral falloff of earthquakes, fractal dimension of complex rupture, *b* value, and the scaling strength on fault, *J. Geophys. Res.*, **96**, 6291-6302, 1991) showed that a flat acceleration spectrum could be explained by a variation of stress drop that was independent of length scale on a fault. Such a scale-independent stress drop is consistent with observations of stress drop being independent of seismic moment. Interestingly, this same variation in stress drop on the fault produced a population of subevents with *b* values of 1. This *b* value is similar to those reported for earthquakes in most regions, suggesting that the population statistics of earthquakes may be related to the same stress drop variation responsible for the generation of high-frequency ground motion.

207. For example, N.A. Abrahamson, J.F. Schneider, and J.C. Stepp, Empirical coherency functions for applications to soil-structure interaction, *Earthquake Spectra*, **7**, 1-27, 1992; M.I. Todorovska and M.D. Trifunac, Amplitudes, polarity and time of peaks of strong ground motion during the 1994 Northridge, California, earthquake, *Soil Dyn. Earthquake Engr.*, **16**, 235-258, 1997; P. Bodin, S.K. Singh, M. Santoyo, and J. Gomberg, Dynamic defor-

mations and shallow sediments in the Valley of Mexico, I: Three-dimensional strains and rotations recorded on a seismic array, *Bull. Seis. Soc. Am.*, **87**, 540-550, 1997.

208. Such short-range incoherence may be attributed to spatially varying site conditions, which cause seismic waves to be distorted from the plane waves. In the case of Pinyon Flat, F. Vernon, J. Fletcher, L. Carroll, A. Chave, and E. Sembera (Coherence of seismic body waves from local events as measured by a small-aperture array, *J. Geophys. Res.*, **96**, 11,981-11,996, 1991) concluded that coherence was reduced due to slight irregularities in depth of the weathered layer of granodiorite at the site.

209. For example, see Working Group on California Earthquake Probabilities, Seismic hazards in southern California: Probable earthquakes, 1994-2024, *Bull. Seis. Soc. Am.*, **85**, 379-439, 1995.

210. The Northridge earthquake had more than 20,000 recorded aftershocks.

211. Y. Okada, Internal deformation due to shear and tensile faults in a half-space, *Bull. Seis. Soc. Am.*, **82**, 1018-1040, 1992.

212. Several different probability distributions can be used to characterize earthquake recurrence, although the most common for time-dependent hazard analysis is the log-normal distribution. The most critical parameter in these calculations is the ratio of the standard deviation (σ) of the interoccurrence times and the average of these times (T_{ave}), or the coefficient of variation σ/T_{ave}. For low values of σ/T_{ave}, earthquake occurrence is almost periodic; high values indicate large variability in recurrence times. Present research is focused on determining the magnitude and physical origin of this coefficient for different regions of the world.

213. See Working Group on California Earthquake Probabilities, Seismic hazards in southern California: Probable earthquakes, 1994-2024, *Bull. Seis. Soc. Am.*, **85**, 379-439, 1995.

214. For a summary of the contents of this issue, see the introductory paper by N.A. Abrahamson and K.M. Shedlock, Overview of ground motion attenuation models, *Seis. Res. Lett.*, **68**, 9-23, 1997.

215. E.H. Field and the SCEC Phase III Working Group, Accounting for site effects in probabilistic seismic hazard analyses of southern California: Overview of the SCEC Phase III report, *Bull. Seis. Soc. Am.*, **90**, S1-S31, 2000.

216. Senior Seismic Hazard Analysis Committee, *Recommendations for Probabilistic Seismic Hazard Analysis: Guidance on Uncertainty and Use of Experts*, U.S. Nuclear Regulatory Commission, NUREG/CR-6372, Washington, D.C., 1997.

217. International Conference of Building Officials, *Uniform Building Code*, Whittier, Calif., 3 volumes, 1997; Building Seismic Safety Council, *NEHRP Recommended Provisions for Seismic Regulations for New Buildings and Other Structures, Part 1—Provisions*, FEMA 302, Washington, D.C., 336 pp., 1997; Building Seismic Safety Council, *The 2000 NEHRP Recommended Provisions for New Buildings and Other Structures*, FEMA 368, Washington, D.C., 374 pp., 2000; International Building Council, *International Building Code*, Falls Church, Va., 756 pp., 2000.

218. At present, there is debate among proponents of this stochastic method as to whether the source spectrum is best represented by a Brune spectrum with a single corner frequency or by a model having two corner frequencies. See D.M. Boore, Stochastic simulation of high-frequency ground motions based on seismological models of the radiated spectra, *Bull. Seis. Soc. Am.*, **73**, 1865-1894, 1983; G.M. Atkinson and D.M. Boore, Evaluation of models for earthquake source spectra in eastern North America, *Bull. Seis. Soc. Am.*, **88**, 917-934, 1998.

219. For example, see R.W. Graves (Simulating seismic wave propagation in 3D elastic media using staggered-grid finite-differences, *Bull. Seis. Soc. Am.*, **86**, 1091-1106, 1996) and K.B. Olsen (Site amplification in the Los Angeles Basin from 3D modeling of ground motions, *Bull. Seis. Soc. Am.*, **90**, S77-S94, 2000).

6

Research Opportunities and Requirements

Much remains to be learned about the physics and geology of earthquakes. Few problems are more challenging to science or strategically relevant to the nation, and few have a greater potential for mobilizing efforts to elucidate the fundamental geological processes that shape our planet. Over the 25 years of the National Earthquake Hazard Reduction Program (NEHRP), scientists in the United States and elsewhere have made substantial strides in understanding seismic phenomena, but many key questions remain to be answered. It is fair to say that a comprehensive theory of earthquakes—one that adequately describes the dynamical interactions among faults, as well as the basic features of rupture nucleation, propagation, and arrest—does not yet exist. The quest for such a theory is the principal driver for basic research in the field and the primary motivation for the geosystems approach.

Basic research on this problem connects closely with the practical issues in earthquake risk reduction through the common aim of forecasting earthquakes and anticipating their effects. Economic losses from earthquakes are escalating owing to rapid urbanization in tectonically active areas (Figure 1.2). As a practical matter, the nation must counter the threat by redoubling its efforts to prepare for and respond to earthquakes. The Committee on the Science of Earthquakes posed three questions regarding the Earth science component of this national effort: (1) What are the primary goals for the next decade of earthquake research? (2) What resources will be needed to achieve these goals? (3) In what areas can technological investments accelerate progress? This chapter presents the committee consensus on these issues in nine key areas where better knowledge is needed. The order of topics reflects the organization of the earthquake research effort laid out in Chapters 3 through 5 and is not intended to imply a prioritization.

6.1 FAULT CHARACTERIZATION

Almost all destructive earthquakes are generated by the sudden slippage of faults near the Earth's surface. Most dangerous faults in the continental crust (i.e., those that slip at average rates greater than a few millimeters per year) can be identified through a combination of geologic, geodetic, and seismologic measurements. New technologies in all three fields have enhanced the ability to locate active faults and assess their seismogenic potential (Chapter 4). These advances now warrant a substantial expansion in regional data gathering, including the densification of seismic and geodetic monitoring systems and more intense efforts to mine the geological record of fault activity. The goal should be a comprehensive catalog of fault information.

Goal: Document the location, slip rates, and earthquake history of dangerous faults throughout the United States.

Fault characterization at the detail required for comprehensive seismic hazard analysis will require nationwide efforts to improve capabilities in the three main observational areas of seismology, geodesy, and geology:

- a national seismic network capable of recording all earthquakes down to moment magnitude (M) 3 with fidelity across the entire seismic bandwidth and with sufficient density to determine the source parameters, including focal mechanisms, of these events; the location threshold for regional networks should reach M 1.5 in areas of high seismic risk;
- geodetic instrumentation for observing crustal deformation within active fault systems with enough spatial and temporal resolution to measure all significant motions, including aseismic events and the transients before, during, and after large earthquakes; and
- programs of geologic field study to quantify fault slip rates and determine the history of fault rupture over many earthquake cycles.

Major programs for augmenting seismic and geodetic instrumentation have been proposed by the NEHRP science agencies. The Advanced National Seismic System (ANSS), to be deployed by the U.S. Geological

BOX 6.2 EarthScope Initiative

EarthScope is an initiative to build a network of multipurpose instruments and observatories that will significantly expand capabilities to observe the structure and active tectonics of the North American continent.[1] The initiative, which was proposed by NSF with participation from NASA, USGS, and the Department of Energy, will deploy four new observational facilities:

1. **USArray** will improve the resolution of seismic images of the continental lithosphere and deeper mantle below the United States and adjacent regions. USArray will have three components: (1) a transportable telemetered array of 400 broadband seismometers designed to provide real-time data from a regular grid; (2) a flexible array of about 2400 portable seismometers (using natural and explosive sources) for high-density, shorter-term observations of key targets within the footprint of the larger transportable array; and (3) a fixed network of seismometers to provide continuous long-term observations and augment the USGS National Seismic Network.
2. **San Andreas Fault Observatory at Depth (SAFOD)**, an NSF and USGS project, will directly sample fault-zone materials (rock and fluids), measure a wide variety of fault-zone properties, and monitor a creeping and seismically active fault at depth. A 4-kilometer-deep hole will be drilled through the San Andreas fault zone at seismogenic depth above the hypocenter of the 1966 Parkfield earthquake. Fault-zone rock and fluid will be retrieved for laboratory analysis, and geophysical parameters, including seismicity, pore pressure, temperature, and strain, will be measured and monitored downhole and in adjacent areas. Instruments will be emplaced for long-term (20-year) observations of fluid activity, seismicity, and deformation.
3. **Plate Boundary Observatory (PBO)** will permit study of the three-dimensional strain field resulting from deformation along the Pacific-North American plate boundary. PBO will comprise (1) a backbone network of continuously recording, telemetered GPS receivers with 100- to 200-kilometer spacing to provide a long-wavelength, long-period synoptic view of the entire plate boundary zone from Alaska to Mexico, and (2) clusters of borehole strainmeters and seismometers and GPS receivers in tectonically active areas such as major faults and magmatic systems.
4. **Interferometric Synthetic Aperture Radar (InSAR)** will provide spatially continuous, intermittent strain measurements over wide geographic areas via a dedicated satellite mission to be carried out jointly between NASA, NSF, and USGS. InSAR images will complement the continuous GPS point measurements of PBO. The ideal mission will provide dense spatial (100 meters) and temporal (every eight days) coverage with vector solutions accurate to 2 millimeters over all terrain types.

A recent National Research Council report endorsed all four components of EarthScope and recommended that they be implemented as quickly as possible.[2] Deployment costs for the EarthScope instrumental systems are estimated to be $64 million for USArray, $17.4 million for SAFOD, $91.3 million for PBO, and $245 million for InSAR. Data analysis and management expenditures are expected to be $15 million to $20 million per year.

---

[1] *EarthScope: A New View into the Earth,* EarthScope Project Plan, EarthScope Working Group, 36 pp., October 2001.

[2] National Research Council, *Review of EarthScope Integrated Science,* National Academy Press, Washington, D.C., 61 pp., 2001.

SOURCE: National Research Council, *Basic Research Opportunities in Earth Science,* National Academy Press, Washington, D.C., 154 pp., 2001.

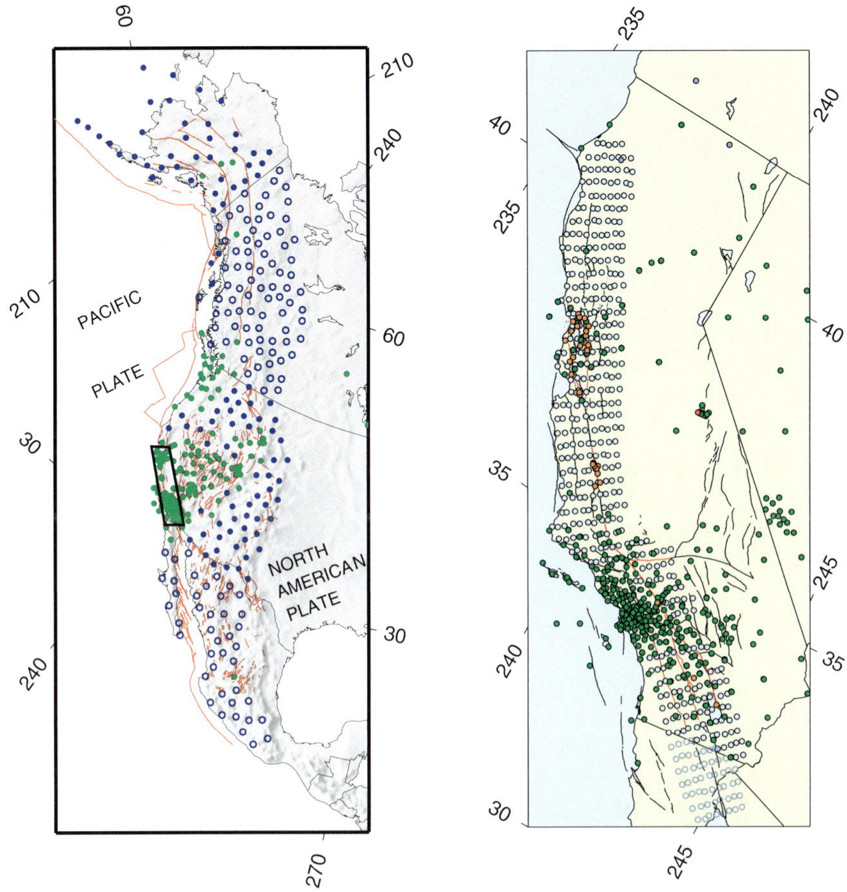

FIGURE 6.1 Map showing approximate configuration of the Plate Boundary Observatory. *Left panel*: Backbone array of continuous GPS receivers will capture the long-wavelength decadal field; instrument spacing varies between 100 and 200 kilometers. Filled blue circles give presently planned locations of stations in the United States; open blue circles, in Canada and Mexico. Existing sites are shown in green. *Right panel*: Strawman distribution of continuous GPS and strainmeters for an instrument cluster to cover the San Andreas fault system, comprising 400 new GPS receivers and 175 new strainmeters. Filled blue circles give presently planned locations of GPS stations in United States; open blue circles, in Mexico. Existing sites are shown in green. Existing strainmeters are shown in red. New strainmeters would be deployed along the most seismogenic portions of San Andreas fault system, highlighted in red. SOURCE: PBO Steering Committee, The Plate Boundary Observatory: Creating a four-dimensional image of the deformation of western North America, White paper providing the scientific rationale and deployment strategy for a Plate Boundary Observatory based on a workshop held October 3-5, 1999. Available at <http://www.earthscope.org>.

The long-term slip rates of most major faults in North America are either unknown or, at best, constrained by geologic measurements at only one or two sites. Geologic field work, combined with precise accelerator mass spectrometer dating using ^{14}C and cosmogenic isotopes (^{36}Cl, ^{10}Be, ^{26}Al), is necessary to quantify late-Pleistocene and Holocene slip rates on major faults. Similarly, deformation rates at longer (Quaternary to Tertiary) time scales, as delineated by $^{40}Ar/^{39}Ar$, fission-track, and uranium-thorium-helium dating, are required to resolve the evolution of slip and rock-uplift rates. The geologic mapping of faults and the measurement of fault slip rates and prehistoric events is coordinated by the USGS, state geological surveys, and multi-institutional research organizations, such as the Southern California Earthquake Center (SCEC). However, additional programmatic resources will be needed to characterize the active faulting at the comprehensive level envisaged in this report.

Slip-rate data are especially lacking in contractional provinces, where many questions still remain about how strain is partitioned among the major faults and between seismogenic faults and aseismic folding. New techniques in tectonic geomorphology could play a major role in addressing these issues. The evolving geomorphic character of former depositional surfaces inferred by combining detailed field mapping and geochronology with precise digital elevation models is particularly powerful in assessing the patterns of deformation associated with blind thrust faults, where the absence of surface ruptures confounds the traditional paleoseismic approaches. Laser altimetry from aircraft using light detection and ranging (LIDAR) systems can be used to investigate faulting and the surface deformations caused by buried faults. Shaded-relief maps generated from the Shuttle Radar Topography Mission's 30-meter data could be the basis for mapping Earth's active faults, folds, and seismically induced landforms between 60 degrees north and 60 degrees south. These data could be the topographic component of global seismic slope stability maps, created at scales that would be useful for long-term regional land-use planning. Significant impediments to the use of these new technologies for earthquake science include the cost of data, national security restrictions on availability, inadequate training, and the lack of cooperative programs.

Topographic data and analyses are necessary but not sufficient to understand the actively deforming lithosphere. In many cases, seismic reflection, deep and shallow boring, and other technologies are required to investigate the subsurface. Remote-sensing geophysical techniques, such as active source seismic reflection and refraction and gravity maps, are valuable for understanding how surface maps of the strain field from GPS, geologic mapping, and geomorphology continue into the subsurface. Heretofore, subsurface data for regional neotectonic studies have been scavenged principally from collections made by the resource extraction industry. Programs designed specifically to collect high-resolution subsurface images using the three-dimensional techniques developed in the search for petroleum could advance earthquake science significantly.

Earthquakes occurring beneath the oceans pose significant hazards to the world population, which is becoming increasingly concentrated along continental coastlines. The main risk in the coastal zones of Cascadia, Alaska, Japan, and Indonesia, for example, arises from thrust faults that intersect the surface many kilometers offshore but are fully capable of generating destructive ground motions and tsunamis. To date, most earthquake research has relied on data from arrays of seismometers, geodetic positioning of benchmarks, and geologic mapping confined to land areas. A major objective of earthquake science should therefore be to extend observational systems and data bases into the offshore environments (Box 6.3). For example, bathymetric maps by various new technologies have revealed ancient slumps and the traces of active faults on the seafloor. Systematic bathymetric mapping in active regions could reveal structures with significant seismogenic or tsunamigenic potential.

BOX 6.3 Marine-Based Earthquake Studies

The oceans offer natural earthquake laboratories and research opportunities not available on land. The geologic structure and history of the oceanic lithosphere, which is considerably different and usually much simpler than the continents, provide an excellent starting place for understanding the fundamentals of fault-zone processes. The seismogenic zone for major thrust faults can be sampled directly only by drilling in the offshore region. The international Ocean Drilling Program has already drilled a number of accretionary prisms to provide ground truth for three-dimensional structural models and sample the active décollement. Ship-based acoustic mapping and seismic reflection and refraction surveys can characterize the three-dimensional structure of marine fault systems at a fraction of the cost of terrestrial surveys. The NSF-sponsored Margins Program has placed a high priority on imaging plate boundaries in the offshore region, and it would be extremely advantageous to follow up such studies with earthquake observatories.

Although offshore earthquake observatories are considerably more expensive than land-based stations, there are no longer major technological impediments to installing them as an integral part of plate boundary networks. Such installations address all of the major aspects of the earthquake problem, but they are especially useful in determining how lithospheric deformation is controlled by lithospheric architecture in relatively simple and well-imaged geologic environments, understanding the details of rupture nucleation and propagation in major thrust systems, and characterizing the strain cycle in regions with the highest rates of relative plate motion. Project NEP-

(continued on following page)

BOX 6.3 *(continued)*

TUNE, a proposed fiber-optic observatory offshore the Pacific Northwest,[1] provides an ideal opportunity to establish a dense network of submarine seismic and geodetic stations in a region of high seismic risk. To be most effective, NEPTUNE data should be integrated with observations from the proposed Plate Boundary Observatory.[2] Deployment of state-of-the-art broadband stations on the ocean floor would also provide better coverage and resolution of seismic sources worldwide, as well as more complete tomographic coverage of the Earth's interior structure, which is a high priority of the International Ocean Network program. Much like the Program for the Array Seismic Studies of the Continental Lithosphere (PASSCAL) deployments on the continents, the semipermanent or temporary deployment of dense arrays of broadband ocean bottom seismometers can complement the sparse permanent stations for the investigation of regional seismicity on oceanic plate boundaries and in intraplate settings. Plans for the development of a PASSCAL-type facility for broadband deployments on the ocean floor are under way.

Seafloor geodesy is an especially challenging but potentially rewarding area for new research. Absolute gravimeters promise precise vertical positioning on the seafloor.[3] Horizontal positioning relies on acoustic ranging to surface floats precisely positioned by GPS.[4] The acoustic link is the largest source of error, due to uncertainties in sound velocity and currents, but despite its lower overall accuracy, such data would still be quite useful for submarine thrust faults with high rates of convergence.

---

[1] NEPTUNE Phase 1 Partners, Real-time, long-term ocean and Earth studies at the scale of a tectonic plate: NEPTUNE feasibility study, prepared for the National Oceanographic Partnership Program, June 2000. Available at <http://www.neptune.washington.edu/pub/documents/hi-qual_feas_study/hi-res_whole.pdf>.

[2] National Research Council, *Review of EarthScope Integrated Science*, National Academy Press, Washington, D.C., 61 pp., 2001.

[3] M.A. Zumberge, E.L. Canuteson, and J.A. Hildebrand, The utility of absolute gravity measurements on the seafloor, *Proceedings of the International Symposium on Marine Positioning*, M. Kumar, G.A. Maul, and G.S. Seeber, eds., Hanover, Germany, pp. 87-94, 1994.

[4] F.N. Speiss, C.D. Chadwell, J.A. Hildebrand, L.E. Young, G.H. Purcell Jr., and H. Dragert, Precise GPS/acoustic positioning of seafloor reference points for tectonic studies, *Phys. Earth Planet. Int.*, **108**, 101-112, 1998.

6.2 GLOBAL EARTHQUAKE FORECASTING

By combining studies of the geological record with seismic and geodetic monitoring, it is possible to forecast which fault systems will produce large earthquakes over long periods of time (decades to centuries). This type of long-range forecasting is essential for seismic hazard analysis, and further work on the problem should receive a high priority. Such research should address the dynamics of rupture propagation over multiple fault segments, the irregularity of the earthquake cycle, and the tendency of earthquakes to cluster in space and time. The diversity of earthquakes and their geological environments necessitates a global approach. Great earthquakes ($M \geq 8$) are infrequent in the United States, but events of this magnitude occur about once per year on a worldwide basis. The experience collected through global studies of seismicity is therefore crucial to earthquake forecasting at the upper limits of the magnitude range. For example, although the Cascadia subduction zone has not produced a great earthquake in historical times, much can be inferred about the prospects of future events from the study of subduction-zone seismicity in other regions. Likewise, an improved understanding of the San Andreas fault will likely come from new knowledge about analogous strike-slip faulting in New Zealand, Turkey, and China, or perhaps from the study of the anomalous, slow earthquakes that characterize oceanic transform faults (Figure 6.2).

Goal: Forecast earthquakes on a global basis by specifying accurately the probability of earthquake occurrence as a function of location, time, and magnitude, as well as the magnitude limits and other characteristics of likely earthquakes in a given place.

Current forecasting schemes rely on controversial assumptions about stress, deformation, and earthquakes (e.g., characteristic earthquake behavior), so the immediate objective should be to formulate rigorous statistical tests of such forecasts. The evaluation of various forecasting methods can then proceed as new earthquakes occur and past ones are discovered, and the results can be used to improve the ideas and assumptions behind the forecasts. Global scope is important for capturing enough earthquakes to test hypotheses rapidly and to sample different tectonic and stress conditions.

Earthquake forecasting requires the basic information on fault structure and kinematics described in the previous section. The Global Seismic Network (GSN) deployed by the Incorporated Research Institutions for Seismology and the USGS just recently reached its design goal of 125 stations; in conjunction with other networks, it is furnishing unprecedented data on the source processes during major earthquakes in remote areas (see Section 4.1). Another monitoring network of obvious importance to global forecasting is the International GPS Service, which furnishes a well-defined global reference frame for regional studies of tectonic deformation (*1*). These facilities deserve long-term support and, in the case of the GSN, extension to include seafloor observatories (Box 6.3).

Fault and deformation data are not yet adequate in many seismically active parts of the world. However, new and recently expanded programs are beginning to provide these data on a global scale, including

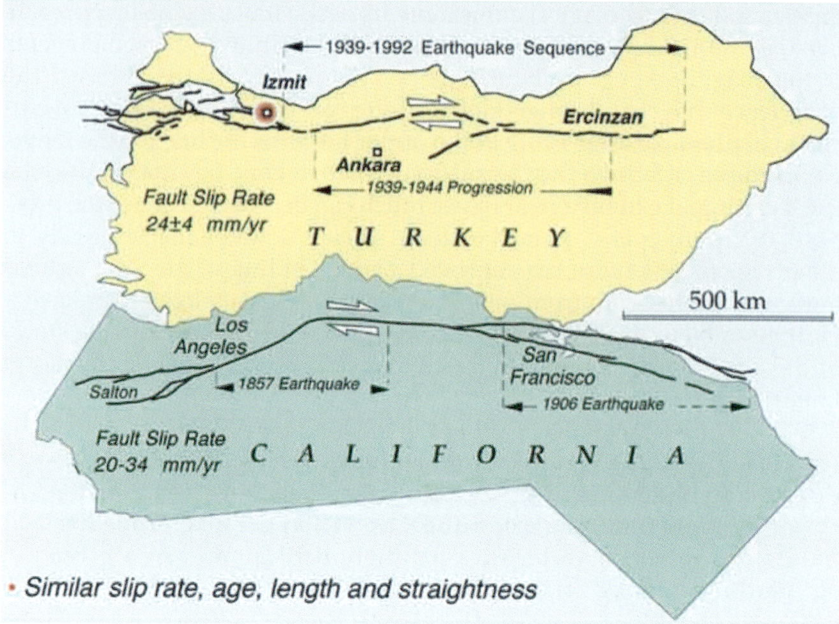

FIGURE 6.2 Comparison between major, plate-bounding active strike-slip faults in California (San Andreas fault) and Turkey (North Anatolian fault) showing similarities in length, slip rate, and general geometry of the fault zones. Both faults have a history of frequent earthquakes of M 7 and higher, with irregular time intervals of decades to centuries between major events. The westward progression of earthquake epicenters in northern Turkey, beginning in 1939 and culminating in the Izmit earthquake of 1999, is reason for concern because the city of Istanbul (population 15 million) lies only 100 kilometers to the west of the Izmit epicenter. Major devastation and loss of life are to be expected if earthquake activity migrates westward into this urban area. It is not known if similar patterns in earthquake migration might occur along the San Andreas fault, but this question could be addressed through paleoseismological studies of San Andreas earthquakes over the last 10,000 years. SOURCE: R. Stein, U.S. Geological Survey.

global and regional earthquake monitoring, regional strain-rate measurements on many active plate boundaries, paleoseismic investigations of earthquake series on major faults throughout the world, neotectonic studies and satellite topography to map and characterize faults on land and near shore, and marine geophysical surveys to study faults on continental margins and in ocean basins.

Given the scope of the problem, broad-based international collaboration is essential. Pioneering efforts such as the Global Seismic Hazard

Assessment Program (Section 3.3), the Risk Assessment Tools for Diagnosis of Urban Areas Against Seismic Disasters, the Global Strain Rate Map Project of the International Lithosphere Project (ILP), the Global Fault Mapping Project, and the ILP Working Group on Earthquake Recurrence Through Time are providing uniform standards and data access, and they should be expanded with aggressive data-gathering efforts that exploit the new technologies described in this report.

6.3 FAULT-SYSTEM DYNAMICS

Much less is known about the feasibility of earthquake prediction over intervals of years to decades than about long-term forecasting. No algorithm for intermediate-term prediction has unequivocally demonstrated predictive skill at a statistically reliable level. However, there are both observational and theoretical reasons to believe that large-scale failures within some fault systems may be predictable on intermediate time scales, provided that adequate knowledge of the history and present state of the system can be obtained. It is not yet clear whether probabilistic forecasting methods can be devised that take advantage of this potential predictability, but such methods could contribute significantly to the reduction of earthquake losses. Therefore, basic research on the issue should be vigorously pursued, including a broad spectrum of research directed toward gaining a better fundamental understanding of fault-system dynamics.

Goal: Understand the kinematics and dynamics of active fault systems on interseismic time scales, and apply this understanding in constructing probabilities of earthquake occurrence, including time-dependent (non-Poissonian) earthquake forecasting.

The study of fault systems relies heavily on the information supplied by earthquake geology, particularly paleoseismology and tectonic geomorphology. A good example is the long-standing issue of regional earthquake clustering, in which periods of high earthquake activity ("seismic storms") are separated by periods of relative quiescence. Clustering is clearly crucial to earthquake forecasting that is poorly constrained by the short catalogs of instrumental seismology. Paleoseismic techniques can be used to identify sequences of slip events at particular points on a fault, but these events must be precisely correlated in time and space to investigate clustering. The dense sampling and precise dating needed for this task are still lacking even along well-studied faults such as the San Andreas.

Seismologic data are also essential for testing hypotheses regarding earthquake clustering, including foreshocks and aftershocks. Modeling seismicity on a fault network relies on accurate and complete seismic catalogs for the recognition of regional patterns in seismicity and for detailed studies of specific earthquake sequences. Upgrading the regional networks proposed as part of the ANSS will greatly facilitate seismological studies of fault systems in the United States. These instrumental improvements will enhance earthquake information and encourage the development of new seismological products, such as the cataloging of fault planes, rupture lengths, and slip propagation directions for moderate-size earthquakes.

Better structural data are needed on fault segmentation, along with an improved mechanical understanding of the role of segmentation in fault rupture. Work on fault-zone complexity suggests fundamental differences in behavior between mature and immature faults. If segment boundaries play a key role in the termination of ruptures, then highly segmented faults may tend to be more characteristic in their behavior or at least more predictable in the lateral extent of future ruptures. A long, smooth fault such as the San Andreas may not have any "hard" segment boundaries, making the size of ruptures more sensitive to time-dependent stress heterogeneities.

Changes in Coulomb stress parameters resulting from large earthquakes have been invoked as a quasi-static mechanism for modifying seismicity rates. If such dynamical interactions significantly affect the timing of earthquakes on nearby faults, they must be accounted for in any model of earthquake recurrence. An important issue is the role of time-dependent phenomena, such as transient fault creep, fault healing, poroelasticity, and viscoelasticity, in stress transfer. The best constraints on these processes come from near-fault deformations following large earthquakes, which can be measured using GPS and InSAR geodesy.

A true understanding of fault-system dynamics will be reached only by integrating the disparate observations of stress, strain, and rheology into self-consistent models that can be tested against observations not yet collected. Simulations of earthquake occurrence on fault networks are required to understand the behavior of natural fault systems and to address fundamental questions relating to earthquake occurrence, such as the effects of stress interactions and prior earthquakes in determining earthquake probability (Figure 6.3). The practical objective of this research is to develop procedures that can assimilate all information on fault-system behaviors into probabilistic forecasts and update these forecasts consistently based on seismic activity and other new information. A proper interpretation of fault systems must begin with a detailed representation of the active structural elements; in particular, it will be necessary to

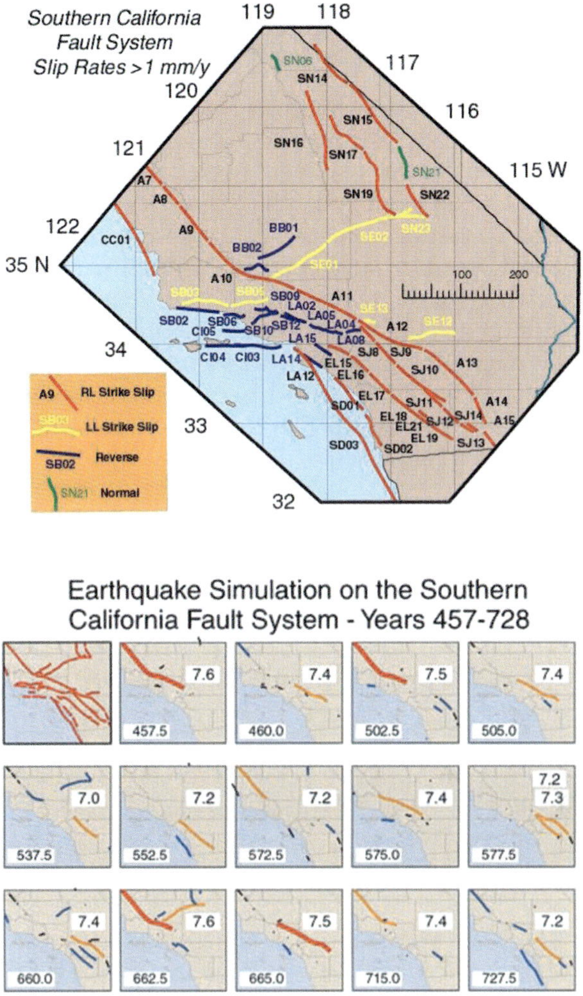

FIGURE 6.3 Prototype "earthquake simulator" for southern California, in which fault driving stress is balanced against fault frictional resistance using a two-dimensional, quasi-static approximation. Top panel shows faults included in the simulation; colors indicate faulting style. Bottom panel plots large (M > 6) earthquakes calculated for a 270-year interval. Line color corresponds to different magnitudes. The numbers at the lower left of each panel indicate time in years, and the frames update on the occurrence of an M > 7 event. Physically based simulations such as these serve as useful platforms for hazard analysis and data assimilation. SOURCE: S.N. Ward, A synthetic seismicity model for southern California: Cycles, probabilities, and hazard, *J. Geophys. Res.*, **101**, 22,393-22,418, 1996. Copyright 1996 American Geophysical Union. Reproduced by permission of American Geophysical Union.

quantify the representation of major active faults in all three spatial dimensions. Further objectives should include the following:

- the integration of other information, such as three-dimensional seismic velocities and attenuation parameters, surface topography, surface geology, and subsurface geologic horizons, into unified structural representations of active fault systems;
- the use of these three-dimensional representations as the basis for combining all available data on geodetic velocities and fault slip rates into kinematically consistent models of deformation zones; and
- the extension of these kinematical representations to fully dynamical models that incorporate realistic rheologies, boundary tractions, and body forces. The latter can be inferred by combining surface topography and gravity with density variations measured at the surface and inferred from seismic tomography.

A dynamical description of fault system behavior must include the state of stress and how it changes with time. Stress can be measured directly in the near-surface environment accessible by mining and drilling, but the principal way to infer the stress field at depth is through modeling. The four observational facilities of the EarthScope program will provide critical data for this purpose, including precise measurements of surface deformation gradients and their temporal evolution (PBO and InSAR), detailed images of subsurface structures that control stress and strain heterogeneity (USArray), and better knowledge of deformation processes at depth (the San Andreas Fault Observatory at Depth [SAFOD]).

The research objectives outlined above will require (1) the ability to manipulate large data sets from geology, geodesy, and seismology; (2) the development of novel techniques for analyzing and interpreting these data sets; and (3) substantial information technology (IT) resources for developing, verifying, and maintaining community models and making them available to a heterogeneous, widely distributed group of users. The considerable experience accumulated by the petroleum industry over the last decade has shown that the construction of faithful and flexible structural representations is a difficult task, requiring significant resources and substantial reliance on advanced IT tools.

6.4 FAULT-ZONE PROCESSES

The considerations detailed in Section 5.3 highlight the importance of determining the composition, structure, and physical state of fault-zone materials and damaged border zones. At larger scales there is a need for better characterization of fault junctions and the structure and mechanical properties of fault-jog materials over which, or through which, rupture jumps in transferring slip from one fault segment to another.

Goal: Characterize the three-dimensional material properties of fault systems and their response to deformation through a combination of laboratory measurement, high-resolution structural studies, and in situ sampling and experimentation.

More information on microscale processes is needed to formulate realistic macroscopic representations of the strength variations and the dynamic response of fault materials. Laboratory research has already provided promising representations of fault friction for understanding the nucleation of slip instability and other processes taking place at low strain rates (Section 4.4). Major advances in experimental techniques and facilities will be needed to elucidate the dynamic phase of fault response, in which rapid large slip rates and large slips may cause extreme temperature excursions and weakening by pressurization of pore fluids, melt generation, or other processes that are yet to be well documented. These phenomena can be partly addressed by high-speed rotary shear experiments, but challenges remain to study sliding under realistic triaxial stressing, to determine the effects of abrupt changes in sliding rate, and to confine rapidly heated fluids including melt. Advanced experimental techniques (e.g., shock and impact loading, stored energy bars, optical sensors of motion and thermal radiation, high-speed photography) in current use by other scientists concerned with dynamic response of materials have already contributed to understanding certain issues in dynamic friction and faulting (see Chapter 5) and should be employed more widely in laboratory earthquake studies.

Detailed investigations of fault properties and processes, in particular fluid-dominated processes, are needed on spatial dimensions from centimeters to kilometers to fill the scale gap between laboratory studies of rock mechanics and regional geophysical studies of active faulting. Such localized studies might best be done by careful investments in a set of "natural laboratories," where high-resolution methods focused on imaging fault structure and deformation processes could be combined with systematic programs for in situ sampling and experimentation. Examination of exhumed faults explores the mechanical importance of shear localization structures and the evidence of fluids and local melting. Sample extraction by ultradeep drilling is feasible, at least at down to shallower seismogenic depths (2 to 6 kilometers) for many surface-breaking faults. Borehole data are needed to extrapolate laboratory results on laboratory-scale samples to natural faults, to elucidate the generation of fault-zone structure (ultracataclastic core, cracked and damaged border zone), and to clarify why some faults creep whereas others slip in large earthquakes. Deep drilling of the Nojima fault in Japan has delivered samples from a depth of 2 kilometers. The EarthScope initiative proposes to construct SAFOD, which will sample the fault, monitor its seismicity and strain, and perform in situ experiments to 4-kilometer depth at Parkfield, California (Figure 6.4), where the USGS maintains a long-term, multidisciplinary program for the study of earthquake processes.

Focusing observations on carefully chosen, well-instrumented areas facilitates the coordination of activities across multiple groups of investigators, encouraging the types of multidisciplinary studies that are essential to understanding earthquake processes and fault-system behaviors. It also provides a long-term basis for capitalizing on field-based research. If

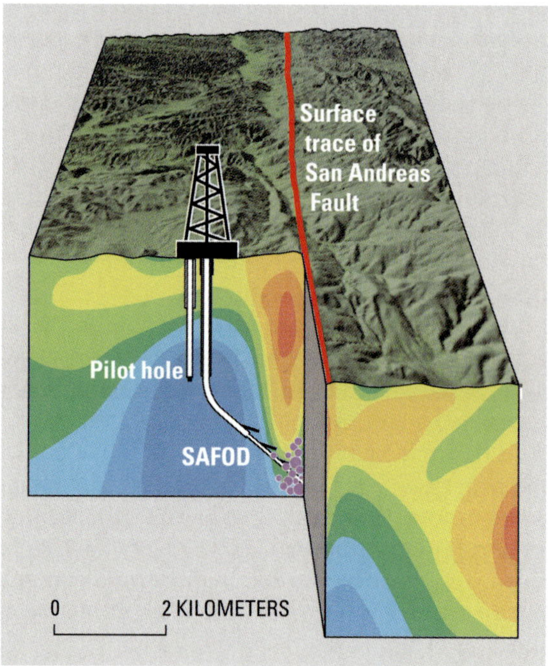

FIGURE 6.4 Schematic cross section of the San Andreas fault zone at Parkfield, showing the SAFOD drill hole proposed as part of the EarthScope project, and the pilot hole being drilled in 2002. Violet dots represent areas of persistent minor seismicity at depths of 3 to 4 kilometers. The colors in the subsurface show electrical resistivity of the rocks as determined from surface surveys; the lowest-resistivity rocks (red) above the area of minor earthquakes may represent a fluid-rich zone. SOURCE: EarthScope Working Group.

the investigations are well directed, and the data properly analyzed and archived, then the returns on previous research investments can be compounded as more data are collected. Each observational study within the natural laboratory adds to the database, improving the context for future work. Natural laboratories are well suited to field operations that are logistically complicated and expensive, as in the collection of spatially dense data sets and the monitoring of phenomena over extended time intervals. The committee endorses a recommendation in the recent National Research Council report *Basic Research Opportunities in Earth Science*, to establish an Earth Science National Laboratory Program within the NSF (2). Many types of earthquake natural laboratories would be able to deliver new data on fault-zone processes. Pore-fluid pressurization in deep boreholes could be used to induce seismic events, which could be recorded by borehole seismometers to characterize nucleation and the early stages of rupture dynamics (a modern-day update of the Rangely oil field experiments conducted in the 1960s; see Chapter 2). Data could be collected from seismic networks in deep mines where small earthquakes are rapidly, and to some extent controllably, generated by the advance of mine faces and other underground developments.

Synoptic studies in natural laboratories could furnish an important observational base for developing theoretical and numerical models of fault systems, and they could yield the essential data by which these models are ultimately validated. Fault zones are complex structures at all scales, and numerical simulations are required to integrate laboratory observations, field data on fault-zone structure and composition, and pore-fluid interactions to obtain constitutive representations of macroscale faulting. Simulations of fault-zone properties must account for many nonlinear phenomena, such as thermal transients during large earthquakes, that may directly alter fault strength (or even result in melting) and may couple with pore-pressure effects. Rigorous physical modeling begins with testable microscale processes and carries out the appropriate analyses that scale up through the geometric complexities of fault networks to understand the implications for natural events. Dynamical simulations at various scales will be needed to assess the discrepancies among laboratory-based friction laws, observed fault-system behaviors (e.g., earthquake productivity, postseismic response), and seismological data on large earthquakes (e.g., fracture energies, particle velocities and accelerations).

6.5 EARTHQUAKE SOURCE PHYSICS

Better knowledge of earthquake source physics on the short time scales of fault rupture will improve the understanding of how strong ground motions are generated, as well as the processes that lead up to the fault ruptures that cause earthquakes. Earthquake scientists are not optimistic about the prospects for short-term earthquake prediction schemes that are based on observations of precursory behavior. Owing to the chaotic behavior of fault nucleation and rupture, accurate prediction of the time, place, and size of specific large earthquakes may not be possible. Nevertheless, near-field observations before and during large earthquakes are too few and too limited to rule out categorically the feasibility of short-term earthquake prediction (3), and further work on the problem is warranted. The key scientific issues involve the relationships between the dynamical processes that govern the nucleation, propagation, and arrest of rupture. More certainly, improvements in long-term forecasting and the still-bright prospects for intermediate-term prediction will depend on understanding the dynamical connection between the evolution of the stress field on interseismic time scales of decades to centuries and the stress heterogeneities created and destroyed during the few seconds of an earthquake.

Goal: Understand the physics of earthquake nucleation, propagation, and arrest in realistic fault systems and the generation of strong ground motions by fault rupture.

The most important data for constraining rupture processes are close-in seismographic and strainmeter recordings during earthquake ruptures, combined with geologic measurements, GPS measurements, and InSAR images of the near-fault deformation field. Effective use of this information in the study of rupture physics will depend on the following:

- detailed three-dimensional models of seismic wave-speed structure needed to account for propagation effects in the waveform data;
- fine structure of fault zones from geologic mapping and remote sensing, seismicity information, and trapped-wave studies of low-velocity zones and anisotropy;
- development of new techniques to obtain experimental data on rock friction at high sliding velocities and large slips;
- better understanding of fault-zone fluid processes from geologic observations on exhumed faults and in situ observations from SAFOD and other drilling experiments; and
- constraints on the state and evolution of stress, including a better characterization of past earthquake history and stress-transfer mechanisms.

Numerical simulations are central to research in rupture dynamics, because they provide the physical basis for linking laboratory experiments and field data on small-scale fault-zone processes with large-scale observations of seismic waveforms and geodetic deformation fields. Full three-dimensional dynamical simulations are needed to address issues regarding fault-zone complexities (e.g., nonplanarity of fault surfaces, stepovers, and branches in fault networks) and the selection among competitive rupture paths. Because of the wide separation between the inner scale of faulting (e.g., frictional breakdown at the rupture front, as small as 10 meters) and its outer scale (e.g., total rupture length up to hundreds of kilometers), the computational difficulties are truly enormous, requiring terascale resources from the national computational grid.

Rupture modeling involves nonlinear processes and geometrical complexities on various length scales, and there is no consensus methodology optimal for all aspects of the problem. A desirable simulation framework, therefore, would allow user-supplied rupture modules to be embedded in, and coupled to, a fast, simple, three-dimensional wave propagation model (e.g., a finite-difference code). With appropriate links to community structural models, this approach would furnish a framework for comparing waveform computations with recorded seismic data from past events and, thus, for improving the models, as well as the predictive simulations, for future earthquake scenarios. An important research objective is the validation of rupture-dynamics simulations using a set of "reference earthquakes" as a basis for comparison. Establishing these reference events will require (1) improved analysis of the geologic, geodetic and seismologic observations, especially strong-motion data, from these events; and (2) the collection of additional data on the three-dimensional structure and properties of the faults that ruptured.

Although many of the basic scaling properties of shallow earthquakes are shared by intermediate- and deep-focus earthquakes, it would appear unlikely that the same microscopic physics pertains to fault ruptures at all depths. A truly comprehensive theory of earthquakes must be able to explain the similarities and differences of earthquakes at different depths. Deep-focus events are difficult to study because the closest observations are necessarily hundreds of kilometers above the hypocenters. Global seismic networks and temporary deployments of portable seismometer arrays above active subduction zones are yielding the most direct data on natural events, while laboratory studies of deformation are furnishing the information on the microscale physics of earthquake instabilities at the high pressures and temperatures of descending slabs.

6.6 GROUND-MOTION PREDICTION

Seismic shaking is influenced heavily by the details of how seismic waves propagate through complex geological structures. Strong ground motions can be enhanced by resonances in sedimentary basins and wave multipathing along sharp geologic boundaries at basin edges, as well as by amplification due to near-site properties. Although near-site effects such as liquefaction can be strongly nonlinear, most aspects of seismic-wave propagation are linear phenomena described by well-understood physics. Therefore, if the seismic source could be specified precisely and the wave velocities, density, and intrinsic attenuation were sufficiently well known, it would be possible to predict the time history of strong motions by a numerical calculation. The research goal is to use this physics-based approach to go beyond empirical attenuation relationships in characterizing strong ground motions and their secondary effects.

Goal: Predict the strong ground motions caused by earthquakes and the nonlinear responses of surface layers to these motions—including fault rupture, landsliding, and liquefaction—with enough spatial and temporal detail to assess seismic risk accurately.

Research in this field should focus on urban areas where the consequences of large earthquakes are most severe (Box 6.4). In particular, past earthquakes have demonstrated that areas of damage are often localized in highly populated sedimentary basins near active faults. Site-specific information about the time histories of shaking will be needed for performance-based design of structures in such settings. The challenge of urban hazard mapping is to predict ground-motion effects over an extended region with an acceptable level of reliability. Ground-motion maps for the 1989 Loma Prieta and 1994 Northridge events demonstrated that while the major urban regions of California were sufficiently instrumented to determine a first-order distribution of ground motions, the networks were not dense enough to provide a direct correlation of local damage patterns with ground-motion levels. Ground-motion simulations can potentially be used to interpolate the recorded data for more detailed seismic zonation, provided that the subsurface structure is adequately characterized.

Plausible objectives for the next 10 years are (1) to determine the structure of high-risk areas well enough to model the surface motions from a specified seismic source at all frequencies up to at least 1 hertz and (2) to formulate useful, consistent, stochastic representations of surface motions up to at least 10 hertz. At present, computer simulations in areas where the three-dimensional structure of the crust is best known, such as the Los Angeles region, can model the peak amplitudes only below about 0.3 hertz (Figure 6.5). To extend these calculations to the higher frequencies needed for engineering applications, a much greater volume of seismological data will be needed to map the three-dimensional structure of the crust. Future densification of urban seismic networks should strive for

BOX 6.4 Urban Hazard Characterization

An important objective of earthquake research is to develop the capability to produce detailed maps of earthquake shaking and other seismic hazards for urban areas with high seismic risk, as well as site-specific time histories of ground shaking in these urban areas expected during large earthquakes. These hazard maps and time histories should incorporate local site response and basin effects, rupture directivity and dynamics, and time-dependent probabilities of large earthquakes on all relevant faults. The steps necessary for improving the prediction of ground motions in urban areas include the following:

- characterization of the recurrence times and time-dependent probabilities of large earthquakes on faults capable of producing strong motions, as described in Section 6.1;
- instrumentation of urbanized sedimentary basins with dense seismograph arrays that can record both weak and strong ground motions and sensors installed within structures to evaluate building performance during strong shaking, as envisaged under the 1999 ANSS program plan; these arrays should be supplemented with borehole arrays that measure how motions change as waves propagate upward from rock through the sedimentary section;
- development of detailed three-dimensional models of seismic velocities, density, and attenuation using seismic reflection-refraction, gravity, travel times of seismic waves, surface-wave dispersion, and waveform modeling; detailed measurements of shear-wave velocity and nonlinear properties are required in the upper 300 meters, data on the latter would come from borehole arrays; recordings of small and moderate earthquakes can be used to validate the three-dimensional models;
- simulations of ground motions for potential large earthquakes using validated three-dimensional models of seismic velocity and attenuation to make detailed maps of various shaking intensity parameters; the most advanced simulations would include dynamic rupture, three-dimensional basin effects, and nonlinear soil response—the latter would be used to assess liquefaction potential (with surficial geology and water table information) and landslide probability (with slope data and material properties); and
- implementation of the research results through cooperative efforts with engineers, risk evaluators, urban planners, and emergency management planners and education of the general public.

station spacing as short as 1 kilometer, or even less in some areas. These dense arrays should have sufficient dynamic range to record the strong shaking from large events as well as the weaker ground motions from small and moderate earthquakes. Small earthquakes recorded on such networks will furnish the dense data coverage needed for constructing three-dimensional models of sedimentary basins, while the strong-motion data will be essential for validating the ability of numerical models to predict ground shaking in future large earthquakes.

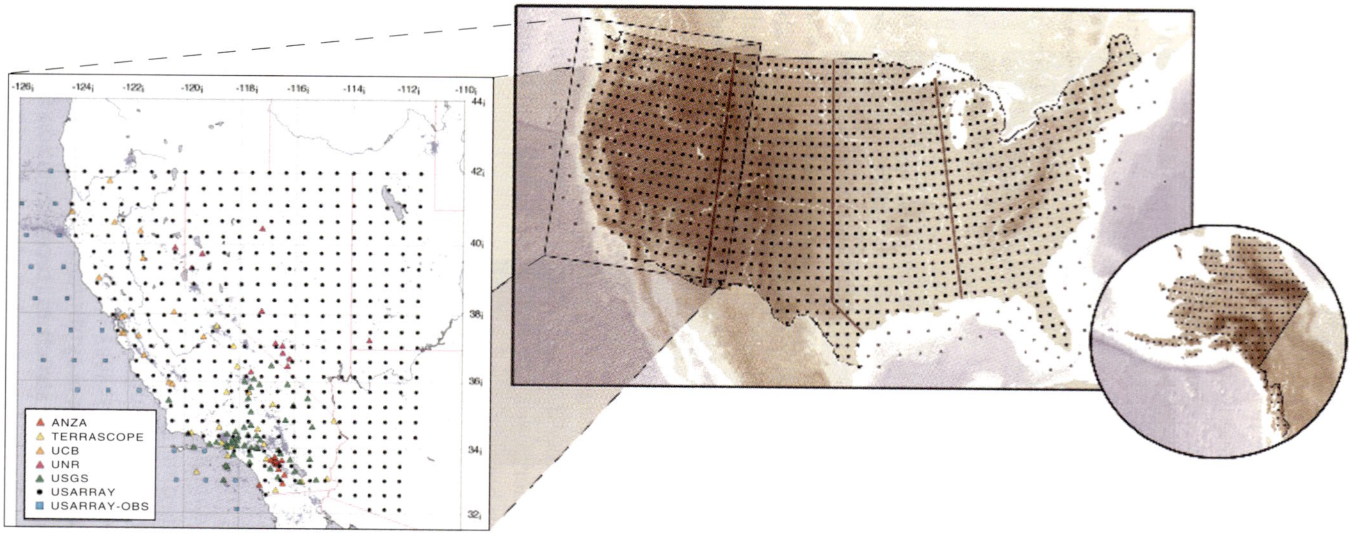

FIGURE 6.6 Configuration of the transportable Big Foot seismic array proposed as a part of the USArray component of the EarthScope initiative. The underlying grid shows a regular spacing of 70 kilometers, resulting in approximately 1600 sites in the lower 48 states (actual deployment patterns will depend on site availability and therefore be less regular). USArray will begin operations in California, encompassing the SAFOD site at Parkfield. The western United States detail (*left panel*) shows the nominal coverage of the first full deployment of the 400 portable stations in the Big Foot array. Existing broadband stations are shown as colored symbols. SOURCE: EarthScope Working Group.

other information on major features, particularly faults. Numerical modeling methods for three-dimensional elastic wave propagation are now fairly mature, with several efficient, capable codes in regular use; a framework thus exists for the construction of community wave propagation models that would comprise many of the functional features needed for regional-scale ground-motion simulation. However, the ground-motion problem is further complicated by the fact that source excitation and wave propagation are intimately coupled, both as forward problems (inertial effects are important in rupture dynamics) and as inverse problems (excitation and propagation must be untangled to interpret seismograms). The coupling becomes stronger as the ground-motion frequencies become higher. A better understanding of rupture dynamics, the subject of the previous section, is therefore crucial to ground-motion modeling.

Developing flexible and feasible methods for assimilating wavefield observations (e.g., from USArray and ANSS) into updated three-dimensional structural models—the structural inverse problem—is both a conceptual and a computational challenge. In principle, the inversion can be set up as a linearized perturbation to an initial three-dimensional structure, and the sensitivity of the model parameters to various data values can be computed numerically. In practice, the huge computations will require substantial theoretical work and an enhanced information infrastructure.

Improved regional models will allow a better separation of source and propagation effects, facilitating more accurate hypocenter locations, focal mechanisms, and higher-order source parameters, including better imaging of rupture kinematics. The latter will be significant for initializing the stress field in fully dynamic rupture calculations. As the three-dimensional structures used in waveform simulations improve, the need for "site effects" to correct for incomplete descriptions of wave propagation should be reduced, which will help to isolate the true near-surface site response. Numerical codes that can account for soil nonlinearity are numerous, but computational approaches continue to evolve. The modeling framework must therefore accommodate a range of soil-modeling methodologies by developing a consistent interface for coupling ground-motion calculations to nonlinear codes.

Ground-motion prediction has the potential to enhance greatly local seismic zonation by including effects of rupture directivity, the orientation of the fault (e.g., the hanging-wall effect), and structures such as sedimentary basins, basin edges, and buried folds and faults. The development of procedures that include these effects ultimately will make earthquake scenario maps and seismic zonation more tractable, although it will greatly increase the complexity of their production. Owing to these complexities, nonscientists who could employ this information in performance-based design, risk evaluation, and emergency management need to be educated about the capabilities and implementations of research results. In particular, the procedures described above can be applied in near real time immediately following an earthquake to improve the ground-motion maps now available (Figure 1.5) for guiding emergency response.

6.7 SEISMIC HAZARD ANALYSIS

Seismic hazard analysis provides the methodology for combining all information on seismic hazards into probabilistic and scenario-based predictions of the ground motions. This input is critical to earthquake engineering and design, as well as to the predictions of human casualties, damage to the built environment, and economic losses that can be expected from future earthquakes. Much applied research is still needed to improve the earthquake forecasts, attenuation relations, and site-response factors needed to apply seismic hazard analysis techniques. The methodology of seismic hazard analysis is now fairly mature, but it is highly empirical, based on many simplifications and approximations. Current attenuation relationships fail to capture much of the variance observed in the ground motions from actual earthquakes (Section 5.6). A recent study by the Southern California Earthquake Center concluded that "any model that attempts to predict ground motion with only a few parameters will have substantial intrinsic variability. Our best hope for reducing such uncertainties is via waveform modeling based on the first principles of physics" (4). The time is right to integrate physics-based models of earthquake processes into an improved scientific framework for seismic hazard analysis and risk management that explicitly considers the time dependence of seismic phenomena.

Goal: Incorporate time dependence into the framework of seismic hazard analysis in two ways: (1) by using rupture dynamics and wave propagation in realistic geological structures to predict strong-motion seismograms (time histories) for anticipated earthquakes, and (2) by using fault-system dynamics to forecast the time-dependent perturbations to average earthquake probabilities.

Seismic hazard analysis employs many of the data sets described under the other major science issues. These databases should be available as part of the community modeling framework, since alternative earthquake source models cannot be compared rigorously unless they are based on common sets of data such as earthquake catalogs, GPS velocity vectors, and fault geometries and slip rates. Similar requirements exist for the strong-motion data sets and seismic velocity models that are used in the development and validation of ground-motion prediction models. These community data sets and models should be readily accessible on-line and have audit trails that allow them to be traced back to the original sources. A computational infrastructure is needed to allow real-time, dynamic access to current data that reside at community data centers. The challenge will be to construct data bases and models that can represent uncertainties and accommodate a wide variety of potential uses, while still encouraging and incorporating creative science.

Few recordings of strong ground motions from earthquakes greater than M 7 are available for earthquake engineering research (5). Consequently, ground-motion models may not adequately represent the damage potential of large earthquakes. For example, the response spectrum, which is the most common representation of ground motion for engineering design, is not very sensitive to the duration of strong motion, which is significant for large earthquakes. Also, the response spectrum may not provide an adequate representation of the damage potential of long-period ground motions from large earthquakes. Near-fault displacement pulses can place very large deformation demands on flexible frame buildings. These buildings are sometimes designed to withstand deformation that extends as much as 10 times beyond the elastic limit. The design of these buildings for nonlinear response is critically dependent on the reliability of the ground-motion level used for design, especially at the longer periods that scale strongly with earthquake magnitude. In particular, there is an urgent need for an improved representation of the near-fault rupture directivity pulse for use in earthquake engineering, because the response spectrum does not provide an adequate representation of pulse-type motions.

Probabilistic seismic hazard analysis (PSHA) and waveform modeling are complementary approaches to seismic hazard analysis. For example, the composite PSHA hazard estimate can be disaggregated to find the most menacing scenarios for a given site (e.g., Figure 3.10), and ground-motion simulations can be performed to generate time histories for those events. The notion of full waveform modeling of scenario earthquakes is not new to seismic hazard analysis of course, but opportunities exist for a greater level of interdisciplinary research, including collaboration with earthquake engineers, to develop a methodology appropriate for design and risk management applications. Over the long term, the most effective strategy for reducing the economic losses in earthquakes will be through the design of structures to withstand seismic shaking at specified levels of performance (see Section 3.5). Performance-based design is based on prediction of the deformations and failures of structural systems and their likelihood. For severe earthquake ground motions, such prediction requires representative time histories for dynamic analysis of these complex, highly nonlinear systems. Reliable procedures for simulating ground-motion time histories have to be developed, tested rigorously against the available strong-motion data, and then applied in earthquake engineering research and practice. Significant research is needed on how to use waveform modeling to characterize the probability distributions of ground-motion time histories (or parameters derived from those time histories) in a way that properly accounts for both aleatory and epistemic uncertainties.

Constructing the appropriate probability distributions using brute-force Monte Carlo techniques requires multiple runs for many potential earthquake sources and geologic structures and thus increases demands on the resources for numerical simulation and the analysis of simulation output. Efforts to analyze the problem this way have required in excess of 1000 accelerograms and nonlinear structural analyses. Such solutions are unacceptable in standard practice. It will be more efficient and effective to combine seismic hazard analysis for a structure-specific, ground-motion intensity measure (or vector of such indicators) with a limited number (order of 10) of nonlinear analyses to produce the desired product—which might be, for example, the annual probability that a building's maximum interstory drift (relative displacement) exceeds a specified level. Similar approaches should be feasible for other engineered facilities, such as earth dams, transportation networks, and urban infrastructure. The implementation of this engineering interface can be facilitated through multidisciplinary research centers such as the SCEC, Pacific Earthquake Engineering Research Center, Mid-America Earthquake Center, and Multidisciplinary Center for Earthquake Engineering Research.

Substantial opportunities for collaboration with the earthquake engineering community are presented by the NSF Network for Earthquake Engineering Simulation (NEES), a nationwide, distributed collaboratory comprising a variety of engineering testing facilities and simulation capabilities, currently under construction. Ground-motion data and simulations are critical inputs to the NEES project, and an active collaboration between NEES and earthquake science will be required.

6.8 SEISMIC INFORMATION SYSTEMS

A major advance in seismic monitoring and ground-motion recording is the integration of high-gain regional seismic networks with strong-motion recording networks to form comprehensive seismic information systems. On regional scales, such information systems provide essential information for guiding the emergency response to earthquakes, especially in urban settings. Seismic data from a regional network can be processed immediately following an event and broadcast to users, which include emergency response agencies and responsible government officials, utility and transportation companies, and other commercial interests. The parameters include traditional estimates of origin time, hypocenter location, and magnitude, as well as maps of ground motions pertinent to damage assessments. The need for such systems became clear after the 1994 Northridge, California, and 1995 Kobe, Japan, earthquakes. The lack of accurate information on ground shaking in Kobe, for example, delayed the emergency response and resulted in needless damage and casualties (Box 2.4).

Goal: Develop reliable seismic information systems capable of providing (1) time-critical information about earthquakes needed for rapid alert and assessment of impact, including strong-motion maps and damage estimates, and (2) early warning of impending strong ground motions and tsunamis outside the epicentral zones of major earthquakes.

The ANSS, with its planned nationwide network of on-line broadband seismographs and strong-motion sensors, can fulfill some of these objectives. It can provide time-critical information for immediate public safety and emergency response when the dangers of an earthquake, tsunami, or volcanic eruption arise. The delivery of this information will require a coordinated information infrastructure that feeds directly into emergency management systems through robust, secure connections and can take advantage of new communication pathways, including the Internet and the World Wide Web.

A fully implemented ANSS will provide the framework for upgrading regional seismic information systems to real-time warning systems that, in favorable situations, can automatically notify critical facilities of impending shaking tens of seconds before the seismic waves arrive at the site. The implementation of seismic warning systems will depend critically on automated broadcasting and decision making, and the construction and testing of such systems will require extensive collaborations between seismologists and end users. Although the operational responsibilities for seismic information and warning systems correctly reside with the USGS, the participation of regional network operators, whose missions now include public information and education, is essential.

6.9 PARTNERSHIPS FOR PUBLIC EDUCATION AND OUTREACH

In a rapidly expanding society, it is difficult to balance earthquake risk against the economic expenses and social strictures (regulations and enforcement) required to reduce risk. It is also difficult to prescribe what roles scientists should play in this intrinsically political process. Most observers support the proposition that scientists should increase their involvement in two critical areas: (1) implementing new knowledge gained from research to improve mitigation technologies and strategies, and (2) educating people about the nature of earthquake hazards and how to use earthquake information in seismic preparation and response. Experience from the NEHRP makes it clear that the efforts of individual scientists and research organizations can be amplified through partnerships with other technical groups, such as engineers and social scientists, as well as with the end users of earthquake science—emergency responders, disaster managers, insurance agencies, and public officials (see Section 3.6).

Goal: Establish effective partnerships between earthquake scientists and other communities to reduce earthquake risk through research implementation and public education.

Interdisciplinary collaborations are critical in establishing a firm technical basis for civic action and strengthening the resolve of public officials to improve mitigation strategies. Adapting probabilistic seismic hazard analysis to the needs of performance-based design is an example of where more cooperation between scientists and engineers could pay off in a big way. However, government agencies have not been particularly effective in sponsoring these types of interactions. For example, NSF manages research in earthquake science and earthquake engineering out of separate directorates. Moreover, while the USGS holds federal responsibility for the operational aspects of earthquake monitoring and hazard assessments, there is no federal equivalent for earthquake engineering. Owing to such organizational deficiencies, the most successful interactions have tended to involve the professional engineering societies and regional research centers.

Successful partnerships between those who conduct research and those who use research results usually require active and continuous collaborations, sustained through two-way communication. They profit from the involvement of people who have knowledge of implementation processes, have a tolerance for ambiguity, accept the high transaction costs associated with interdisciplinary activities, and are willing to overcome communication problems by developing a common language. An iterative process, with repeated opportunities for researchers and end users to educate each other, advances the concept of joint ownership. This in turn can lead to consensus and implementation of mutually identified priorities for earthquake hazard awareness, mitigation product development, and information dissemination. Workshops, short courses, and field excursions that bring scientists together with end users can be effective mechanisms for promoting two-way communication and stimulating new approaches to solving practical problems. These activities can be enhanced by using technical briefs on research results in a form ready for application by professionals.

When it comes to the general public, the interpretation of scientific research—reducing results to understandable, usable products that improve hazard awareness, public safety, and mitigation efforts—is an essential part of the educational process. User-friendly information, distributed widely in print and over the Internet, can advance public understanding of the severity of an earthquake threat and motivate vulnerable populations to take protective measures. People are obviously most attuned to the issues of earthquake risk just after widely felt earthquakes, and scientists must be ready to engage the public through the news and other media during these teachable moments. The *ShakeMap* and *Did You Feel It?* web sites (6) set up by the USGS have received a large number of hits immediately following recent tremors, indicating that people are willing to invest considerable time and effort to understand what is happening in the ground beneath their feet.

The primary educational role for many earthquake scientists is at the university level, where teaching and research go hand-in-hand in developing the careers of young scientists. A healthy national program in research-based graduate education is essential to attract bright students and overcome the manpower limitations that now constrain some fields in earthquake science. Two trends in undergraduate education that deserve more support are the incorporation of information technology and computer science in geophysics training and an emphasis on environmental processes and histories in geologic training. Other educational objectives include

- studies of earthquakes in the laboratory and field to enrich the educational experiences of students from all backgrounds and help them appreciate the excitement of basic and applied science;
- scientist-mentored summer internships for undergraduates, such as those funded through the NSF Research Experiences for Undergraduates program;
- work with museums to create novel and interactive learning environments;
- development of K-12 earthquake curricula in accordance with the National Science Education Standards and state-sponsored efforts, such as California's Earthquake Loss Reduction Plan; guiding these efforts should be specific objectives to (1) structure these Earth science curricula

in ways that appeal to students from underrepresented groups, and (2) achieve better meshing between K-12 and college-level Earth science education; and

- workshops for K-12 and college-level educators to demonstrate and encourage the use of educational resources, curricula, and field-based experiences, in accordance with established career development standards.

6.10 RESOURCE REQUIREMENTS

The programmatic support required for earthquake research during the next 10 years will outstrip the resources currently available through NEHRP and other federal programs. The ANSS and EarthScope initiatives, for example, would greatly improve the observational capabilities for earthquake science in the United States and would contribute substantially to the objectives outlined in this report. To bring ANSS into full operation will require capital investments of approximately $170 million, and its annual operational costs are estimated to be about $47 million (Box 6.1). In comparison, the congressional appropriation for the entire USGS component of the NEHRP budget was only $50 million in FY 2001. Deployment costs for the EarthScope instrumental systems are estimated to be $91.3 million for PBO, $245 million for InSAR, $64 million for USArray, and $17.4 million for SAFOD (Box 6.2). Data analysis and management activities will require an additional $15 million to $20 million per year during the first decade of EarthScope operations. The total FY 2001 geoscience expenditures by the NSF in support of NEHRP were only about $12 million. The research opportunities for characterizing the structure and history of active fault systems warrant a severalfold increase in the neotectonic and paleoseismic studies currently supported by the USGS and the NSF. Work in this area is limited by the small number of earthquake geologists engaged in this type of research, underlining the need for increasing efforts in geoscience education at both the undergraduate and the graduate levels.

Research to understand earthquakes and their effects is central to continuing efforts to decrease earthquake risk. The technological and conceptual developments documented in this report have positioned the field of earthquake science for major advances. Investments made now will eventually pay off in terms of saved lives and reduced damage. These returns can be realized sooner by encouraging unconventional lines of research; coordinating scientific activities across disciplines and organizations, especially between scientists and engineers; and supporting international programs to investigate the global diversity of earthquake behavior. The transition to a systems-oriented science has important ramifications for the types of cooperative research activities and organizational structures that are most effective in addressing the basic and applied problems of earthquake research. In particular, there is a critical need to maintain and expand scientific centers where the disciplinary activities of many research organizations can be coordinated, evaluated, and synthesized into system-level models of regionalized earthquake behavior. In addition to their key role in basic earthquake science, such multidisciplinary centers have proven to be effective organizations for the dissemination of earthquake information and research results, the formulation of science-based strategies for loss reduction, and the education of the general public and nonscientist professionals concerned with disaster mitigation and loss reduction.

NOTES

1. The International GPS Service (IGS) is an International Service under the Federation of Astronomical and Geophysical Data Analysis Services. IGS has developed a worldwide system to put high-quality GPS data on-line within one day and data products within two weeks. The system consists of about 125 satellite tracking stations, 3 data centers, and 7 analysis centers. GPS data are used to generate GPS satellite ephemerides, Earth rotation parameters, IGS tracking station coordinates and velocities, and GPS satellite and IGS tracking station clock information. See < http://igscb.jpl.nasa.gov/>.

2. National Research Council, *Basic Research Opportunities in Earth Science*, National Academy Press, Washington, D.C., pp. 101-105, 2001.

3. For example, aseismic "silent earthquakes" have recently been observed on the thrust interface of subduction zones by geodetic networks in Japan and Cascadia, but the interplay between these events and major earthquakes in subduction zones is not understood; see Section 4.2.

4. E.H. Field and the SCEC Phase III Working Group, Accounting for site effects in probabilistic seismic hazard analyses of Southern California: Overview of the SCEC Phase III report, *Bull. Seis. Soc. Am.*, **90**, S1-S31, 2000.

5. A notable exception is the excellent data set collected by the Taiwan Central Weather Bureau from the 1999 Chi-Chi, Taiwan, earthquake (M 7.6).

6. See <http://www.trinet.org/shake/index.html>; <http://pasadena.wr.usgs.gov/shake/>.

7

Summary

Earthquakes pose the most dangerous natural threat to the built environment. The estimated economic loss from earthquakes exceeds $4 billion per year in the United States alone, and the risks are rising throughout the world as nations expand their urban centers and infrastructure in geologically active regions. In 1977, Congress established the National Earthquake Hazard Reduction Program (NEHRP) to counter this threat through scientific and engineering research. NEHRP-sponsored studies have significantly improved long-term forecasts of earthquakes and their site-specific effects, as well as the ability to respond rapidly to earthquake disasters. This research has clearly paid off in the formulation of more effective risk reduction strategies. However, there has been considerable controversy about how government and the private sector can best implement loss reduction measures through regulatory policies, economic incentives, long-term investments, and public education. Part of this debate concerns the role of scientific research in earthquake mitigation.

The Committee on the Science of Earthquakes, sponsored in part by a grant from the National Academy of Sciences, has conducted this study to appraise the needs for future research from four complementary perspectives: (1) the need to improve seismic safety and performance of the built environment, especially in highly exposed urban areas; (2) the requirements for disseminating information rapidly during earthquake crises; (3) the opportunities for exciting basic science, particularly in the context of current research on complex natural systems; and (4) the responsibility for educating people at all levels of society about the causes and effects of earthquakes. The study comprises five elements, each presented as a chapter of the report:

1. Survey of basic and applied earthquake science from ancient times to the present day, with a discussion of lessons drawn from past research
2. Evaluation of the current status of seismic hazard analysis and its connections to earthquake engineering, loss estimation, and risk mitigation
3. Examination of the new technologies in the main observational disciplines of seismology, geodesy, geology, and rock mechanics
4. Technical assessment of the key issues for future earthquake science, including the application of a dynamical systems approach to integrate observations
5. Analysis of research opportunities and requirements.

7.1 CURRENT CAPABILITIES

The study of earthquakes, like the science of many other complex natural systems, is still in its juvenile stages of exploration and discovery. Research has been focused on two primary problems: (1) earthquake complexity and how it arises from the brittle response of the lithosphere to deep-seated forces, and (2) the forecasting of earthquakes and their site-specific effects. Investigations of the first problem began with attempts to place earthquake occurrence in a global framework and contributed to the discovery of plate tectonics, while work on the second addressed the needs of earthquake engineering and led to the development of seismic hazard analysis. The historical separation between these two lines of inquiry has been narrowed by recent progress on dynamical models of earthquake occurrence and strong ground motion. This research has transformed the field from a haphazard collection of disciplinary activities into a more coordinated system-level science that seeks to describe seismic activity not just in terms of individual events, but as an evolutionary process involving dynamical interactions within networks of interconnected faults. The bright prospect for "earthquake system science" is a major theme of this report.

Experience shows that much can be learned from multidisciplinary investigations coordinated in the aftermath of large earthquakes, and it makes clear the importance of standardized instrumental data and geologic field work. During the last decade, research has been accelerated through the development of new observational and computational technologies. Subsurface imaging can now be applied with sufficient resolution to delineate the deep, three-dimensional architecture of fault systems. Neotectonic studies are improving constraints on fault geometries and long-term slip rates, and paleoseismology is furnishing an extended record of past earthquakes, revealing evidence for the clustering of large events in "seismic storms." The Global Positioning System (GPS) and interferometric synthetic aperture radar (InSAR) satellites are mapping with unprecedented resolution the crustal deformations associated with individual earthquakes, long-term tectonic loading, and the stress interactions among nearby faults. Networks of broadband seismometers have been deployed to record earthquake ground motions faithfully at all frequencies and amplitudes. By using high-performance computing and communications, scientists now have the means to process massive streams of observations in real time and, through numerical simulation, to quantify the many aspects of earthquake physics that have been resistant to standard analysis.

Large earthquakes can be forecast on time scales of decades to centuries by combining the information from the geological record with data from seismic and geodetic monitoring. Earthquake scientists have begun to understand how geological complexity controls the strong ground motion during large earthquakes and, working with engineers, how to predict the site-specific response of buildings, lifelines, and critical facilities to seismic excitation. The long-term expectations for potentially destructive shaking have been quantified in the form of seismic hazard maps, which display estimates of the maximum shaking intensities expected at each locality in the United States. Once a large earthquake has occurred, automated systems can rapidly and accurately compute hypocenter location, fault-plane orientation, and other source parameters. Predicted distributions of the extent of strong ground motions can be broadcast in near real time, helping to anticipate damage and guide emergency response. In the case of distant, suboceanic earthquakes, post-event predictions of the earthquake-generated sea waves (tsunamis) can warn coastal communities with sufficient lead times to permit evacuation. Similarly, seismic activity can be used to warn about impending volcanic eruptions, as illustrated by the successful prediction of the Mt. Pinatubo eruption in 1991.

7.2 SCIENCE GOALS

Despite these successes, many scientific questions about earthquakes remain to be answered. No available theory adequately describes the dynamical interactions among faults or the basic features of rupture nucleation, propagation, and arrest. From a practical perspective, the search for a comprehensive theory is motivated by the need to understand (1) active fault systems on time scales of days to centuries for the purpose of improving earthquake forecasting and (2) fault ruptures on time scales of seconds to minutes for the purpose of predicting strong ground motions.

Extending the capabilities for long-term forecasting through better fault-system models is critical to improving seismic hazard analysis. On short time scales (hours to days), no method for event-specific earthquake prediction has yet demonstrated skill at a statistically reliable level; indeed, the chaotic nature of brittle deformation may imply that useful short-term prediction cannot be achieved, even with substantial improvements in the ability to detect precursory signals. Near-field observations before and during large earthquakes are too few and too limited, however, to rule out categorically the feasibility of short-term earthquake prediction. For example, aseismic "silent earthquakes" have recently been observed on the thrust interface of subduction zones by geodetic networks in Japan and Cascadia, but the interplay between these events and major earthquakes in subduction zones is not understood. Moreover, there are both observational and theoretical reasons to believe that large-scale failures within some fault systems may be predictable on intermediate time scales (years to decades), provided that adequate knowledge about the history and present state of the system can be obtained. It is not yet clear whether probabilistic forecasting methods can be devised that take advantage of this potential predictability, but such methods could contribute significantly to the reduction of earthquake losses. Fundamental understanding of earthquake predictability will likely come through a broad research program with the goals of improving knowledge of fault-zone processes; the nucleation, propagation, and arrest of fault ruptures; and stress interactions within fault networks.

Better ground-motion prediction will depend on the ability to model fault ruptures, wave propagation, and the near-surface response to wave excitation. Three-dimensional simulation of the ground motions generated by dynamic fault ruptures is a challenging problem, owing to the complex physics of rock failure, the wide separation between the inner and outer scales of faulting, and the computational requirements for representing realistic geologic structures. Accurate prediction of strong ground motions requires detailed information about the heterogeneities in material properties and the stress field that govern high-frequency wave propagation.

The committee identified specific long-term goals in nine areas of interdisciplinary research that offer exceptional opportunities to further the national effort in earthquake science:

1. *Fault Characterization*: Document the location, slip rates, and earthquake history of dangerous faults throughout the United States.
2. *Global Earthquake Forecasting*: Forecast earthquakes on a global basis by specifying accurately the probability of earthquake occurrence as a function of location, time, and magnitude, as well as the magnitude limits and other characteristics of likely earthquakes in a given place.

3. *Fault-System Dynamics*: Understand the kinematics and dynamics of active fault systems on interseismic time scales, and apply this understanding in constructing probabilities of earthquake occurrence, including time-dependent earthquake forecasting.

4. *Fault-Zone Processes*: Characterize the three-dimensional material properties of fault systems and their response to deformation through a combination of laboratory measurement, high-resolution structural studies, and in situ sampling and experimentation.

5. *Earthquake Source Physics*: Understand the physics of earthquake nucleation, propagation, and arrest in realistic fault systems and the generation of strong ground motions by fault rupture.

6. *Ground-Motion Prediction*: Predict the strong ground motions caused by earthquakes and the nonlinear responses of surface layers to these motions—including fault rupture, landsliding, and liquefaction—with enough spatial and temporal detail to assess seismic risk accurately.

7. *Seismic Hazard Analysis*: Incorporate time dependence into the framework of seismic hazard analysis in two ways: (1) by using rupture dynamics and wave propagation in realistic geological structures to predict strong-motion seismograms (time histories) for anticipated earthquakes, and (2) by using fault-system dynamics to forecast the time-dependent perturbations to average earthquake probabilities.

8. *Seismic Information Systems*: Develop reliable seismic information systems capable of providing (1) time-critical information about earthquakes needed for rapid alert and assessment of impact, including strong-motion maps and damage estimates, and (2) early warning of impending strong ground motions and tsunamis outside the epicentral zones of major earthquakes.

9. *Education and Outreach*: Establish effective partnerships between earthquake scientists and other communities to reduce earthquake risk through research implementation and public education.

7.3 RESEARCH OPPORTUNITIES AND REQUIREMENTS

In order to realize these goals, substantial investments will have to be made in the main observational disciplines of seismology, geodesy, geology, and rock mechanics. The seismological requirements include national and regional seismic networks capable of recording all earthquakes down to magnitude 3 (1.5 in urban areas of high seismic risk) with sufficient fidelity and density to determine the focal mechanisms and other source parameters of the smaller events. The recently completed Global Seismic Network (GSN) must be maintained to furnish data on source processes of major earthquakes around the world. Portable arrays of seismometers are needed to study aftershocks and other forms of earthquake clustering and to provide the high-resolution structural information necessary for investigating the architecture of fault systems and predicting ground motions. Probing the structure of sedimentary basins will require the extensive use of artificial-source reflection and refraction seismology, as well as seismographic data from deep boreholes to calibrate the effects of near-surface layering. Plausible 10-year objectives are to determine the structure of high-risk urban areas well enough to model the surface motions from deterministic seismic sources at all frequencies up to at least 1 hertz and to formulate useful, consistent, stochastic representations of surface motions up to at least 10 hertz.

Geodetic instrumentation should be deployed for observing crustal deformations within active fault systems with enough spatial and temporal resolution to measure all significant motions, including aseismic events and the transients before, during, and after large earthquakes. This endeavor will require combining pointwise measurements using strainmeters and GPS with continuous deformation images from a dedicated U.S. InSAR satellite mission. Laser and radar altimetry are needed to produce the precise digital elevation models for investigating surface faulting and the deformations caused by buried faults.

The determination of fault slip rates and rupture histories over many earthquake cycles will require the combination of geologic field study and high-precision age dating. At present, the long-term slip rates of most major faults in North America are either unknown or, at best, constrained by geologic measurements at only one or two sites. Slip-rate data are especially lacking in contractional provinces, where many questions still remain about how strain is partitioned among the major faults and between seismogenic faults and aseismic folding. New techniques of tectonic geomorphology can address these issues. To investigate the important problem of earthquake clustering, paleoseismologists must date slip episodes at particular points on a fault well enough to establish event sequences. The objective should be catalogs of large earthquakes spanning thousands of years on all major faults where such paleoseismic investigations are feasible.

Better information on microscale processes is needed to formulate realistic macroscopic representations of the strength variations and the dynamic response of fault materials. Advances in laboratory-based techniques will be required to elucidate the dynamic phase of fault response, in which rapid large slips may cause large temperature excursions and weakening by pressurization of pore fluids and/or melt generation. Field examination of exhumed faults should explore the mechanical importance of shear localization structures and the evidence for fluids and/or local melting. Borehole data are needed to extrapolate laboratory results on laboratory-scale samples to natural faults, elucidate the generation of fault-zone structure, and clarify why some faults creep while others slip in large earthquakes. High-resolution methods focused on imaging fault structure and deformation processes should be combined with systematic programs for in situ sampling and experimentation at an established set of long-term natural laboratories.

The diversity of earthquakes and their geological environments necessitates a global approach. Global scope is particularly important for capturing enough earthquakes to test the controversial assumptions of earthquake forecasting schemes under different tectonic and stress conditions. A truly comprehensive theory of earthquakes must be able to explain the similarities and differences of earthquakes at different depths, including the perplexing occurrence of events to depths of nearly 700 kilometers in descending lithospheric slabs. The GSN and temporary deployments of portable seismometer arrays above active subduction zones will yield the requisite data on intermediate-focus and deep-focus earthquakes, while laboratory studies of deformation will furnish information on the microscale physics of earthquake instabilities at the high pressures and temperatures of descending slabs. Pioneering international efforts such as the Global Seismic Hazard Assessment Program, the Global Strain Rate Map Project, the Global Fault Mapping Project, and the Working Group on Earthquake Recurrence Through Time are providing uniform standards and data access for seismic hazard analysis on a global scale, and they should be expanded with aggressive data-gathering efforts that exploit the new technologies described in this report. A major objective of earthquake science should be to extend observational systems and data bases into the oceans to understand the distribution of offshore faulting and its seismogenic and tsunamigenic potential.

Progress toward understanding fault-system dynamics will depend on the ability to integrate the disparate observations regarding stress, strain, and rheology into self-consistent models that can be tested against observations not yet collected. Dynamical simulations at various scales are needed to assess the discrepancies among laboratory-based friction laws, observed fault-system behaviors, and seismological data on large earthquakes. Simulations of earthquake occurrence on fault networks are required to understand the behavior of natural fault systems and to address fundamental questions relating to earthquake occurrence, such as the effects of stress interactions and prior earthquakes in determining earthquake probability. The practical objective of this research is to develop procedures that can assimilate all information on fault-system behaviors into probabilistic forecasts and can update these forecasts consistently based on seismic activity and other new information.

Reliable procedures for simulating ground-motion time histories have to be developed, rigorously tested against the available strong-motion data, and then applied in earthquake engineering research and practice. As richer data sets are collected, a primary challenge will be to set up a computational framework for the systematic refinement of three-dimensional wave-speed and attenuation models and the use of these models in the calculation of synthetic seismograms. A consistent interface will be needed for coupling ground-motion calculations to nonlinear soil-modeling codes. Significant research is needed on how to use waveform modeling to characterize the probability distributions of ground-motion time histories, or parameters derived from those time histories, in a way that properly accounts for both aleatory and epistemic uncertainties.

The transition of earthquake science to a systems-oriented, physics-based approach has important ramifications for the types of cooperative research activities and organizational structures that will be most effective in addressing the basic and applied problems of earthquake research. In particular, additional support is needed for scientific centers and distributed collaboratories with advanced information technology infrastructures, where the disciplinary activities of many research groups can be coordinated, evaluated, and synthesized into system-level models of earthquake behavior. In addition to their key role in basic earthquake science, such centers have proven to be effective in disseminating earthquake information and research results, formulating science-based strategies for loss reduction, and educating groups at all levels about the role of science in disaster mitigation and loss reduction.

7.4 RESOURCE REQUIREMENTS

The technological investments and programmatic support required for earthquake research during the next 10 years will outstrip the resources currently available through NEHRP and other federal programs. Major initiatives by the two NEHRP science agencies illustrate this situation. The U.S. Geological Survey (USGS) has proposed the deployment of an Advanced National Seismic System (ANSS), which would upgrade the U.S. National Seismographic Network, modernize regional networks, and deploy 6000 strong-motion stations in high-risk urban areas. A fully implemented ANSS would upgrade regional networks to modern seismic information systems and provide the framework for developing real-time warning systems. The ANSS plan, if brought into full operation, would greatly improve seismological instrumentation in the United States and would contribute substantially to the objectives outlined in this report. This system will require capital investments of approximately $170 million, and its annual operational costs are estimated to be about $47 million. In comparison, the congressional appropriation for the entire USGS component of the NEHRP budget was only $50 million for FY 2001.

A second example is the EarthScope initiative, proposed by the National Science Foundation (NSF). This facility-oriented program includes the Plate Boundary Observatory (PBO), which would expand existing geodetic networks with additional permanent GPS stations and campaign-style observations and fill major gaps in measurements of plate boundary deformation in the western United States. The second geodetic component—a satellite-based InSAR imaging system—would map decimeter-level deformations of fault ruptures continuously over areas tens to hundreds of kilometers wide, as well as a range of nonseismic phenomena such as volcano inflation, glacial flow, and ground subsidence. USArray, the seismological component of EarthScope, would map lithospheric structure nationwide on scales of tens of kilometers and provide new capabilities for active-source imaging of specific features, including sedimentary basins where seismic risk is often high. EarthScope would also construct a San Andreas Fault Observatory at Depth (SAFOD) that would, for the first time, sample the fault by deep drilling, monitor its seismicity and strain, and perform in situ experiments to depths of 4 kilometers. Deployment costs for the EarthScope instrument systems are estimated to be \$91.3 million for PBO, \$245 million for InSAR, \$64 million for USArray, and \$17.4 million for SAFOD. Data analysis and management will require an additional \$15 million to \$20 million per year during the first decade of EarthScope operations. In comparison, total FY 2001 geoscience expenditures by the NSF in support of NEHRP were about \$12 million.

Geologic field work will be an important part of EarthScope; yet even if this NSF initiative were fully funded, it would not boost resources for earthquake geology to the levels envisaged in this report. The research opportunities for characterizing the structure and history of active fault systems warrant a severalfold increase in the neotectonic and paleoseismic studies currently supported by the USGS, as well as the NSF. Work in this area is limited by the small number of earthquake geologists engaged in this type of research, underlining the need for increasing efforts in geoscience education at both the undergraduate and the graduate levels.

Experience gained during NEHRP demonstrates that the willingness of society to invest in risk reduction is best achieved through an active collaboration among scientists, engineers, government officials, and business leaders, working together among an informed populace. A corollary is that earthquake research will contribute to risk reduction more effectively when it is carried out in a context that recognizes the problem's engineering, economic, and political dimensions. No agency is responsible for ensuring an integrated approach to research problems in earthquake science and engineering, and better mechanisms should be developed for bringing the two fields together to exploit potential synergies. Cooperation among the NSF and USGS earthquake science centers and the NSF earthquake engineering centers will be critical to this goal. Science participation in engineering programs such as the Network for Earthquake Engineering Simulation should be encouraged.

Research to understand earthquakes and their effects is central to continuing efforts to decrease earthquake risk. The technological and conceptual developments documented in this report have positioned the field of earthquake science for major advances. Investments made now will eventually pay off in terms of saved lives and reduced damage. These returns can be realized sooner by encouraging unconventional lines of research; coordinating scientific activities across disciplines and organizations, especially between scientists and engineers; and supporting international programs to investigate the global diversity of earthquake behavior. Few problems are more challenging to science or strategically relevant to the nation, and few have a greater potential for elucidating the fundamental geological processes that shape the face of the planet.

APPENDIX A

Major Federal Earthquake Programs

Federal funding of earthquake research began in 1930 with the expenditure of about $10,000 for support of seismology, increasing to approximately $500,000 by 1957. The Department of Defense initiated Project Vela Uniform in 1959 to obtain a better understanding of seismic phenomena needed for nuclear test detection and treaty verification, and by 1961 the total federal expenditures for seismological research had risen to almost $30 million, much of it for deployment of the World Wide Standardized Seismographic Network. Support in this area began to decline in 1964 as Vela Uniform achieved its research goals, and the research programs sponsored by the Department of Commerce, Department of the Interior, and National Science Foundation (NSF) were reoriented toward the investigation of earthquake hazards. In 1967, the total federal support for earthquake research was $7.4 million, with this amount split approximately equally between engineering and geoscience studies.

The first official call for a comprehensive program of earthquake research in the United States came in 1965 from a select committee convened by the President's Office of Science and Technology, chaired by Frank Press (*1*). Although its major recommendations were not implemented immediately, the committee's report initiated a series of assessments in earthquake science and engineering that continued into the next decade (*2*). Motivated by these studies and spurred by the disastrous 1971 San Fernando earthquake in southern California, Congress established the National Earthquake Hazard Reduction Program (NEHRP) in 1977 with the authorization of $55 million dollars for the U.S. Geological Survey (USGS) and NSF. NEHRP continues to be the mainstay of federal support for earthquake studies.

NEHRP was established by the Earthquake Hazards Reduction Act of 1977 (Public Law 95-124) to "reduce the risks of life and property from future earthquakes in the U.S. . . ." The act authorized additional funds for the USGS and NSF to conduct wide-ranging studies on the fundamental causes of earthquakes with several practical objectives, including the identification of earthquake hazards; the development of an earthquake prediction capability; the preparation of plans for mitigation, preparedness, and response activities; the development of seismic design and construction standards; and the education of the public about earthquakes hazards. In 1980, the act was amended to include the National Institute of Standards and Technology (NIST, then the National Bureau of Standards) and to designate the newly created Federal Emergency Management Agency (FEMA) as the lead agency. The roles of the four agencies were further clarified in the 1990 NEHRP Reauthorization Act, which cast their primary responsibilities as follows:

- FEMA coordinates the NEHRP program, plans and manages the federal response to earthquakes, funds state and local preparedness exercises, and supports the development of improved seismic design and construction techniques for new buildings and retrofit guidelines for existing buildings.
- USGS conducts and supports Earth science investigations into the origins of earthquakes, predicts earthquake effects, characterizes earthquake hazards, and disseminates Earth science information.
- NSF funds earthquake engineering research, basic Earth science research, and earthquake-related social science research.
- NIST conducts and supports engineering studies to improve seismic provisions of building codes, standards, and practices for buildings and lifelines.

Congress reauthorizes the program at intervals of one to three years, most recently for FY 2001 ($101.5 million), FY 2002 ($105.8 million), and FY 2003 ($110.3 million) (3). The split among the agencies for FY 2001 is 48 percent (USGS), 30 percent (NSF), 20 percent (FEMA), and 2 percent (NIST). The President's budget request for FY 2003 is $117.9 million. Appropriations for NEHRP have declined significantly in constant dollars since the late 1970s.

In addition to funding ongoing earthquake hazards activities of the four NEHRP agencies, the Earthquake Hazard Reduction Authorization Act of 2000 authorizes the establishment and operation of the Advanced National Seismic Research and Monitoring System (ANSS) and the Network for Earthquake Engineering Simulation (NEES). The purpose of the ANSS is to "organize, modernize, standardize, and stabilize the national, regional, and urban seismic monitoring systems in the United States, including sensors, recorders, and data analysis centers, into a coordinated system that will measure and record the full range of frequencies and amplitudes exhibited by seismic waves, in order to enhance earthquake research and warning capabilities." The act authorizes five years of USGS funding to establish the system ($33 million to $35 million per year) and two years of operating funds ($4.5 million in FY 2002 and $10.3 million in FY 2003).

NEES was established at NSF to "upgrade, link, and integrate a system of geographically distributed experimental facilities for earthquake engineering testing of full-sized structures and their components and partial-scale physical models." The program was authorized for four years, including $28.2 million in FY 2001, $24.4 million in FY 2002, $4.5 million in FY 2003, and $17 million in FY 2004.

In addition to the four designated NEHRP agencies, a number of federal agencies conduct earthquake loss reduction activities, including the Department of Energy (seismic regulations for nuclear and conventional power plants), Department of Transportation (seismic regulations for bridges and highways), Environmental Protection Agency (control of hazardous materials), General Services Administration (standards for federal construction), Department of Veterans Affairs (standards for hospital construction), Nuclear Regulatory Commission (seismic and earthquake engineering criteria for nuclear power plants), and Department of Housing and Urban Development (repair and rehabilitation of residential structures). Many of these agencies also engage in earthquake research and development, including the following:

- *Department of Defense:* The Army Corps of Engineers conducts seismic research at its laboratories, and the Defense Threat Reduction Agency funds university research related to seismic monitoring of the Comprehensive Test Ban Treaty.
- *Department of Energy* (DOE): Los Alamos National Laboratory programs include basic research on the physical properties of Earth materials, reservoir microearthquakes, and earthquake strong motion; the development of seismic imaging methods and microborehole tools; and operation of the Los Alamos Seismographic Network. DOE's Nuclear Explosion Monitoring Program is responsible for providing the United States National Data Center with analytic tools to monitor underground nuclear explosions.
- *National Oceanic and Atmospheric Administration*: The National Hazard Data Center collects post-event data and images, and assists in the

detection, location, and evaluation of the extent of certain hazards using satellite data.

• *National Aeronautics and Space Administration*: The Solid Earth and Natural Hazards Program sponsors research and disaster reduction based on remote-sensing technologies such as the Global Positioning System and synthetic aperture radar.

The National Earthquake Loss Reduction Program was proposed in 1996 to coordinate these activities (4), but it has never been fully implemented.

NOTES

1. Office of Science and Technology, *Earthquake Prediction: A Proposal for a Ten Year Program of Research*, White House, Washington, D.C., 134 pp., 1965.

2. Among the reports that led to congressional enactment of NEHRP were the following: Ad Hoc Interagency Working Group, *Proposal for a Ten-Year National Earthquake Hazard Program*, Federal Council for Science and Technology, Washington, D.C., 1968; National Research Council, *Earthquake Engineering Research*, National Academy Press, Washington, D.C., 1969; General Accounting Office, *Need for a National Earthquake Research Program*, GAO report B-175621, Washington, D.C., 1972; *Earthquake Prediction and Hazard Mitigation, Options for USGS and NSF Programs*, National Science Foundation and U.S. Geological Survey, Washington, D.C., 1976. An interesting account of the programmatic development of earthquake science in the United States is given by R.E. Wallace in *Earthquakes, Minerals, and Me, Oral History Interviews with S. Scott*, U.S. Geological Survey Open-File Report 96-260, Menlo Park, Calif., pp. 60-84, 1996.

3. Earthquake Hazards Reduction Authorization Act of 2000, Public Law 106-503.

4. National Earthquake Strategy Working Group, *Strategy for National Earthquake Loss Reduction*, prepared for the National Science and Technology Council, April 1996, <http://www.ostp.gov/NSTC/html/USGS/>.

APPENDIX B

Acronyms and Abbreviations

AEL	annualized earthquake losses
AMS	accelerator mass spectrometry
ANSS	Advanced National Seismic System
ARPA	Advanced Research Projects Agency
ASCE	American Society of Civil Engineers
ATC	Applied Technology Council
BSE	Basic Safety Earthquake
BSSC	Building Seismic Safety Council
CDMG	California Division of Mines and Geology
CGS	California Geological Survey
CLVD	compensated linear vector dipole
CMT	centroid moment tensor
CTBT	comprehensive test ban treaty
DEM	digital elevation model
DMC	Data Management Center (IRIS)
DOE	Department of Energy
ECHO	Earth Change and Hazard Observatory
EPEDAT	Early Post-Earthquake Damage Assessment Tool
FDSN	Federation of Digital Seismographic Networks
FEMA	Federal Emergency Management Agency
GPS	Global Positioning System
GSHAP	Global Seismic Hazard Assessment Program
GSN	Global Seismic Network
IBC	International Building Code

IDC	International Data Center
IDNDR	International Decade for Natural Disaster Reduction
IGS	International GPS Service for Geodynamics
ILP	International Lithosphere Project
IMS	International Monitoring System
InSAR	interferometric synthetic aperture radar
IRIS	Incorporated Research Institutions for Seismology
ISC	International Seismological Centre
ISS	International Seismological Summary
IT	information technology
JMA	Japan Meteorological Agency
LARSE	Los Angeles Region Seismic Experiment
LIDAR	light detection and ranging
M	moment magnitude
MMI	Modified Mercalli Intensity
NASA	National Aeronautics and Space Administration
NEES	Network for Earthquake Engineering Simulation
NEHRP	National Earthquake Hazard Reduction Program
NEIS	National Earthquake Information Service
NIBS	National Institutes for Building Standards
NIST	National Institute of Standards and Technology
NRC	National Research Council
NSF	National Science Foundation
NSMP	National Strong-Motion Program
OHP	Ocean Hemisphere Project
PASSCAL	Program for the Array Seismic Studies of the Continental Lithosphere
PBO	Plate Boundary Observatory
PDE	Preliminary Determination of Epicenters
PEER	Pacific Earthquake Engineering Research
PGA	peak ground acceleration
PGV	peak ground velocity
PHIVOLCS	Philippine Institute of Volcanology and Seismology
PSHA	probabilistic seismic hazard analysis
RAMP	Rapid Array Mobilization Program
Sa	spectral acceleration
SAFOD	San Andreas Fault Observatory at Depth
SCEC	Southern California Earthquake Center
SCIGN	Southern California Integrated GPS Network
SHA	seismic hazard analysis
SLR	Satellite Laser Ranging
SRTM	Shuttle Radar Topography Mission
Sv	spectral velocity

UBC	Uniform Building Code
USCGS	U.S. Coast and Geodetic Survey
USGS	U.S. Geological Survey
USNSN	U.S. National Seismic Network
VLBI	Very Long Baseline Interferometry
WWSSN	World Wide Standardized Seismographic Network

Index

F

H

I

J

K

L

M

N

O

P

R

S

T

U